M. S. Welling

Dictionary of
Paints, Adhesives and Sealants
German/English

Wörterbuch
Farben, Klebstoffe und Dichtungsmittel
Deutsch/Englisch

Vertrieb:
Bundesrepublik Deutschland, Österreich: VCH, Postfach 101161,
D-69451 Weinheim (Bundesrepublik Deutschland)
Schweiz: VCH, Postfach, CH-4020 Basel (Schweiz)
übrige Länder: Pentech Press Ltd., 3 Graham Lodge, Graham Road,
London NW4 3DG (England)

ISBN 3-527-29348-5 (VCH, Weinheim)
ISBN 0-7273-0706-1 (Pentech Press)
ISSN 0930-6862

M. S. Welling

parat

Dictionary of Paints, Adhesives and Sealants German/English

Wörterbuch Farben, Klebstoffe und Dichtungsmittel Deutsch/Englisch

Weinheim · New York
Basel · Cambridge · Tokyo

Titel der Originalausgabe: German – English Dictionary of Paints,
Adhesives and Sealants
erschienen im Verlag Pentech Press Ltd., London, UK

Dipl.-Chem. M. S. Welling
5 Campbell Croft
Edgeware, Middlesex HA8 8DS
United Kingdom

Herausgeber der Reihe „parat“
Dr.-Ing. H.-D. Junge
Cavaillonstraße 78/I
69469 Weinheim
Bundesrepublik Deutschland

1. Auflage 1995

Lektorat: Roland Wengenmayr

Die Deutsche Bibliothek – CIP-Einheitsaufnahme
Welling, Manfred S.:
Dictionary of paints, adhesives and sealants : english, german = Wörterbuch Farben, Klebstoffe und Dichtungsmittel / M. S. Welling. – Weinheim ; New York ; Basel ; Cambridge ; Tokyo : VCH, 1995
(Parat)
Einheitssacht.: German-english dictionary of paints, adhesives and sealants
ISBN 3-527-29348-5
NE: HST

Preface

This dictionary has been compiled for technical translators who have to render German texts dealing with paints, adhesives and sealants into English, as well as for those who have to read such texts in the course of their work. It contains more than 18,000 words dealing with every aspect of paint, printing ink, adhesives and sealants technology. The subjects covered include polymer chemistry and technology, properties and testing, additives, equipment and applications. It is the first dictionary to treat this complex subject in depth and covers the latest developments.

The purpose of the dictionary is to give translators a reliable tool which will help them produce technically correct work. Many of the English terms have been researched and verified by reference to technical articles, books and company literature published in Britain and the United States. None of the English equivalents given have been taken from existing dictionaries or published translations of technical articles, specifications, recommendations and similar documents which are very often unreliable. Translation examples and technical explanations have been given where it was thought that they might be helpful.

There were instances where the meaning of a word was in doubt, or where a word could not be confirmed by reference to published literature. Here I was fortunate to have the help of a number of friends and colleagues in the industry. My special thanks go to John Bellchamber, Grant Edwards and Chris Thomlinson of the translation department of Hoechst UK Ltd., to Prof.Dr. Gerd Habenicht, Dr. John Haim and Herbert Stern who, between them, helped to solve and clarify a great many technical and linguistic problems.

Suggestions for amendments or improvements of the dictionary will be gratefully received and should be sent to me at 5 Campbell Croft, Edgware, Middlesex HA8 8DS, England. Suggested additional German words should be given in the original context if possible.

M.S. Welling

General information

Abbreviations used in this dictionary:

e.g. for example
i.e. that is
q.v. which see

Use of typefaces and symbols
Bold type has been used for German keywords and phrases, e.g. **festigkeitsbestimmend** and **die festigkeitsbestimmenden Größen der Klebstoffe wie z.B...**

Italics have been used
1. for comments and explanations, e.g. **Lufteintrag** air introduced *(e.g. into a paint by stirring)*
2. for explanations of words which cannot be translated literally, e.g. **sicherheitsrelevant** *important for safety*

Brackets have been used
1. to enclose English words, or parts thereof, which can be omitted without altering the meaning, e.g. (degree of) corrosion resistance, photoresist (material), rain(fall), primer coat(ing) etc.
2. to enclose explanatory English words printed in italics, e.g. **Fülle** 1. body *(of a paint)*. 2. build *(of a paint film)*

Oblique strokes have been used to indicate that any of the two words separated by the stroke can be used, e.g. jointing filler/compound, wood glaze/varnish/lacquer. Hyphenated words here count as one word.

Note:
Where several English equivalents are given for a German word, the order in which these are printed does **not** indicate an order of preference. Only the context - and the translator's approach to the subject - can determine which of several alternatives is the most suitable.

Sources of reference
The German keywords and phrases which form the basis of this dictionary were extracted from the technical and sales literature published by the major German and Swiss paint, adhesive and sealant manufacturers, German paint and adhesives journals and various German technical books. The English equivalents were verified, where necessary, by reference to the following books, technical journals and literature by the companies listed.

BOOKS
Berendsen, A.M., *Marine Painting Manual,* Graham and Trotman, London (1989)
Booth, K. (editor), *Industrial Packaging Adhesives,* Blackie, Glasgow (1990)
Brewis, D.M., D. Briggs (editors), *Industrial Adhesion Problems,* Orbital Press, Oxford (1985)
Calbo, L.J. (editor), *Handbook of Coatings Additives,* Marcel Dekker, New York (1992)
Cohen, E.D. and E.B. Gutoff (editors), *Modern Coating and Drying Technology,* VCH Publishers, New York (1992)
Crighton, A., *Adhesives in Packaging* Pira Information Centre, Leatherhead (1991)

Grainger, S. (editor), *Engineering Coatings - Design and Applications*, Abington Publishing, Cambridge (1989)
Hartshorn, S.R. (editor), *Structural Adhesives - Chemistry and Technology,* Plenum Press, New York (1986)
Hemingway, R.W., A.H. Corner and S.J. Branham (editors), *Adhesives from Renewable Resources,* American Chemical Society, Washington D.C. (1989)
Holmberg, K. and A. Matthews, *Coatings Tribology,* Elsevier, Amsterdam (1994)
Kinloch, A.J., *Adhesion and Adhesives - Science and Technology,* Chapman & Hall, London (1987)
Kinloch, A.J. (editor), *Structural Adhesives - Developments in Resins and Primers* Elsevier, London (1986)
Klosowski, J.M., *Sealants in Construction,* Marcel Dekker Inc. New York (1989)
Lambourne, R. (editor), *Paints and Surface Coatings: Theory and Practice,* Ellis Horwood, Chichester (1987)
Landrock, A.H., *Adhesives Technology Handbook,* Noyes Publications, New Jersey (1985)
Leach, R.H. (editor), *The Printing Ink Manual,* van Nostrand Reinhold, Wokingham (1988)
Lee, L.-H. (editor), *Adhesives, Sealants and Coatings for Space and Harsh Environments,* Plenum Press, New York (1988)
Lee, L.-H. (editor), *Adhesive Chemistry,* Plenum Press, New York (1984)
Lees, W.A., *Adhesives in Engineering Design,* The Design Council, London (1984)
Lehr, W.D., *Powder Coating Systems,* McGraw-Hill, New York (1991)
Misev, T.A., *Powder Coatings - Chemistry and Technology,* John Wiley, Chichester (1991)
Mittal, K.L., (editor), *Adhesive Joints - Formation, Characteristics and Testing,* Plenum Press, New York (1984)
Morgans, W.M., *Outlines of Paint Technology,* Edward Arnold, London (1990)
Morgans, W.M., *Resins and Varnishes for Paints and Inks,* Selection and Industrial Training Administration, Manchester (1982)
Munger, C.G., *Corrosion Prevention by Protective Coatings,* National Association of Corrosion Engineers, Houston, Texas (1984)
Panek, J.R. and J.P. Cook, *Construction Sealants and Additives,* John Wiley, New York (1991)
Parsons, P.(editor), *Surface Coatings: Raw Materials and their Usage,* Chapman and Hall, London (1993)
Patrick, R.L. (editor), *Treatise on Adhesion and Adhesives,* (vols. 1-6), Marcel Dekker, New York (1967-1089)
Paul, S., *Surface Coatings Science and Technology,* John Wiley, Chichester (1985)
Piron, D.L., *The Electrochemistry of Corrosion,* National Association of Corrosion Engineers, Houston, Texas (19910
Pizzi, A., K.L. Mittal (editors), *Handbook of Adhesive Technology,* Marcel Dekker, New York (1994)
Pizzi, A. (editor), *Wood Adhesives: Chemistry and Technology,* Marcel Dekker, New York (1989)
Sadek, M.M. (editor), *Industrial Applications of Adhesive Bonding,* Elsevier, London (1987)
Satas, D.(editor), *Coatings Technology Handbook,* Marcel Dekker Inc., New York (1991)
Schmid, E.V., *Exterior Durability of Organic Coatings,* FMJ International Publications, Redhill (1988)
Sharma, M.K. (editor), *Surface Phenomena and Additives in Water- Based Coatings and Printing Technology,* Plenum Press, New York (1991)
Shields, J., *Adhesives Handbook,* Butterworth, London (1984)
Skeist, I. (editor), *Handbook of Adhesives,* van Nostrand Reinhold, New York (1977)
Strafford, K.N., P.K. Datta, C.G. Googan (editors), *Coatings and Surface Treatment for Corrosion and Wear Resistance,* Ellis Horwood, Chichester (1984)
Strafford, K.N., P.K. Datta and J.S. Gray (editors), *Surface Engineering Practice: Processes, Fundamentals and Applications in Corrosion and Wear,* Ellis Horwood, New York (1990)
Suzuki, I., *Corrosion Resistant Coatings Technology,* Marcel Dekker Inc. New York (1989)
Tank, G.F., *Industrial Paint Finishing Techniques and Processes,* Ellis Horwood, New York (1991)
Wake, W.C. (editor), *Synthetic Adhesives and Sealants,* John Wiley, Chichester (1987)
Ward, J. (editor), *Thermoset Powder Coatings,* FMJ International Publications, Redhill (1989)
Wicks, Z.W., F.N. Jones and S.P. Pappas, *Organic Coatings, Science and Technology Vol.1: Film Formation, Components and Appearance* and *Vol.2: Applications, Properties and Performance,* John Wiley, New York (1992 and 1994)
Woodbridge, R. (editor), *Principles of Paint Formulation,* Blackie, Glasgow (1991)
Wu, Souheng, *Polymer Interface and Adhesion,* Marcel Dekker Inc. New York (1982)
Collins Dictionary of the English Language, Collins, London & Glasgow (1979)
The New Collins Thesaurus, Collins, London & Glasgow (1984)
The Concise Oxford Dictionary, Oxford University Press (1976)

TECHNICAL JOURNALS AND OTHER PUBLICATIONS

Adhesive Age
American Ink Maker
American Paint and Coatings Journal
Coatings Comet
Corrosion
Corrosion Science
Dyes and Pigments
European Adhesives and Sealants
European Polymer Paint Colour Journal
Focus on Pigments
Industrial Corrosion
Industrial Finishing
Ink and Print International
International Journal of Adhesion and Adhesives
Journal of Adhesion
Journal of Adhesion Science and Technology
Journal of Coatings Technology
Journal of Protective Coatings and Linings
Journal of the Oil and Colour Chemists' Association
Journal of Thermal Spray Technology
Journal of Waterborne Coatings
Modern Paints and Coatings
Oxidation of Metals
Paint and Ink International
Paint and Resin
Pigment and Resin Technology
Polymer Paints Colour Journal
Radnews
Review of Progress in Coloration and Related Topics
Surface Coating International
Waterborne and High-solids Coatings

LITERATURE PUBLISHED BY THE FOLLOWING COMPANIES

Allied Colloids Ltd.
Anchor Chemical Ltd.
Bactria Biocides Ltd.
BIP Speciality Resins Ltd.
Bohlin Instruments Ltd.
Brookhaven Instruments Ltd.
Ciba Polymers Ltd.
Cornelius Chemical Company Ltd.
Croda Colours Ltd.
Croda Resins Ltd.
Croxton + Garry Ltd.
Datacolour International Ltd.
Dunwood Polymer Services Ltd.
Elcometer Instruments Ltd.
Foscolor Ltd.
Glen Creston Ltd.
Harcros Durham Chemicals Ltd.
Hays Colours Ltd.
Johnson Matthey Ltd.
Petrochem UK Ltd.
Redland Minerals Ltd.
Scott Bader Co. Ltd.
Sheen Instruments Ltd.
Vinamul Ltd.
Yorkshire Chemicals Ltd.

A

AAS atom absorption spectroscopy
Abbau degradation
Abbau, biologischer biodegradation
abbaubar degradable
abbaubar, biologisch biodegradable
abbaubar, photochemisch photodegradable
Abbaubarkeit degradability
Abbaubarkeit, biologische biodegradability
abbauen 1. to degrade, to disintegrate, to break down. 2. to relieve, to dissipate *(e.g. stresses)*
Abbauerscheinungen signs of degradation
Abbaumechanismus degradation mechanism
Abbauprodukt degradation product
Abbauprozeß degradation process
Abbaurate degradation rate
Abbaureaktion degradation reaction
Abbauverhalten degradation behaviour
Abbeizpaste paint stripping paste
Abbindegeschwindigkeit setting speed
Abbindemechanismus setting/curing mechanism
abbinden to set, to harden *(adhesives, cement etc.)*
abbindend, hydraulisch hydraulic
abbindend, physikalisch physically setting
Abbindeverhalten setting characteristics
Abbindevermögen setting characteristics
Abbindezeit setting speed
Abbindung setting, curing
Abblätterungen peeling, flaking *(of paint film)*
Abbremsen braking
Abbröckeln crumbling, peeling, flaking *(of paint)*
Abdeckpapier release paper
abdestillieren to distil off
Abdichtmasse sealant (compound)
Abdichtung sealing
Abdunsten evaporation
Abdunstkinetik evaporation kinetics
Abdunstrate evaporation rate
Abdunstverhalten evaporation properties/characteristics
Abfall 1. waste. 2. decrease, reduction
Abfallbeseitigungsanlage waste disposal plant
Abfallentsorgung waste disposal
Abfallmenge (amount of) waste
Abfallminimierung waste minimisation
Abfallprodukt waste product
Abfallreduzierung waste reduction
Abfallvermeidung preventing waste
Abfallverminderung reducing the amount of waste produced
Abfallverwertung waste recycling
Abfallverwertung, thermische energy recovery/recycling
abfiltrieren to filter off
Abfüllen filling
Abgas flue/waste gas
Abgasreinigung waste/flue gas scrubbing
abgebunden set, cured
abgelöst detached *(e.g. paint film)*
abgesättigt saturated
abgeschieden 1. separated. 2. deposited *(e.g. a paint film on a metal surface by electrophoresis)*
abgeschrägt tapered
abgeschrägte einschnittige Überlappung bevelled/tapered lap joint
abgeschrägte Nut-Federverbindung scarf tongue and groove joint
abgeschrägte Überlappung tapered lap joint
abgeschrägte zweischnittige Laschung tapered double strap lap joint
abgeschrägte zweischnittige Überlappung bevelled double lap joint **abgeschwächte Totalreflexion** attenuated total reflection
abgesetzt 1. settled out. 2. recessed
abgesetzte einschnittige Laschung recessed single strap joint
abgesetzte Rohrsteckverbindung landed lap tubular joint
abgesetzte Überlappung double butt lap joint
abgesetzte Überlappung mit Abschrägung double scarf lap joint
abgesetzte zweischnittige Laschung recessed double strap joint
abgetönt tinted
abgewittert weathered
Abheben lifting *(of a paint film)*
Abietinsäure abietic acid
Abkühlbedingungen cooling conditions
Abkühldauer cooling time
abkühlen to cool down
Abkühlgeschwindigkeit cooling rate
Abkühlkurve cooling curve
Abkühlphase cooling phase
Abkühlprozeß cooling process
Abkühlspannungen cooling stresses
Abkühlung cooling
Abkühlungsverhältnisse cooling conditions
Abkühlvorgang cooling process
Abkühlzeit cooling time
ablagern to deposit
Ablagerung 1. deposit, sediment. 2. deposition
Ablaufbeständigkeit resistance to running/sagging

Ablaufdiagramm flow diagram
Ablaufeigenschaften running/sagging properties
ablaufen to run, to sag
ablauffest sag resistant
Ablaufneigung tendency to run/sag
Ablaufschema flow diagram/sheet
Ablaufsicherheit sag resistance
Ablaufstabilisator antisag agent
Ablaufverhalten sagging properties
Ablaufwiderstand sagging resistance
Ablenkfrequenz sweep frequency
ablösbar detachable
ablösen to lift, to become detached *(paint film)*
Ablösung lifting, detachment *(of paint film)*
Abluft waste/outgoing air, fumes
Ablüftdauer drying time
ablüften to allow to evaporate *(e.g. solvent)*
Ablüftezeit drying time, time required for solvent to evaporate
Abluftreinigung waste air purification
Ablüftungsbedingungen drying conditions
abmessen to measure out
Abmessungen dimensions, measurements
Abminderungsfaktor reducing factor
Abmischung 1. mixing, blending. 2. mix: **in Abmischng mit** mixed with
Abmusterung colour matching
Abnahme decrease
abnehmend decreasing
Abperleffekt water repellency
Abplatzfestigkeit resistance to spalling
Abplatzungen spalling
abrasiv abrasive
Abrasivität abrasiveness
Abrasivkörner abrasive particles
Abrasivkörper abrasive substance
Abrasivstoff abrasive
Abrasivverschleiß abrasive wear
Abrasivwirkung abrasive effect
abreiben 1. to rub (off). 2. to grind, to pass through the mill *(e.g. pigment paste)*
Abreicherung drop in concentration
Abreißversuch pull-off test
Abrieb abrasion
abriebarm abrasion resistant
abriebbeständig abrasion resistant
Abriebbeständigkeit abrasion resistance
Abriebeigenschaft abrasion resistance
abriebfest abrasion resistant
Abriebfestigkeit abrasion resistance
Abriebgeschwindigkeit abrasion/wear rate
Abriebprüfmaschine abrasion tester
Abriebprüfung abrasion test
Abriebspuren traces of abrasion
Abriebswert abrasion resistance
Abriebverhalten abrasion characteristics/resistance
Abriebwiderstand abrasion resistance
Abriebwirkung abrasive effect
Absacken curtaining, sagging
Absackerscheinungen curtaining, sagging
Abschälen peeling
Abschälfestigkeit peel strength
abschalten to switch off
Abschältest peel test
Abschätzung estimate
abscheidbar, elektrophoretisch capable of being deposited cataphoretically/electrophoretically/by electrodeposition
abscheidbar, kataphoretisch capable of being deposited electrophoretically/cataphoretically/by electrodeposition
Abscheidemethode method of separation
abscheiden 1. to separate. 2. to deposit *(e.g. a paint film on a metal surface by electrophoresis)*
Abscheidung 1. separation. 2. deposition *(e.g. of a paint film on a metal surface by electrophoresis)*
Abschirmeigenschaften screening/shielding properties
Abschirmmedium screening medium
Abschirmung screening, shielding
Abschirmwirkung screening/shielding effect
Absenkung lowering
Absetzbeständigkeit resistance to sedimentation/settling out
Absetzen settling out, sedimentation
Absetzerscheinungen settling out, (signs of) sedimentation
Absetzneigung tendency to settle (out)
Absetzverhalten sedimentation behaviour
Absetzverhinderungsmittel anti-settling agent
Absinken decrease, drop
Absolutdruck absolute pressure
Absolutmessung absolute determination
Absoluttemperatur absolute temperature
Absolutwert absolute value
absorbierend absorbent
absorbiert absorbed
Absorption absorption
Absorptionsband absorption band
Absorptionsfähigkeit absorptive capacity
Absorptionsfrequenz absorption frequency
Absorptionsgeschwindigkeit absorption rate
Absorptionskoeffizient absorption coefficient
Absorptionskurve absorption curve
Absorptionsmaximum maximum absorption
Absorptionspigment absorption pigment
Absorptionsrate absorption rate
Absorptionsspektroskopie absorption spectroscopy
Absorptionsspektrum absorption spectrum
Absorptionsverhalten absorption behaviour
Abspaltung separation *(e.g. of water or volatile compounds during a chemical reaction)*
Abspaltungsprodukt decomposition product
Abstoßung repulsion
Abstoßungskräfte repulsive forces
Abstrahlung radiation

Abtastkopf scanning head
Abtastlaserstrahl scanning laser beam
Abtastung scanning
Abtönen tinting
Abtönfarbe tinter
Abtönpaste tinting paste
Abtönsystem tinting system
Abtrag abrasion
Abtragtiefe depth of abrasion
Abtrennung separation
Abtrocknen 1. drying. 2. evaporation *(e.g. of solvent)*
Abwasser waste water, effluent
Abwasseraufbereitung sewage/effluent treatment/purification
Abwasserreinigungsanlage sewage purification plant
Abweichung deviation
Abwickelaggregat unwind unit
abwiegen to weigh out
abzentrifugieren to centrifuge off, to separate by centrifugation
abziehbar strippable
Abziehlack strip lacquer
Acetal acetal
Acetalcopolymerisat acetal copolymer
Acetaldehyd acetaldehyde
Acetalgruppe acetal group
Acetalharz acetal resin
Acetalisierung acetalisation
Acetalisierungsgrad degree of acetalisation
Acetalmischharz acetal copolymer
acetalvernetzt acetal cured
Acetat acetate
Acetation acetate ion
Acetessigsäure acetoacetic acid
Acetessigsäurerest acetoacetate group
Acetoacetatester acetoacetate ester
Acetoacetatgruppe acetoacetate group
acetoacetyliert acetoacetylated
Aceton acetone
acetonbeständig acetone resistant
Acetonitril acetonitrile
acetonlöslich acetone-soluble
Acetophenon acetophenone
Acetoxygruppe acetoxy group/radical
Acetylacetonperoxid acetylacetone peroxide
Acetylcellulose acetyl cellulose
Acetylen acetylene
Acetylenalkohol acetylene alcohol
Acetylengruppe acetylene group
Acetylensauerstoff oxyacetylene
Acetylgruppe acetyl group/radical
Acetylierung acetylation
Acrylamid acrylamide
Acrylamidcopolymer acrylamide copolymer
Acrylamidgruppe acrylamide group
Acrylamidmonomer acrylamide monomer
Acrylamidosulfonsäure acrylamidosulphonic acid
Acrylat acrylate
Acrylat-Polyolharz acrylate-polyol resin
acrylatbasiert acrylate-based
Acrylatbasis, auf acrylate-based
Acrylatcopolymer acrylic copolymer
Acrylatcopolymerisat acrylate copolymer
Acrylatdispersion acrylic dispersion
Acrylatdispersionskleber acrylic dispersion adhesive
Acrylatester acrylate, acrylic ester
Acrylatfarbe acrylic paint
Acrylatgruppe acrylic group
Acrylathaftkleber acrylic pressure sensitive adhesive
Acrylathaftkleberlösung acrylic pressure sensitive adhesive solution
Acrylathaftklebermonomer acrylic monomer for making pressure sensitive adhesives
Acrylatharz acrylic resin
Acrylatharzbasis, auf acrylic resin-based
Acrylatharzlack acrylic paint/enamel
Acrylatklebstoff acrylic adhesive
Acrylatlatexteilchen acrylic dispersion particle
acrylatmodifiziert acrylate-modified
Acrylatmonomer acrylic monomer
Acrylatoligomer acrylic oligomer
Acrylatpolymer acrylic polymer
Acrylatpolymerisat acrylic polymer, polyacrylate
Acrylatverdicker acrylic thickener
acrylbeschichtet acrylate-coated
Acrylbindemittel acrylic binder
Acrylcopolymer acrylic copolymer
Acrylcopolymerisat acrylic copolymer
Acryldecklack acrylic paint/enamel
Acryldispersion acrylic emulsion/dispersion
Acryldispersionsharz acrylic dispersion resin
Acrylemulsion acrylic emulsion/dispersion
Acrylester acrylate
Acrylesterbasis, auf acrylate-based
Acrylestercopolymerisat acrylic copolymer
Acrylesteremulsion acrylic emulsion
Acrylestergruppe acrylic group
Acrylesterverbindung acrylate/acrylic compound
Acryletheramid acrylic ether amide
Acrylglanzfarbe acrylic gloss paint
Acrylgruppe acrylic group
acrylhaltig containing acrylate
Acrylharz acrylic resin
Acrylharzaußenanstrich exterior acrylic finish
Acrylharzbindemittel acrylic binder
Acrylharzfarbe acrylic paint
Acrylharzkunststoff acrylic polymer
acrylharzmodifiziert acrylate-modified
Acrylharzpolymer acrylic polymer
acrylisch acrylic
Acrylklarlack clear acrylic lacquer
Acrylklebstoff acrylic adhesive
Acryllack acrylic paint/enamel
Acryllackschicht acrylic coating/film
Acryllatex acrylic dispersion/emulsion
Acrylmonomer acrylic monomer
Acrylnitril acrylonitrile

Acrylnitrilcopolymerisat acrylonitrile copolymer
Acrylnitrilgehalt acrylonitrile content
Acrylnitrilgruppe acrylonitrile group
Acrylnitrilrestgehalt residual acrylonitrile content
Acryloligomer acrylic/acrylate oligomer
Acryloylgruppe acryloyl group
Acrylpolymer polyacrylate
Acrylrest acrylic group
Acrylsäure acrylic acid
Acrylsäurealkylester alkyl acrylate
Acrylsäurebutylester butyl acrylate
Acrylsäureester acrylate, acrylic acid ester
Acrylsäureethylester ethyl acrylate
Acrylsäureharz acrylic resin
Acrylsäuremethylester methyl acrylate
Acrylschicht acrylic film/coating
Acrylsilanharz acrylic silane resin
Acrylverbindung acrylic compound
Acrylverdicker acrylic thickener
acyclisch acyclic
acyliert acylated
Acylierung acylation
Acylisocyanate acyl isocyanate
Acylphosphinoxid acyl phosphine oxide
Acylphosphinsäure acylphosphinic acid
Acylurethan acyl urethane
Addition addition
Additionsreaktion addition reaction
Additionsvernetzung addition crosslinking/curing
Additiv additive
additiventhaltend containing additives
additivfrei free from additives
Addukt adduct
Adduktbindemittel adduct binder
Adduktharz adduct resin
Adhäsion adhesion
Adhäsionsadditiv adhesion promoting agent/additive
Adhäsionsarbeit adhesive energy
Adhäsionsart type of adhesion
Adhäsionsbereich glueline
Adhäsionsbruch adhesive failure/fracture
Adhäsionseffekt adhesive effect
Adhäsionseigenschaften adhesive properties
adhäsionserhöhend adhesion-improving/-enhancing
Adhäsionsfestigkeit adhesive strength
adhäsionshemmend adhesion-inhibiting
Adhäsionsklebung adhesive bonding
Adhäsionskraft adhesive force
Adhäsionsmechanismus adhesion mechanism
Adhäsionsprozeß adhesive process
Adhäsionsriß adhesive failure
Adhäsionstheorie theory of adhesion
Adhäsionsverhalten adhesive properties
Adhäsionsversagen adhesive failure
Adhäsionswert adhesion
adhäsiv adhesive
Adhäsivbruch adhesive failure
Adhäsivfestigkeit adhesive/bond strength
Adhäsivversagen adhesive failure
Adhäsivzusammensetzung adhesive composition
ADI-Wert acceptable daily intake, ADI
Adipinsäure adipic acid
Adipinsäureester adipate
Adsorbens adsorbent
Adsorber adsorption unit
adsorbierend adsorbent
adsorbiert adsorbed
Adsorption adsorption
Adsorptions-Desorptionsgleichgewicht adsorption-desorption equilibrium
Adsorptionsenthalpie adsorption enthalpy
Adsorptionsfähigkeit adsorptive capacity
Adsorptionsgleichgewicht adsorption equilibrium
Adsorptionsgrad degree of adsorption
Adsorptionsisotherme adsorption isotherm
Adsorptionskapazität adsorptive capacity
Adsorptionskoeffizient adsorption coefficient
Adsorptionsprozeß adsorption process
Adsorptionsrate adsorption rate
Adsorptionsschicht adsorption layer
Adsorptionsverhalten adsorption behaviour/characteristics
Adsorptionsvorgang adsorption process
Adsorptionswärme heat of adsorption
aerob aerobic
Aerosol 1. aerosol. 2. mist
Aerosolbehälter aerosol (container)
Aerosoldose aerosol can
Aerosoldosenlack aerosol paint
Aerosollack aerosol paint
Aerosolverpackung aerosol can
Affinität affinity
Agar agar
Agglomerat agglomerate
Agglomeration agglomeration
Agglomerationsgrad degree of agglomeration
agglomerieren to agglomerate
Agglomerierung agglomeration
Aggregat 1. aggregate. 2. unit *(of machinery or equipment)*
Aggregatzustand state of aggregation
aggressiv aggressive, corrosive
Aggressivität aggressiveness, corrosiveness
Airless-Spritzen airless spraying
Airless-Spritzlackieren airless spraying
Airlessverarbeitung airless spraying
AKM *abbr. of* **Alkyd-Melaminharz**, alkyd-melamine resin
AKM-Beschichtung alkyd-melamine coating
aktinisch actinic
aktiv active
Aktivator activator
Aktivatorzusatz addition of activator
aktivierbar capable of being activated
aktiviert activated
Aktivierung activation
Aktivierungsenergie activation energy
Aktivierungsmittel activating agent

Aktivität activity
Aktivitätskoeffizient activity coefficient
Aktivkohle activated charcoal
Aktivkohlebett activated charcoal layer
Aktivkohlefilter activated charcoal filter
Aktivpigment reactive pigment
Aktivsauerstoff active oxygen
Aktivsauerstoffgehalt active oxygen content
Aktivsubstanz active substance
aktuell topical
akustisch acoustic
akusto-optisch opto-acoustic
akut acute
Akzeptanz acceptance
Al-Legierung aluminium alloy
Alanin alanine
Aldehyd aldehyde
Aldehydgruppe aldehyde group
Aldimin aldimine
Aldoximogruppe aldoximo group/radical
Algen algae
Algenbewuchs algal fouling
Alginat alginate
algizid algicidal
Algizid algicide
alicyclisch alicyclic
Aliphat aliphatic solvent
aliphatisch aliphatic
Alizarin alizarin
Alkali alkali
alkaliadsorbierend alkali adsorbent
Alkalialkoholat alkali alcoholate
Alkalibelastung exposure to alkaline conditions
...wenn keine allzu großen Alkalibelastungen der Untergünde vorherrschen ...if the surfaces are not too alkaline
alkalibeständig alkali resistant
Alkalibeständigkeit alkali resistance
alkaliempfindlich alkali-sensitive, sensitive to alkalis
alkalifrei alkali-free
Alkalihydroxid alkali hydroxide
Alkaliion alkali ion
alkalilöslich alkali soluble
Alkalimetall alkali metal
alkaliresistent alkali resistant, resistant to alkalis
Alkaliresistenz alkali resistance
alkalisch alkaline
alkalisch härtend alkali-curing
Alkalistabilität alkali resistance
Alkalisulfit alkali sulphite
Alkalität alkalinity
alkaliunlöslich alkali-insoluble
Alkan alkane
Alkanolamin alkanolamine
Alkansulfonsäure alkanesulphonic acid
Alkohol alcohol
Alkoholabspaltung separation of alcohol
Alkoholat alcoholate
alkoholbeständig alcohol resistant
Alkoholbeständigkeit alcohol resistance
alkoholfrei non-alcoholic
alkoholgelöst dissolved in alcohol
alkoholhaltig containing alcohol
alkoholisch alcoholic
Alkoholkette alcohol chain
alkohollöslich alcohol soluble
Alkoholverträglichkeit compatibility with alcohol
Alkoholyse alcoholysis
alkoholysiert alcoholysed
Alkoxyalkylgruppe alkoxyalkyl group
alkoxyfunktionell alkoxyfunctional
Alkoxygruppe alkoxy group
Alkoxymethylengruppe alkoxymethylene group
Alkoxymethylgruppe alkoxymethyl group
Alkoxymethylol alkoxymethylol
Alkoxyradikal alkoxy radical/group
Alkoxyrest alkoxy radical/group
Alkoxysilan alkoxysilane
Alkoxysilanpropylester alkoxysilane propyl ester
Alkoxysilylamin alkoxysilylamine
Alkoxysilylgruppe alkoxysilyl group
Alkoxythiophen alkoxythiophen
Alkyd alkyd
Alkyd-Melaminharz alkyd-melamine resin
Alkydacrylat alkyd acrylate
Alkydbindemittel alkyd binder (resin)
Alkydeinbrennlack alkyd stoving paint/enamel
Alkydemulsionsfarbe alkyd emulsion paint
Alkydestergruppe alkyd ester group
Alkydharz alkyd resin
Alkydharz, fettes long-oil alkyd (resin)
Alkydharz, mittelfettes medium-oil alkyd (resin)
Alkydharzanstrich alkyd finish
Alkydharzdeckanstrich alkyd finish
Alkydharzdeckbeschichtung alkyd finish
Alkydharzdeckbeschichtungsstoff alkyd paint
Alkydharzdecklack alkyd topcoat/finish
Alkydharzemulsion alkyd emulsion
Alkydharzfarbe alkyd paint
Alkydharzgrundierung 1. alkyd primer. 2. alkyd primer coat
Alkydharzklarlack clear alkyd varnish/lacquer
Alkydharzlack alkyd paint
Alkydharzlackierung alkyd finish
Alkydharzlösung alkyd resin solution
Alkydharzwasserlack water-based alkyd paint
Alkydlack alkyd paint
Alkydlasur alkyd glaze
Alkydsystem alkyd system/formulation
Alkylacetat alkyl acetate
Alkylalkohol alkyl alcohol
Alkylalkoxysilan alkylalkoxysilane
Alkylamin alkylamine
Alkylaminogruppe alkylamino group
Alkylammoniumsalz alkyl ammonium salt
Alkylaromat aromatic alkyl compound
Alkylarylsulfonat alkyl aryl sulphonate
Alkylbenzol alkyl benzene
Alkylbenzylsulfonat alkyl benzyl sulphonate
Alkylbernsteinsäure alkyl succinic acid

Alkylbromid alkyl bromide
Alkylcarbonsäure alkyl carboxylic acid
Alkylderivat alkyl derivative
Alkyldiamin alkyl diamine
Alkylenglykol alkylene glycol
Alkylenkette alkylene chain
Alkylenoxid alkylene oxide
Alkylester alkyl ester
Alkylgruppe alkyl group
Alkylhydroperoxid alkyl hydroperoxide
alkyliert alkylated
Alkylkette alkyl chain
Alkylkomponente alkyl component
Alkylmethylpolysiloxan alkyl methyl polysiloxane
Alkylolverbindung alkylol compound
Alkylorganosilan alkyl organosilane
Alkylperester alkyl per-ester
Alkylperoxidradikal alkylperoxy radical
Alkylphenol alkyl phenol
Alkylphenolharz alkyl phenol resin
Alkylphenolpolyglykolether alkylphenol polyglycol ether
Alkylphenoxyrest alkyl phenoxy radical/group
Alkylphosphit alkyl phosphite
Alkylpolysiloxanpolymer alkylpolysiloxane polymer
Alkylpolysulfid alkyl polysulphide
Alkylrest alkyl group
Alkylseitenkette alkyl side chain
Alkylsilan alkyl silane
Alkylsiliconharz alkyl silicone resin
Alkylsubstituent alkyl substituent
Alkylsulfid alkyl sulphide
Alkylsulfonat alkyl sulphonate
Alkylsulfonsäureester alkyl sulphonate
Alkyltrialkoxysilan alkyl trialkoxysilane
Alkyltrichlorsilan alkyl trichlorosilane
Alkylverbindung alkyl compound
Alkylvinylether alkyl vinyl ether
Alkylvinylpolysilan alkyl vinyl polysilane
Alkylzinncarboxylat alkyl tin carboxylate
Alleinbindemittel sole binder
Alleinweichmacher sole plasticiser
Allergie allergy
allergieneutral *not producing an allergic reaction*
Allergiker allergic person
Alleskleber general-purpose adhesive
Allgemeineigenschaften general properties
Allophanat allophanate
Allophanatbindung allophanate linkage
Allophanatbindungsanteil allophanate linkage content
Allophanatstruktur allophanate structure
Allylalkohol allyl alcohol
Allylamin allylamine
Allylchlorid allyl chloride
Allyldiolether allyldiol ether
Allylester allyl ester
Allylether allyl ether
Allylgruppe allyl group
Allylharz allyl resin
allyliert allylated
Allylmonomer allyl monomer
Allylverbindung allyl compound
Allzweckkautschuk general purpose rubber
Allzweckvernetzungsmittel general purpose catalyst
Altanstrich old paint film
Altbausanierung restoration/refurbishment of old buildings
Altbauten old buildings
Altbeton old concrete
alternierend alternating
Alterung ageing
Alterung, künstliche articifial ageing
Alterung, thermische heat ageing
Alterungsanfälligkeit susceptibility to ageing
Alterungsauswirkungen effects of ageing
Alterungsbeanspruchung ageing
alterungsbedingt due to ageing
Alterungsbedingungen ageing conditions
alterungsbeständig ageing resistant
Alterungsbeständigkeit ageing resistance
Alterungseigenschaften ageing properties/characteristics
Alterungseinfluß effect of ageing
alterungsempfindlich susceptible to ageing
Alterungserscheinungen signs of ageing
Alterungsgeschwindigkeit ageing rate/speed
Alterungskennwert ageing properties *(for translation example see under* ***Alterungswerte****)*
Alterungsmodul ageing modulus
Alterungsprozeß ageing process
Alterungsrißbildung cracking due to ageing
Alterungsschutzmittel antioxidant
Alterungstemperatur ageing temperature
Alterungsuntersuchung ageing test
Alterungsursache cause of ageing
Alterungsverhalten ageing properties
Alterungsvorgang ageing process
Alterungswerte ageing properties: **verbesserte Alterungswerte** improved ageing resistance
Alterungszeit ageing period
Altgummi waste rubber
Altlasten waste residues
Altlastensanierung removing/cleaning up waste residues
Aluminium aluminium
Aluminiumacetylacetonat aluminium acetyl acetonate
Aluminiumammoniumsulfat aluminium ammonium sulphate
Aluminiumblech aluminium sheet
Aluminiumbronzepulver aluminium bronze powder
Aluminiumchelat aluminium chelate
Aluminiumchlorid aluminium chloride
Aluminiumguß cast aluminium
Aluminiumhydroxid aluminium hydroxide
Aluminiumion aluminium ion
Aluminiumklebung 1. bonding of aluminium. 2.

bonded aluminium joint
Aluminiumlackierung 1. aluminium finish. 2. painting of aluminium
Aluminiumlegierung aluminium alloy
Aluminiumnitrat aluminium nitrate
Aluminiumoberfläche aluminium surface
aluminiumorganisch organo-aluminium
Aluminiumoxid aluminium oxide, alumina
Aluminiumphosphat aluminium phosphate
Aluminiumpigment aluminium pigment
Aluminiumsilikat aluminium silicate
Aluminiumstearat aluminium stearate
Aluminiumsulfat aluminium sulphate
Aluminiumtriphosphat aluminium triphosphate
Aluminiumtripolyphosphat aluminium tripolyphosphate
Aluminiumverbindung aluminium compound
Aluminiumverklebung 1. bonding of aluminium. 2. bonded aluminium joint
Aluminiumwerkstoff aluminium material
Alupigment aluminium pigment
Aluplatte aluminium sheet
Ameisensäure formic acid
Amidgruppe amide group
Amidharz amide resin
Amidin amidine
Amin amine
Aminacrylat aminoacrylate
Aminaddukt amine adduct
Aminaldehydharz aminoaldehyde resin
Aminalkylsilan aminoalkyl silane
Aminanteil amine content
Aminbeschleuniger amine accelerator
aminblockiert amine-blocked
Aminepoxidharz aminoepoxy-resin
aminfunktionell aminofunctional
Amingehalt amine content
amingehärtet amine-cured
Amingruppe amino group
Aminhärter amine hardener/catalyst
Aminhärtung amine cure
Aminharz amino resin
Aminhydrochlorid aminohydrochloride
aminkatalysiert amine-catalysed
aminisch amine-type
Aminkatalysator amine catalyst
Aminlichtschutzmittel amine light stabiliser
aminmodifiziert amine-modified
Aminoacrylat aminoacrylate
Aminoacrylatcopolymerisat aminoacrylate copolymer
Aminoacrylester aminoacrylic ester
Aminoalkohol aminoalcohol
Aminoalkylacrylat aminoalkyl acrylate
Aminoalkylester aminoalkyl ester
Aminoamid aminoamide
Aminobenzoesäure aminobenzoic acid
Aminocapronsäure aminocaproic acid
Aminocarbonsäure aminocarboxylic acid
Aminocarboxylat aminocarboxylate
Aminochinolin aminoquinoline
Aminodiphenylamin aminodiphenylamine
Aminoendgruppe terminal amino group
Aminoessigsäure aminoacetic acid
aminofunktionell aminofunctional
Aminogruppe amino group
Aminoharz amino resin
Aminophenol aminophenol
Aminoplast aminoplastic
Aminoplastharz amino resin
Aminopolyether aminopolyether
Aminopropanol aminopropanol
Aminopropylgruppe aminopropyl group
Aminorest amino radical
Aminosäure aminoacid
Aminosäurebaustein aminoacid unit
Aminosilan aminosilane
Aminourethanharz aminourethane resin
Aminoverbindung amino compound
Aminsalz amine salt
Aminurethane aminourethane
aminvernetzend amine-curing
Aminvernetzer amine curing agent
Aminzahl amine value
Ammoniak ammonia
Ammoniakplasma ammonia plasma
Ammoniakwasser ammonia
Ammoniumacrylat ammonium acrylate
Ammoniumbenzoat ammonium benzoate
Ammoniumgruppe ammonium group
Ammoniumnitrat ammonium nitrate
Ammoniumpolyphosphat ammonium polyphosphate
Ammoniumsalz ammonium salt
Ammoniumverbindung ammonium compound
amorph amorphous
amphiphil amphiphilic
amphoionisch amphoionic
Amphotensid amphosurfactant
amphoter amphoteric
Amphoterharz amphoteric resin
amphoterisch amphoteric
Amplitude amplitude
Amplitudendämpfung amplitude damping
Amplitudenverhältnis amplitude ratio
Amplitudenverteilung amplitude distribution
amtlich zugelassen officially approved
amtliche Prüfanstalt official test establishment
amtliche Zulassung official approval/authorisation
anaerob anaerobic
Analyse analysis
Analyse, dynamisch-mechanische dynamic-mechanical analysis
Analyse, dynamisch-thermomechanische dynamic-thermomechanical analysis
Analyse, thermische thermoanalysis
Analyse, thermogravimetrische thermogravimetric analysis
Analysedaten analytical data
Analysemethode analytical method/technique
Analysenergebnis analytical result

Analysenverfahren analytical procedure/technique
Analysenwaage analytical balance
Analysenqualität analytical grade
Analytik analysis
Analytiker analyst
analytisch analytical
anaphoretisch anaphoretic
Anatas anatase
Anätzen etching
anfällig susceptible
Anfälligkeit susceptibility
Anfangsfarbstärke initial/original colour intensity
Anfangsfestigkeit initial/original strength
Anfangsfeuchte initial/original moisture content
Anfangsflexibilität initial/original flexibility
Anfangsglanz initial/original gloss
Anfangshaftung initial/original adhesion
Anfangsphase initial stage
Anfangstack initial tack
Anfangstemperatur initial/starting temperature
Anfangstrocknung initial drying
Anfangsviskosität initial/original viscosity
Anfangsvolumen initial/original volume
Anfangswert initial/original figure
Anfangszustand original state/condition
anfärben to colour
Anforderung demand, requirement
Anforderungskatalog list of requirements
Anforderungsprofil range/list of requirements
angebrochen partly used *(e.g. a tin of paint)*
angefeuchtet moistened
angelöst partly dissolved
angereichert enriched
angesäuert acidified
angetrocknet partly dried
Angriff attack
Anhaftfestigkeit adhesion
Anhaltswert guide value
Anhydrid anhydride
Anhydridanteil anhydride content
Anhydridbasis, auf anhydride-based
anhydridgehärtet anhydride-cured
Anhydridgruppe anhydride group
Anhydridhärtung anhydride cure
Anhydridharz anhydride resin
anhydridvernetzend anhydride-curing
Anhydridvernetzung anhydride cure
Anilin aniline
Anilinpunkt aniline point
Anion anion
anionaktiv anionic
anionisch anionic
anionogen anionic
anisotrop anisotropic
Anisotropie anisotropy
Ankerrührer anchor-type mixer/stirrer
ankondensiert condensed into the molecule
Anlage plant
Anlagenbetreiber plant operator
Anlagenelement (machine) unit
Anlagenkapazität plant capacity
Anlagenkonzept 1. plant design and lay-out. 2. plant, production line
Anlagenprogramm range of machines/equipment
Anlagenspektrum range of machines/equipment
Anlagerung addition *(in a chemical context)*
Anlieferungszustand, im as supplied
anlösen to partly dissolve
Anmachwasser mixing water
Anode anode
anodisch anodic
anodische Tauchlackierung anodic electrodeposition painting
anodisiert anodised
Anodisierung anodisation
anorganisch inorganic
Anpreßdruck contact pressure
anquellbar swellable
Anquellen swelling
Anreibeaggregat rolls, roll mill
Anreibebindemittel pigment paste binder
Anreibeharz pigment paste resin
Anreibung pigment dispersion
Anreicherung enrichment
Anriß initial crack
Anrißschälwiderstand initial peel strength
Ansatz 1. mix, batch. 2. formula, (mathematical) equation
Ansatzgröße batch size
ansäuern to acidify
Ansäuerung acidification
Anschlußfuge connecting joint
Anschmelzen partial melting
Anschmutzbarkeit tendency to pick up dirt
Anstieg increase, rise
Anstrich 1. coat *(of paint, varnish etc.)*, paint film. 2. paint *(not strictly correct, but often used)*
Anstrichaufbau paint film structure
Anstrichbindemittel paint binder
Anstrichdefekt paint film defect
Anstricheigenschaften paint film properties
Anstrichfarbe paint
Anstrichfarbenzusatz paint additive
Anstrichfilm paint film
Anstrichformulierung paint formulation
Anstrichgrund substrate
Anstrichlack paint
Anstrichmasse paint, coating compound
Anstrichmaterial paint
Anstrichmittel paint
Anstrichmittelsystem paint
Anstrichoberfläche paint film surface
Anstrichschäden paint film defects
Anstrichsektor paint sector/industry, coatings industry
Anstrichstoff paint
Anstrichsystem paint
anstrichtechnisch *adjective relating to paint films*
  anstrichtechnische Eigenschaften paint film properties

Anstrichträger substrate, painted surface
Anstrichuntergrund substrate, surface to be painted
Anteil content, amount **flüchtige Anteile** volatiles
Anthranilsäure anthranilic acid, 2-aminobenzoic acid
Antiablaufmittel sag control agent
Antiabsetzeigenschaften anti-settling properties
Antiabsetzmittel anti-sedimentation agent
Antiabsetzverhalten anti-sedimentation characteristics
Antiabsetzwirkung anti-settling effect
antiadhäsiv non-stick
Antialgicid algicide
Antiausschwimmittel anti-flotation agent
antibakteriell anti-bacterial
Antibewuchsanstrich anti-fouling paint
Antibewuchseigenschaften anti-fouling properties
Antibewuchsfarbe anti-fouling paint/composition
Antibewuchszusatz anti-fouling additive
Antiblockeffekt anti-blocking effect
Antiblockeigenschaften anti-blocking properties
Antiblockmittel anti-blocking agent
Antiblocksystem anti-blocking agent
Antiblockwirkung anti-blocking effect
Antidegradans anti-oxidant/-degradant
Antidröhnbeschichtung anti-resonance coating
Antidröhnmasse anti-resonance/resonance-deadening compound
Antidröhnplatte anti-resonance/resonance-deadening panel
Antifoulingagens anti-fouling agent
Antifoulinganstrich anti-fouling coating/finish
Antifoulinganstrichstoff anti-fouling paint/composition
Antifoulingausrüstung anti-fouling finish
Antifoulingbeschichtung antifouling finish
Antifoulingfarbe anti-fouling composition/paint
Antifoulingmittel anti-fouling agent
Antifoulingsystem anti-fouling paint/composition
Antifoulingwirkung anti-fouling effect
Antifungizid fungicide
Antihaftbelag non-stick coating
Antihaftbeschichtung non-stick finish/coating
Antihaftbeschichtungsmaterial non-stick coating compound
Antihafteffekt non-stick effect
Antihafteigenschaften non-stick properties
Antihaftfähigkeit non-stick properties
Antihaftmittel non-stick agent
Antihaftschicht non-stick coating
Antihaftüberzug non-stick coating
Antihaftwirkung non-stick effect
Antihautmittel anti-skinning agent
Antiklebeffekt non-stick effect
Antiklebewirkung non-stick effect
Antikorrosionseigenschaften anti-corrosive properties
Antikorrosionspigment anti-corrosive pigment
Antikorrosionsschicht anti-corrosive coating/film
antikorrosiv anti-corrosive, corrosion resistant
Antikratermittel anti-cratering agent
Antikratzschicht anti-mar finish
Antikratzüberzug anti-scratch finish, anti-mar coating
antimikrobiell antimicrobial
Antimonchlorid antimony chloride
Antimonoxid antimony oxide
Antimontrioxid antimony trioxide
Antioxidans antioxidant
Antioxidationsmittel antioxidant
Antioxidationssystem antioxidant blend
antioxidativ anti-oxidative
Antiozonans anti-ozonant
Antirutschausrüstung non-slip/non-skid finish/coating
Antirutschbeschichtung non-slip/non-skid coating
Antirutschlack non-slip/non-skid paint
Antischaumagens antifoam
Antischaummittel antifoam
Antischimmelmittel fungicide
Antislipmittel non-slip agent
Antistatiklack antistatic varnish/lacquer
Antistatikschicht antistatic coating
Antistatiksystem antistatic agent
Antistatikum antistatic (agent)
Antistatikzusatz antistatic agent/additive
antistatisch antistatic
antrocknen to surface-dry, to become touch dry
Antrocknung (min) touch-dry after (min)
Anwendungsbandbreite range of applications
Anwendungsbedingungen service conditions
Anwendungsbeispiele application examples, typical applications
Anwendungsbereich range of applications
anwendungsbezogen application-related
Anwendungsbreite range of applications
anwendungsfertig ready-to-use
Anwendungsgebiet application area, field of application
Anwendungsgrenztemperatur maximum service/operating temperature
Anwendungsmöglichkeiten application possibilities, possible applications
anwendungsorientiert application-related
Anwendungspalette range of applications/uses
Anwendungsrichtlinien application guidelines
Anwendungsschwerpunkte main/principal applications, most important applications
Anwendungsspektrum range of applications/uses
anwendungsspezifisch application-oriented
Anwendungstechnik 1. application research. 2. application research department. 3. application technology
anwendungstechnisch applicational, application-oriented
Anwendungstemperatur service/working/operating temperature
Anwendungstemperaturgrenze maximum

service/working/operating temperature
Anydridhärter anhydride hardener
Anziehungskräfte forces of attraction
Anzugsgeschwindigkeit curing/setting speed
Apparatebau plant construction
apparativ *adjective relating to equipment*
apparativer Aufwand amount of equipment needed
Apparatur apparatus, equipment
Applikationsanlage 1. paint application unit. 2. adhesive application unit
Applikationsbedingungen application conditions
Applikationseigenschaften application properties/characteristics
Applikationsmethode method of application
Applikationsmethodik method of application
Applikationstechnik method of application
Applikationsverfahren method of application
Applikationsverhalten application properties
Applikationsviskosität application consistency
Applizierung application
äquimolar equimolecular
äquivalent equivalent
Äquivalentgewicht equivalent weight
Arbeiter manual/blue-collar worker
Arbeiterbelegschaft workforce
Arbeitgeber employer
Arbeitnehmer employee
Arbeitnehmer, gewerbliche blue-collar workers
Arbeitsablauf 1. process. 2. process sequence. 3. (operational) procedure
Arbeitsaufnahme energy input/absorption
Arbeitsaufnahmevermögen energy-absorbing capacity
Arbeitsaufwand 1. amount of work (involved). 2. labour costs
arbeitsaufwendig labour-intensive
Arbeitsbedingungen working conditions
Arbeitsfeld field of activity
Arbeitsfolge working/operating sequence
Arbeitsgang operation
Arbeitsgangfolge working/operating sequence
Arbeitsgemeinschaft working party
Arbeitsgerät tool, implement
Arbeitsgeschwindigkeit working/operating speed
Arbeitsgruppe working party
Arbeitshygiene workshop/factory hygiene
arbeitshygienisch *adjective relating to factory or workshop hygiene*
Arbeitsinhalt energy content
arbeitsintensiv labour-intensive
Arbeitskreis working party, study group
Arbeitsleistung 1. performance, efficiency, capacity *(of a machine)*. 2. output *(of power)*. 3. input *(of power)*. 4. rating *(electrical)*. 5. energy, power
Arbeitsplatz workplace
Arbeitsplatzbelastung workplace pollution
Arbeitsplatzkonzentration, maximale maximum allowable concentration, MAC
Arbeitsprinzip operating principle
Arbeitsproduktivität productivity
Arbeitsprozeß process, operation
Arbeitsradius operating radius
Arbeitsraum workroom
Arbeitsschutz works/industrial safety
Arbeitsschutzauflage works safety directive
Arbeitsschutzbekleidung industrial protective clothing
Arbeitsschutzbrille (safety) goggles
Arbeitsschutzkleidung industrial protective clothing
Arbeitsschutzmaßnahmen works safety precautions
Arbeitsschutzrichtlinie works safety guideline
Arbeitssicherheit safety at work
Arbeitsstoff substance, material
Arbeitsstoffverordnung dangerous/hazardous substances directive
Arbeitstag working day
Arbeitstechnik working/production technique
Arbeitstemperatur working/operating temperature
Arbeitsvorgänge operations
Arbeitsweise operation, method of operation
ArbStoffV *abbr. of* **Arbeitsstoffverordnung**, dangerous/hazardous substances directive
Architekturfarbe architectural paint
Argon argon
Argonlaser argon laser
Argonplasma argon plasma
Armaturen fittings
armiert reinforced
Aromat 1. aromatic solvent. 2. aromatic compound
aromatenarm with a low content of aromatic solvents/compounds
aromatenfrei free from aromatic solvents/compounds
Aromatengehalt aromatic content
Aromatengemisch 1. aromatic solvent blend. 2. mixture of aromatic compounds
aromatenreich 1. with a high aromatic solvent content. 2. containing a high percentage of aromatic compounds
Aromatenrest aromatic group
aromatensubstituiert aromatic-substituted
aromatisch aromatic
arylalkylmodifiziert aryl alkyl-modified
Arylalkylphosphat aryl alkyl phosphate
Arylalkylverbindung aryl alkyl compound
Arylcarbonsäure aryl carboxylic acid
Arylgruppe aryl group/radical
Arylphenol aryl phenol
Arylradikal aryl group/radical
Arylrest aryl group/radical
arylsubstituiert aryl-substituted
Arylverbindung aryl compound
Asbest asbestos
Asbestersatz asbestos substitute
Asbestfaser asbestos fibre

asbestfrei asbestos-free
Asbestgewebe asbestos cloth
Asbestplatte asbestos sheet
Asbestzement asbestos cement
Asbestzementplatte asbestos-cement sheet
Asche ash
aschearm with a low ash content
Aschegehalt ash content
Asphalt asphalt
Assoziation association
assoziativ associative
Assoziativverdicker associative thickener
ästhetisch aesthetic
Asthmaanfall attack of asthma
asymmetrisch asymmetric
asymptotisch asymptotic
Atembeschwerden breathing difficulties/problems
Atemmaske face mask
Atemschutz breathing apparatus, respirator
Atemschutzgerät breathing apparatus
Atemschutzhaube smoke hood
Atemschutzmaske face mask
Atemwege airways
ATL *abbr. of* **anodische Tauchlackierung**, anodic electrodeposition painting
Atmosphäre atmosphere
Atmosphärenbedingungen atmospheric conditions
Atmosphärendruck atmospheric pressure
atmosphäril atmospheric
Atmosphärilien atmospheric influences
atmosphärisch atmospheric
atmungsaktiv breathable
Atmungsaktivität breathability
Atmungsorgane respiratory organs
Atom atom
Atomabsorptionsspektroskopie atom absorption spectroscopy
Atomabstand atomic distance
atomar atomic
Atombindung atomic bond
Atomgewicht atomic weight
Atomgruppe atomic grouping
Atomschale atomic shell
Atomverknüpfung atomic linkage
atoxisch non-toxic
ATR attenuated total reflection
ATR-FTIR-Untersuchung ATR-FTIR test, attenuated total reflection Fourier transform infrared spectroscopic test
Ätzbeständigkeit corrosion resistance
Ätzen to pickle, to etch
Ätzmedium pickling solution
Ätzprozeß pickling process
Aufbau, chemischer chemical constitution
Aufbau construction
aufgedampft coated by vapour deposition *(e.g. a thin metal film)*
aufgehellt brightened
aufgeladen charged
aufgepfropft grafted
aufgerakelt applied by knife
aufgetrocknet dried
Aufheizen heating up
Aufheizgeschwindigkeit heating-up rate
Aufheizkurve heating-up curve
Aufheizrate heating-up rate
Aufheizzeit heating-up period
Aufheller brightener
Aufhellung brightening, lightening
Aufhellvermögen brightening power
Aufkonzentrierung concentration *(by evaporation)*
aufladbar chargeable
Aufladbarkeit, elektrostatische electrostatic chargeability
Aufladung charge
Auflage 1. directive, requirement. 2. support. 3. coating
auflösbar soluble
auflösen to dissolve
Auflösung resolution
Auflösungsvermögen resolving power
aufpolymerisiert grafted
Aufrauhen abrading, roughening
Aufschäumung foam height
Aufschlämmung slurry
aufschmelzbar meltable, fusible
Aufschmelzen melting
Aufschmelzklebstoffmasse hot melt adhesive
Aufschwimmen flooding
Aufstieg rise
Auftrag 1. job, project. 2. application *(e.g. of paint or adhesive)*
Auftragsaggregat applicator *(e.g. for adhesives)*
Auftragsart method of application
Auftragsgerät applicator
Auftragsgeschwindigkeit speed of application
Auftragsgewicht rate of application *(e.g. paint or adhesive)*
Auftragsmaschine applicator *(e.g. for adhesives)*
Auftragsmenge amount applied, rate of application *(e.g. paint, adhesive etc.)*
Auftragsmethode method of application
Auftragsstärke application thickness
Auftragstemperatur application temperture
Auftragsverfahren method of application
Auftragswalze applicator roll
Auftragung 1. application *(paint or adhesive)*. 2. plotting *(a curve)*. 3. curve
Auftrocknung drying
Aufwand 1. cost, expense, expenditure. 2. complexity, effort
Aufziehgerät film applicator, film casting instrument
Augenbindehautendzündung conjunctivitis
Augenerkrankung eye infection
Augenkontakt contact with the eyes *(e.g. spilt chemicals)*
Augenreizmittel eye irritant
Augenschleimhaut conjunctiva

Augerspektroskopie Auger spectroscopy
Augerelektronenspektroskopie auger electron spectroscopy
ausbalanciert balanced
Ausbeute yield
Ausbildung training
ausbleichen to fade
Ausblüherscheinungen signs of efflorescence
Ausblühungen efflorescence
ausdampfen to evaporate (from)
Ausdehnung expansion
Ausdehnungskoeffizient coefficient of expansion
Ausdehnungskoeffizient, kubischer coefficient of cubical/volume expansion
Ausdehnungskoeffizient, linearer coefficient of linear expansion
Ausdehnungskoeffizient, thermischer coefficient of (thermal) expansion
Ausdehnungsverhalten expansion characteristics/behaviour
Ausdruck print-out
Ausflockung flocculation
Ausfüllmasse knifing filler, stopper, surfacer
Ausgangshärte original hardness
Ausgangskomponenten original components
Ausgangskonzentration 1. original/initial concentration. 2. outlet concentration
Ausgangslänge original length
Ausgangsmolekül starting molecule
Ausgangsmonomer starting monomer
Ausgangspartikelgrößenverteilung original particle size distribution
Ausgangsprobe original sample
Ausgangsprodukt starting product
Ausgangspunkt starting point
Ausgangsrezeptur starting formulation
Ausgangstemperatur 1. original/starting temperature. 2. outlet temperature
Ausgangsviskosität initial/original viscosity
Ausgangswert initial/original value
Ausgangszustand initial/original condition
ausgefällt precipitated
ausgehärtet cured
ausgerüstet 1. treated. 2. equipped
ausgetrocknet dried (out)
ausgewogen 1. weighed out. 2. balanced *(e.g. properties)*
aushärtbar curable
Aushärte- *see* **Aushärtungs-**
aushärten to cure
Aushärtung curing
Aushärtungsbedingungen curing conditions
Aushärtungsgeschwindigkeit curing rate
Aushärtungsgrad degree of cure
Aushärtungsmechanismus curing mechanism
Aushärtungsofen curing oven
Aushärtungsreaktion curing reaction
Aushärtungstemperatur curing temperature
Aushärtungsverhalten curing characteristics/behaviour
Aushärtungszeit cure time
Aushärtungszyklus curing cycle
Auskleidung lining
Auskreidung chalking
auskristallisieren to crystallise out
Auslagerung ageing
Auslagerungszeit ageing period
Auslaßventil outlet valve
Auslauf outlet
Auslaufbecher flow cup
Auslaufbechermessung flow cup determination
Auslaufzeit flow time
Auslaugung extraction
Ausrichtung alignment
Ausrüstung 1. equipment. 2. application *(of size or finish)*. 3. incorporation *(of plasticiser, antistatic agent etc.)*
Ausschwimmen floating
Ausschwimmen, horizontales floating
Ausschwimmen, vertikales flooding
ausschwitzen to exude
Aussehen appearance
Außenanstrich 1. exterior/outside coating/finish. 2. architectural paint
Außenanwendung exterior/outside use
Außenbeschichtung 1. exterior/outside coating. 2. painting the outside
Außenbeständigkeit outdoor weathering resistance
Außenbewitterung outdoor weathering
Außenbewitterungsbeständigkeit outdoor weathering resistance
Außenbewitterungsprüfung outdoor weathering test
außenbewitterungsstabil resistant to outdoor weathering
Außenbewitterungsversuch outdoor weathering test
Außendurchmesser outside diameter
Außeneinsatz exterior/outdoor use
Außenfarbe exterior paint
Außenfläche outside surface
Außenhaut outer skin
Außenisolierung exterior insulation
Außenlack exterior paint
Außenmuffenverbindung sleeved tubular joint
Außenoberfläche outer/exterior surface
Außenschale outer shell
Außenwände exterior/outside walls
äußerlich weichgemacht externally plasticised
Austauschenthalpie interchange enthalpy
austenitisch austenitic
ausvulkanisiert fully vulcanised
Auswaage amount weighed out
auswandernd migrating
Autodecklack automotive topcoat/finish
Autogrundlack automotive primer
Autohersteller car manufacturer
Autoindustrie car/motor industry
Autoklav autoclave
autoklaviert autoclaved
Autolack automotive paint/enamel

Autolackierung automotive finish
automatisch automatic
automatisierbar capable of being automated
automatisiert automated
Automatisierung automation
automatisierungsfreundlich easily automated, easy to automate
Automatisierungstechnik automation technology/engineering
Automobil car
Automobilbau car construction
Automobilbereich motor industry
Automobildecklackierung automotive finish
Automobilfarbton automotive colour
Automobilindustrie car/motor industry
Automobilkarosserie car body
Automobillack automotive paint/enamel
Automobillackierung automotive finish
Automobilsektor car/motor industry
Automobilzubehör car accessories
Autophoreselack autophoretic paint
Autoreparatur car repair
Autoreparaturhandwerk car repair trade
Autoreparaturlack car refinishing paint
Autoreparaturlackharz car refinishing paint resin
Autoreparaturlackierung car refinishing
Autoscheinwerfer headlamp
Autoxidation autoxidation
autoxidativ autoxidative
autoxidierbar autoxidising
AWETA *abbr. of* **anwendungstechnische Abteilung**, application research department
azeotrop azeotropic
Azeotrop azeotrope
Azidität acidity
Aziridinring aziridine ring
Azofarbstoff azo dye
Azogruppe azo group
Azopigment azo pigment
Azoverbindung azo compound
azovernetzt azo-cured

B

Ba-Salz barium salt
Backlack stoving paint
Bakterien bacteria
Balkendiagramm bar graph
Balkon balcony
Bandbeschichtung coil coating
Bandbreite band width
Bande band
Bandenintensität band intensity
Bandgranulierung strip pelletisation *(of hot melt adhesives)*
Bandlack coil coating paint
Bandlackierung coil coating
Bandlackierverfahren coil coating (process)
Bandmetallackierung coil coating
Bariummetaborat barium metaborate
Bariumoxid barium oxide
Bariumseife barium soap
Bariumstearat barium stearate
Bariumsulfat barium sulphate
Barriere barrier
Barriereeigenschaften barier properties
Barrierefolie barrier film
Barrierepigment barrier pigment
Barriereschicht barrier coating
Barrierewirkung barrier effect
basenempfindlich alkali-sensitive, sensitive to alkalis
basenlöslich alkali-soluble
basenunlöslich alkali-insoluble
basisch basic, alkaline
Basislack 1. base coat, undercoat. 2. base paint
Basismaterial basic material
Basismonomer base monomer
Basisüberzug primer/base coat
Basizität basicity
Bastelklebstoff hobby adhesive
Bastelsektor crafts sector
batteriebetrieben battery operated
Bauchemie *A buzz word intended to convey the idea of synthetic, i.e. chemical products and their relation to the building industry. Since it is untranslatable, it must be treated as synonymous with* **Bauindustrie**, *e.g.* **Diese Produkte werden in zahlreichen Bereichen, wie der Klebstoff- und Lackindustrie, der Bauchemie usw. eingesetzt** These products are used in many fields, e.g. the adhesives and coatings industry, the building sector, etc.
bauchemisch *Like the noun from which it is derived,* (**Bauchemie** *q.v.*), *this word cannot be translated and must therefore be circumscribed, e.g.* **Fast alle modernen bauchemischen Produkte müssen bestimmten Normen entsprechen** Practically all modern synthetic products for the building industry must meet certain standards.
Baudehnungsfuge expansion joint
Baudenkmal historic monument/building
Baudichtstoff structural sealant
Baugenehmigung building permission
Baugewerbe building sector/industry, construction industry
Bauholzleim structural wood adhesive
Bauindustrie building sector/industry, construction industry
baukastenartig modular
Baukastenbauweise modular/unit construction system
Baukastenelement module, unit
Baukastenprinzip modular system/principle
Baukastensystem modular system

Baukleber structural adhesive
Bauklebmasse structural/building adhesive
Bauklebstoff structural/building adhesive
Baukörper building, structure
Baumaterial building material
Baumwollgewebe cotton fabric
bauphysikalisch physical *(the first part of the word is superfluous since it will be clear from the context that building materials are being discussed, e.g.* **die bauphysikalischen Werte des Steines** the physical properties of stone
Bauprodukt building material
Bausektor building sector/industry, construction industry
Baustahl structural steel
Baustein module, building block
Baustelle building site
Baustoff building material
Bautenanstrich 1. exterior/masonry finish. 2. architectural paint
Bautenanstrichmittel architectural paint
Bautenbeschichtung architectural finish
Bautenfarbe architectural paint
Bautenfarbenbereich architectural paints sector
Bautenlack architectural paint
Bautenschutz building conservation
Bautenschutzfarbe architectural/exterior/masonry paint
Bautenschutzmittel masonry water repellent
Bautenschutzsystem masonry water repellent
Bauwerk building
Bauwerksfuge expansion joint
Bauwesen building sector/industry, construction industry
Bauwirtschaft building sector/industry, construction industry
Beanspruchbarkeit load-bearing/-carrying capacity
Beanspruchung stress, load(ing)
Beanspruchung, chemische chemical attack
Beanspruchung, elektrische electrical loading
Beanspruchung, mechanische mechanical stress/loading
Beanspruchung, ruhende static stress/load
Beanspruchung, schlagartige 1. impact stress. 2. sudden stress
Beanspruchung, schwellende fatigue stress
Beanspruchung, statische static stress/load
Beanspruchung, stoßartige impact stress
Beanspruchung, thermische thermal stress
Beanspruchung, wechselnde cyclic stress/loading
Beanspruchung, zügige dynamic stress
Beanspruchungsanalyse stress analysis
Beanspruchungsart type of stress
Beanspruchungsbedingungen conditions of use, service conditions
Beanspruchungsdauer 1. stress duration. 2. time of exposure *(i.e. to certain conditions)*. 3. test period
Beanspruchungsgrenze stress limit, maximum stress
Beanspruchungshäufigkeit stress/loading frequency
Beanspruchungshöhe stress level, amount of stress applied
Beanspruchungsperiode 1. stress duration. 2. time of exposure *(i.e. to certain conditions)*
Beanspruchungsrichtung stress direction
Beanspruchungszeit stress duration, loading time
Beanspruchungszeitraum 1. test period. 2. stress duration
Bearbeitung, mechanische machining
Bearbeitung, spanabhebende machining
Bearbeitung, spanende machining
Bearbeitung, spangebende machining
Bearbeitung, spanlose shaping, forming, thermoforming
Bearbeitungsverfahren machining technique/process
Beaufschlagung subjecting something to *(pressure, high temperatures, impact etc.)*
Becher beaker, cup
Becherglas beaker
Bedachung roof(ing)
Bedampfen 1. metallisation. 2. steam treatment
Bedeckungsgrad hiding power, opacity
bedienerfreundlich operator-/user-friendly
Bedienperson operator
bedienungsfreundlich 1. easy to operate/use. 2. easily accessible
Bedienungspersonal operating personnel, machine operators
bedruckbar printable
Bedruckbarkeit printability
bedrucken to print
bedruckt printed
Beflammen flame treatment
beflammt flame treated
Beflockung flocking
Beflockungsmaschine flock spraying machine
befüllen to fill
Begleiterscheinung side effect
Begleitpapiere accompanying documents
begrenzt limited
begrenzt löslich with limited solubility
begrenzt verträglich with limited compatibility
Behälter container
Behältnis container
behandelt treated
Behandlung treatment
Behandlungsdauer treatment period, duration of treatment
Behandlungszeit treatment period, duration of treatment
beheizbar heatable
beheizt heated
beidseitig on both sides
beigefarbig beige
Beimengung addition, incorporation

beimischen to add, to incorporate
Beistellaggregat ancillary/subsidiary unit
Beizbehälter pickling tank
Beizdauer pickling time/duration
Beize 1. pickling solution. 2. (wood) stain
beizen to pickle *(e.g. metal in acid before painting or bonding)*
Beizlösung pickling solution
Beizmittel pickling solution
Beizprozeß pickling process
Beizsäure pickling acid
Beiztemperatur pickling temperature
Beizverfahren pickling process
Bekleidungsindustrie clothing industry
Belagsstoff covering (material)
Belastbarkeit load-bearing/-carrying capacity
belasted 1. stressed. 2. polluted, contaminated
Belastung 1. stress, load(ing), application of stress, applied load. BUT: **kurzzeitige Belastungen bis zu Temperaturen von 300°C** brief exposure to temperatures of up to 300°C. 2. pollution, contamination
Belastung, korrosive corrosive attack
Belastung, ruhende static stress/load
Belastung, schwellende fatigue stress
Belastung, statische static stress/load
Belastungsart type of (applied) stress
belastungsbezogen stress-related
Belastungsdauer stress duration
Belastungseinrichtung stress/load application device
Belastungsfläche exposed surface
Belastungsfrequenz stress frequency
Belastungsgeschwindigkeit stressing/loading rate
Belastungsgrenze load-bearing/-carrying capacity
Belastungsgröße stress level, amount of stress
Belastungshöhe stress level
Belastungsniveau stress level
Belastungsprüfung loading test
Belastungsrichtung direction of applied stress, direction of loading
Belastungsversuch loading test
Belastungszeit stress duration, loading time
Belastungszyklus stress cycle
belegen to coat
Beleimung coating with glue/adhesive
Beleuchtung illumination
belichtet exposed to light
Belichtung exposure to light
Belüftung ventilation
Belüftungsanlage ventilation plant
Benard'sche Zellen Benard's cells
benetzbar wettable
Benetzbarkeit wettability
benetzt wetted
Benetzung wetting
Benetzungsadditiv wetting agent
Benetzungseigenschaften wetting properties/characteristics
Benetzungsfähigkeit wetting power, wettability
Benetzungsfehler wetting defect
Benetzungsgeschwindigkeit wetting rate
Benetzungshilfsmittel wetting agent
Benetzungsmittel wetting agent
Benetzungsprozeß wetting process
Benetzungsschwierigkeiten wetting problems
Benetzungsstörung wetting defect/fault
Benetzungsverhalten wetting behaviour/characteristics
Benetzungsverhältnisse wetting conditions
Benetzungsvermögen wetting power, wettability
Benetzungsvorgang wetting process
Benetzungswinkel contact angle
benutzerfreundlich user-friendly
Benutzerfreundlichkeit user-friendliness
Benzil benzil
Benzildimethylketal benzil dimethyl ketal
Benzin benzine, petrol, gasoline
Benzinkohlenwasserstoff aliphatic hydrocarbon
Benzoat benzoate
Benzoesäure benzoic acid
Benzoesäuremethylester methyl benzoate
Benzoguanamin benzoguanamine
Benzoguanaminformaldehydharz benzoguanamine formaldehyde resin
Benzoguanaminharz benzoguanamine resin
Benzoin benzoin
Benzoinderivat benzoin derivative
Benzol benzene
Benzolkern benzene ring
Benzolkohlenwasserstoff aromatic hydrocarbon
Benzolsulfonsäure benzene sulphonic acid
Benzophenon benzophenone
Benzoylgruppe benzoyl group
Benzoylperoxid benzoyl peroxide
Benzoylperoxidpaste benzoyl peroxide paste
Benzoylradikal benzoyl radical
Benztriazol benztriazole
Benzylalkohol benzyl alcohol
Benzylcellulose benzyl cellulose
Benzylchlorid benzyl chloride
Beoabachtungsdauer observation/test period
Beobachtungswinkel angle of observation
Beobachtungszeitraum observation period
Berechnung calculation, computation
Berechnungsansatz mathematical equation
Berechnungsbeispiel mathematical example
Berechnungsgleichung mathematical equation
Berechnungsresultat calculated result
Beregnung water spray
Bereich range, region, zone, area
bernsteinfarben amber (coloured)
Bernsteinsäure succinic acid
Bernsteinsäureanhydrid succinic anhydride
Berregnungsvorrichtung water spraying device
Berufskrankheit occupational disease
Berührungsdruck contact pressure
Berührungsfläche contact surface
berührungsfrei 1. contact-less, non-contact. 2. solid-state, electronic

berührungslos 1. contact-less, non-contact. 2. solid-state, electronic
Berührungsstelle point of contact
beschädigt damaged
Beschädigung damage
Beschädigungsgefahr risk of damage
beschichten to coat
Beschichtung coating *(the word is sometimes also used to denote* paint*)*
Beschichtungsanlage coating line/plant
Beschichtungsarbeit painting operation
Beschichtungsaufbau paint film composition
Beschichtungsdicke coating thickness
Beschichtungseigenschaften (paint) film properties
Beschichtungseinheit coating unit
Beschichtungseinrichtung coating machine/unit
Beschichtungsfehler coating defect
Beschichtungsfilm paint film
Beschichtungsfläche surface to be coated/painted
Beschichtungsindustrie paint/coatings industry
Beschichtungsmaschine coating machine
Beschichtungsmasse coating compound
Beschichtungsmaterial 1. coating compound, paint. 2. material to be coated
Beschichtungsmittel coating compound, paint
Beschichtungsoberfläche coating surface, paint film surface
Beschichtungsparameter coating conditions
Beschichtungspulver powder coating
Beschichtungsqualität coating quality, paint film quality
Beschichtungsstation coating station
Beschichtungsstoff coating compound, paint
Beschichtungsstraße coating line
Beschichtungssystem coating compound, paint
Beschichtungstechnologie surface coating technology, paint technology
Beschichtungsüberzug coating
Beschichtungsuntergrund surface *(to be coated or already coated)*, substrate
Beschichtungsverfahren coating process
Beschichtungswerkstoff coating compound, paint
beschleunigen to accelerate, to speed up
Beschleuniger accelerator
beschleunigt accelerated
Beschleunigung acceleration
Beschleunigungsspannung accelerating voltage
beschränkt löslich sparingly soluble
Beständigkeit stability, resistance
Beständigkeit, chemische chemical resistance
Beständigkeit gegen Dauerbelastung creep strength
Beständigkeit, thermische thermal stability, heat resistance
Beständigkeitseigenschaften stability characteristics
Bestandteil constituent, component
Bestandteile, flüchtige volatile matter/content
Bestimmung 1. determination. 2. regulation, order, decree
Bestimmung, naßchemische wet analysis
Bestimmungen, bauaufsichtliche building regulations
Bestimmungen, lebensmittelrechtliche food regulations
Bestimmungsmethode method of determination
Bestimmungsresultat test result
bestrahlt irradiated
Bestrahlung irradiation
Bestrahlungsbedingungen irradiation conditions
Bestrahlungsdosis irradiation dosage
Bestrahlungsintensität radiation intensity
Bestrahlungslichtquelle irradiation light source
Bestrahlungsstärke irradiation intensity
Bestrahlungsvernetzung radiation cure/curing
Bestrahlungsversuch irradiation test
Bestrahlungswinkel irradiation angle
Bestrahlungszeit irradiation time
Betastrahlen beta rays
Beton concrete
Betonanstrich concrete paint
Betonbauwerk concrete structure
Betonbeschichtungsstoff concrete paint
Betonboden concrete floor
Betonersatzmörtel concrete patching mortar
Betonfarbe concrete paint
Betonfertigteil prefabricated concrete unit
Betonfläche concrete surface
Betonimprägnierung concrete impregnation
Betonklebstoff concrete adhesive
Betonoberfläche concrete surface
Betonplatte concrete slab
Betonsanierung concrete repair/restoration
Betonsanierungsfarbe concrete repair paint
Betonschäden concrete cancer
Betonschutz concrete protection
Betonschutzanstrich (protective) concrete finish
Betonschutzfarbe (protective) concrete paint
Betonstruktur concrete structure
Betonversiegelung concrete sealant
Betonwand concrete wall
Betrachtungswinkel angle of observation
Betrieb 1. business, factory, works, establishment. 2. operation
Betriebsanweisung operating instructions
Betriebsaufwand operating costs
Betriebsausfall production breakdown
Betriebsbedingungen operating conditions
Betriebsdruck working/operating pressure
betriebseigen in-plant, in-house
Betriebserfahrung practical experience
betriebsfähig in working order
Betriebsgrößen production data
Betriebshygiene works/factory hygiene
betriebsintern in-house, in-plant
Betriebskontrolle works control
Betriebskontrollprüfung works control test
Betriebskosten operating/running costs
Betriebskostenvergleich operating costs

comparison
Betriebslabor works laboratory
Betriebsmaßstab industrial-/production-scale
betriebssicher dependable, reliable (in operation)
Betriebsstunden hours of operation
Betriebstemperatur operating/working temperature
Betriebstemperaturbereich operating temperature range
Betriebsweise method of operation
Beugung diffraction
Beugungswinkel diffraction angle
Beurteilung assessment, evaluation
beweglich mobile: **bewegliche Teile** moving parts
Beweglichkeit mobility
Bewegungsfuge expansion joint
Bewehrung reinforcement
Bewehrungsstahl reinforcing steel
Bewertung evaluation, assessment
bewittert weathered
Bewitterung weathering
Bewitterung, künstliche artificial weathering
Bewitterung, natürliche natural weathering
Bewitterungsautomat automatic weathering instrument
Bewitterungsbeanspruchung weathering
Bewitterungsbedingungen weathering conditions
Bewitterungsbeständigkeit weathering resistance
Bewitterungsdauer weathering period
Bewitterungseigenschaften weathering properties
Bewitterungseinflüsse weathering influences
Bewitterungsergebnisse results of weathering tests
Bewitterungsexperiment weathering test/experiment
Bewitterungsgerät weathering instrument
Bewitterungsgestell weathering frame
Bewitterungsort weathering location
Bewitterungsplatten weathered panels
Bewitterungsplatz weathering location
Bewitterungsprüfung weathering test
Bewitterungsresistenz weathering resistance
Bewitterungsschäden damage due to (*or:* caused by) weathering
bewitterungsstabil weather resistant
Bewitterungsstabilität weathering resistance
Bewitterungsstandort weathering location
Bewitterungsstation weathering station
Bewitterungsstunden hours of weathering
Bewitterungsverhalten weathering properties
Bewitterungsversuch weathering test
Bewitterungszeit weathering period
Bewitterungszeitraum weathering period
Bewuchs marine fouling
bewuchsabweisend anti-fouling
bewuchsfrei free from marine fouling/growths
bewuchsverhindernd anti-fouling
Beziehung relation(ship)
Bezugshologramm reference hologram
Bezugsmaterial reference material
Bezugsmessung reference determination
Bezugspunkt point of reference
Bezugsquelle source of supply
Bezugstemperatur reference temperature
Bezugszeitraum reference period
bidistilliert double-distilled
Biege-E-Modul flexural modulus of elasticity
Biege-Kriechmodul flexural creep modulus
biegebeansprucht under flexural/bending stress
Biegebeanspruchung flexural/bending stress
Biegebelastbarkeit flexural load-carrying capacity
biegebelastet under flexural/bending stress
Biegebelastung flexural/bending stress
Biegedehnung flexural/bending strain
Biegeelastizitätsmodul flexural modulus of elasticity
Biegefestigkeit flexural/cross-breaking strength
Biegefließspannung flexural yield stress
Biegekraft bending force
Biegekriechversuch flexural creep test
Biegemodul flexural modulus
Biegemoment bending moment
Biegen bending, flexing
Biegeprobe flexural test piece, flexural specimen
Biegeradius bending radius
Biegereißbeständigkeit flex cracking resistance
Biegerißbeständigkeit flex cracking resistance
Biegerißbildung flex cracking
Biegeschälfestigkeit flexural peel strength
Biegeschälversuch flexural peel test
Biegeschälwiderstand flexural peel strength
Biegeschwellfestigkeit flexural fatigue strength
Biegeschwellversuch flexural fatigue test
Biegeschwingfestigkeit flexural fatigue strength
Biegeschwingversuch torsion pendulum test
Biegespannung flexural/bending stress
biegesteif rigid
Biegesteifigkeit 1. rigidity, stiffness. 2. flexural strength
Biegeverhalten flexural behaviour
Biegeversuch flexural/bending test
Biegewechselbeanspruchung flexural fatigue stress
Biegewechselbelastung flexural fatigue stress
Biegewechselfestigkeit flexural fatigue strength
Biegewechselversuch flexural fatigue test
biegeweich flexible, pliable
Biegewinkel bending angle
Biegezug flexural tension
Biegezugfestigkeit tensile strength in bending
biegsam flexible, pliable
Biegsamkeit flexibility, pliability, suppleness
Biegung 1. bending, flexure. 2. deflection
Biegung beim Bruch deflection at break
Biegungs-Verformungsverhalten flexural deformation characteristics

bifunktionell difunctional
Big-Bag-Gebinde big bag
Bildschirm (VDU) screen
Bildschirmgerät monitor
Bildung formation
BImSchG *abbr. of* **Bundesimissionsschutzgesetz**, Federal anti-pollution act, Federal pollution control act
BImSchV *abbr. of* **Bundesimissionsschutzverordnung**, Federal anti-pollution directive
Bimsstein pumice stone
binär binary
Bindefestigkeit bond strength
Bindekraft 1. adhesive force. 2. binding power. 3. bond force
Bindemittel binder
Bindemittel, hydraulisches hydraulic adhesive/cement
bindemittelarm low-solids *(e.g. paint)*
Bindemittelbedarf binder requirement
Bindemitteldispersion binder dispersion
Bindemittelgehalt binder content
Bindemittelkombination binder combination/blend
Bindemittelkonzentration binder concentration
Bindemittellösung 1. (paint) vehicle/medium. 2. binder solution
Bindemittelmenge amount of binder
Bindemittelpolymer polymer binder, binder resin
bindemittelreich high-solids *(paint)*
Bindemittelteilchen binder particle
Bindemitteltyp type of binder
Bindemittelzugabe addition of binder
binderarm with a low binder content
Bindevermögen binding power
Bindung bond, linkage
Bindungsabstand bond distance
Bindungsenergie bond energy
Bindungskraft 1. adhesive force. 2. bond force. 3. binding power
Bindungsmechanismus linkage mechanism
Binnenmarkt home market
Binnenmarkt Europa single European market
Binnenmarkt, europäischer single European market
Binnenmarkt, gemeinsamer single European market
bioabbaubar biodegradable
Biofiltrierung biofiltration
Biokatalysator biocatalyst
biokompatibel biocompatible
Biokompatibilität biocompatibility
Biokunststoff biopolymer
biologisch biological
biologisch abbaubar biodegradable
biologische Abbaubarkeit biodegradability
biologischer Abbau biodegradation
Biomasse biomass
Biopolymer biopolymer
bioreaktiv bioreactive
biozid biocidal
Biozid biocide
Bipolarität bipolarity
Bisacylphosphinoxid bis-acyl phosphine oxide
Bishydroxycarbonsäure bis-hydroxycarboxylic acid
Bisoxazolin bis-oxazoline
Bisphenol-A bisphenol A
Bisphenolharz bisphenol resin
Bitumen bitumen
Bitumenanteil bitumen content
bitumenbeständig bitumen resistant
Bitumenbeständigkeit bitumen resistance
Bitumenemulsion bitumen emulsion
bitumenhaltig bituminous
Bitumenlack bituminous paint
Bitumenmasse bitumen compound
Bitumenpappe roofing felt
Bitumenschmelzmasse hot melt bitumen compound
bituminös bituminous
Biuret biuret
Biuretbindung biuret linkage
Biuretbindungsanteil biuret linkage content
Biuretpolyisocyanat biuret polyisocyanate
Biuretstruktur biuret structure
bivalent divalent
bizyklisch dicyclic
Blanc fixe blanc fixe, precipitated barium sulphate
blank bright *(metal surface)*
blanker Stahl bright steel
blankgestrahlt bright finished
Blasenbildung blistering, formation of bubbles
Blasendurchmesser bubble diameter
Blaseneinschlüsse trapped bubbles
blasenfrei free from bubbles/blisters
blättchenförmig platelet-like
blau blue
Blauer Engel *This is Germany's universally recognised symbol indicating that a product is environment-friendly. There is, as yet, no equivalent term in English, so that translators should write the German expression in inverted commas, followed by something along these lines:* literally "blue angel", a label issued by the Federal Office for the Environment to denote environmentally safe products.
Blauer Engel-Farbe environment-friendly paint
bläulich blueish
Blaupigment blue pigment
Blaustich blue tinge
Blech 1. sheet metal. 2. metal sheet
Blechbandlack coil coating paint/enamel/lacquer
Blechemballage metal container
Blechgebinde metal container
Blechtafel metal panel
Blei lead
Bleiabgabe lead emission
bleibende Bruchdehnung residual/irreversible/

permanent elongation at break
bleibende Dehnung residual/irreversible/permanent elongation
bleibende Haftung permanent adhesion
bleibende Verformung residual/irreversible/permanent deformation
Bleichmittel bleaching agent, bleach
Bleichromat lead chromate
Bleichromatpigment lead chromate pigment
bleifrei lead-free
Bleigehalt lead content
bleihaltig containing lead
Bleimaleat lead maleate
Bleimennige red lead
Bleinaphthenat lead naphthenate
Bleioktoat lead octoate
bleiorganisch organo-lead
Bleiphosphit lead phosphite
Bleiphthalat lead phthalate
Bleipigment lead pigment
Bleiseife lead soap
Bleisilicochromat lead silicochromate
Bleistabilisator lead stabiliser
Bleistearat lead stearate
Bleistifthärte pencil hardness
Bleisulfat lead sulphate
Bleitrockner lead-based drier
Bleiverbindung lead compound
Bleiweiß white lead
Blendsystem blend
Blindprobe 1. blank test. 2. reference sample
Blitzkleber superglue
Blockcopolymer block copolymer
Blockcopolymerisat block copolymer
Blockcopolymerisation block copolymerisation
Blockdarstellung 1. block diagram. 2. bar graph
Blockdiagramm 1. block diagram. 2. bar graph
Blocken blocking
blockfest non-blocking
Blockfestigkeit blocking resistance
blockfrei non-blocking
Blockfreiheit freedom from blocking
blockiert blocked
Blockierung blocking
Blockierungsmittel blocking agent
Blockkraft blocking force
Blockmischpolymer block copolymer
Blockmischpolymerisat block copolymer
Blockneigung blocking tendency, tendency to block
Blockpolymer block polymer
Blockpolymerisat block polymer
Blockpolymerisation bulk polymerisation
Blockpunkt blocking point/temperature
Blocktendenz blocking tendency, tendency to block
Blockungsmittel blocking agent
Blockverhalten blocking behaviour/properties
Bodenabdichtung floor sealant
Bodenausgleichmasse flooring screed, jointless flooring compound
Bodenbelag 1. floorcovering. 2. (floor) screed
Bodenbelagsfirma flooring contractor
Bodenbelagsklebstoff floorcovering adhesive
Bodenbelagsstoff floorcovering
Bodensatz sediment
Bodenverunreinigung soil contamination/pollution
Bogenoffsetdruck sheet-fed offset printing
Bogenwiderstand arc resistance
Bondstellen bonded areas
Borat borate
Bornitrid boron nitride
Borsäure boric acid
Bortrioxid boron trioxide
Brandausbreitung flame spread
Brandbekämpfung fire fighting
Brandentdeckung fire detection
Brandfall, im in case of fire
Brandfortpflanzung flame spread
Brandgefahr danger of fire
brandgeschützt flameproofed, flame resistant
Brandgeschwindigkeit burning rate
Brandgroßversuch large-scale burning test
Brandhemmung flame retardance
Brandklasse flammability rating/classification/group
Brandklasseneinteilung flammability rating/classification/group
Brandklassifizierung flammability rating/classification/group
Brandmeldeanlage fire alarm system
Brandnebenerscheinungen secondary effects of fire
Brandneigung flammability
Brandquelle fire source
Brandrisiko fire risk
Brandschachttest chimney test *(to determine flammability)*
Brandschachtverfahren chimney test *(to determine flammability)*
Brandschutz fire protection
Brandschutzadditiv flame retardant
brandschutzausgerüstet flameproofed, flame resistant
Brandschutzausrüstung 1. making flame retardant/resistant. 2. flame retardant
Brandschutzbeschichtung flameproof coating
Brandschutzgesetze fire regulations
Brandschutzklasse flammability rating/clasification/group
Brandschutzmasse flame resistant composition
Brandschutzmaßnahmen fire precautions
Brandschutzmittel flame retardant
Brandschutznorm fire safety standard
brandschutztechnisch *adjective relating to fire protection:* **brandschutztechnische Anforderungen** fire regulations
Brandschutztür fire door
Brandschutzverhalten fire retardancy
brandsicher flame resistant
Brandsicherheit fire resistance

Brandsicherheitsvorschriften fire regulations
Brandtest flammability test
Brandverhalten fire behaviour
Brandverhaltensklasse fire behaviour rating/classification
brandverhindernd fire-preventing
Brandversuch flammability test
Brandvorbeugung fire prevention
Brandvorschriften fire regulations
Brandweiterleitung flame spread
Brandwidrigkeit fire/flame resistance
braun brown
Braunfärbung brown discolouration
Braunstein pyrolusite
Bräunung brown discolouration
Bräunung, thermische brown discolouration due to overheating
Brechung refraction
Brechungsindex refractive index
Brechungsindexwert refractive index
Brechungszahl refractive index
Brechwert refractive index
Brechzahl refractive index
breitgefächert wide-ranging, widespread
Bremsflüssigkeit brake fluid
brennbar flammable
Brennbarkeit flammability
Brenngeschwindigkeit burning rate
Brennpunkt focal point
Brennstoff fuel
Brennstoff, fossiler fossil fuel
Brennstofftank fuel tank
Brennstrecke burning distance
Brennverhalten fire behaviour
Brenzkatechin pyrocatechol
Briefumschlag envelope
brillant sparkling
Brillanz sparkle
Brinellhärte Brinell hardness
Brom bromine
Bromkresolgrün bromocresol green
Bronze bronze
Brookit brookite
Bruch fracture, failure
Bruch, duktiler ductile fracture/failure
Bruch, Durchbiegung beim deflection at break
Bruch, katastrophaler catastrophic failure
Bruch-Schwingspielzahl number of vibrations to failure
bruchanfällig fragile, liable to break
Bruchanfälligkeit fragility, tendency to crack/break
Brucharbeit fracture energy
Bruchart type of fracture/failure
Bruchausbreitung crack propagation
bruchauslösend fracture-initiating
Bruchbeanspruchung breaking stress
Bruchbelastung breaking stress
Bruchbiegespannung bending/flexural stress at break
Bruchbild fracture photomicrograph
Bruchdehnung elongation at break, ultimate elongation
Bruchdehnung, bleibende residual/irreversible/permanent elongation at break
Brucheinleitung crack initiation
bruchempfindlich fragile, liable to break/crack
Bruchempfindlichkeit fragility, liability to break/crack
Bruchenergie fracture energy
Bruchfestigkeit fracture resistance, breaking strength
Bruchfläche fracture surface
Bruchform type of fracture
Bruchfortschritt crack propagation
Bruchgefahr risk of fracture
Bruchgrenze breaking limit
brüchig brittle
Bruchkante fracture edge
Bruchkraft breaking stress
Bruchkriterium fracture criterion
Bruchkurve fracture curve
Bruchlast breaking stress
Bruchlastspielzahl number of breaking stress cycles
Bruchmechanik fracture mechanics
bruchmechanisch fracture-mechanical
Bruchmechanismus failure mechanism
Bruchmorphologie fracture morphology
Bruchschubfestigkeit shear strength at break
Bruchschubspannung shear stress at break
bruchsicher break resistant
Bruchsicherheit break resistance
Bruchspannung 1. tensile stress at break, ultimate tensile stress, breaking stress. 2. rupture voltage *(in electrophoretic paint deposition)*
Bruchstauchung compression at break
Bruchstück fragment
Bruchverhalten fracture behaviour
Bruchversuch fracture test
Bruchvorgang fracture process
Bruchzähigkeit fracture toughness
Bruchzugscherspannung tensile shear stress at break
Brünierschlamm burnishing sludge
BuAc *abbr. of* **Butylacetat**, butyl acetate
Buchbindeindustrie bookbinding industry
Buchbinderei 1. bookbinding. 2. bookbinding workshop
Buchdruck letterpress printing
Buchdruckerschwärze printer's black
Buchdruckfarbe printing ink
Buche beech
Buchenholz beech
Bundesgesetzblatt Federal Law Gazette
Bundesgesundheitsamt Federal Health Office
Bundesimmissionsschutzgesetz Federal anti-pollution act, Federal pollution control act
Bundesimmissionsschutzverordnung Federal anti-pollution act, Federal pollution control act

Bundesländer, alte the former W.Germany
Bundesländer, neue the former E.Germany
bunt brightly coloured, multi-coloured
Buntlack coloured paint *(as opposed to white paint)*
Buntmetall non-ferrous metal
Buntmetall-Legierung non-ferrous metal alloy
Buntpigment 1. pigment. 2. coloured pigment *(where a distinction is made between* **Buntpigmente** and **Weißpigmente**, *q.v.)*
buntpigmentiert containing coloured pigments *(see explanatory note under* **Buntpigment***)*
Bürette burette
Bürogeräte office equipment
Büromaschinen office equipment
Büromöbel office furniture
Büroraum office
Bürostuhl office chair
Bürste brush

Bürsteffekt brushed surface finish
Bürsten brushing
Bürstenapplikation brush application, application by brush
Butadien butadiene
Butadien-Acrylnitrilpolymer butadiene-acrylonitrile copolymer
Butan butane
Butanflüssiggas liquid butane
Butanol butanol, butyl alcohol
butanolisch butanolic
Butoxyethoxyethanol butoxyethoxyethanol
Butter butter
Buttersäure butyric acid
Buttersäureester butyrate
Butylacetat butyl acetate
Butylacetatlack paint containing butyl acetate
Butylacrylat butyl acrylate
Butylalkohol butyl alcohol
Butylbenzoesäure butyl benzoic acid
Butylbenzol butyl benzene
Butylbenzyladipat butyl benzyl adipate
Butylbenzylphthalat butyl benzyl phthalate
Butylbrenzkatechin butyl pyrocatechol
Butyldiglykol butyl diglycol
Butyldiglykolacetat butyl diglycol acetate
Butylester butyl ester
Butylether butyl ether
Butylglycidylether butyl glycidyl ether
Butylglykol butyl glycol
Butylgruppe butyl group
Butylhydroperoxid butyl hydroperoxide
butyliert butylated
Butylkautschuk butyl rubber
Butylkitt butyl rubber mastic
Butylmethacrylat butyl methacrylate
Butylmethylphenylsiliconharz butyl methyl phenyl silicone resin
Butylmethylsiliconharz butyl methyl silicone resin
Butylperbenzoat butyl perbenzoate
Butylperoctoat butyl peroctoate
Butylperoxid butyl peroxide
Butylphenylglycidylether butyl phenyl glycidyl ether
Butylstearat butyl stearate
Butyltitanat butyl titanate
Butyltrichlorsilan butyl trichlorosilane
butylverethert butyl etherified
Butylzinncarboxylat butyl tin carboxylate
Butylzinnmercaptid butyl tin mercaptide
Butylzinnverbindung butyl tin compound
Butyraldehyd butyraldehyde
Butyrolactam butyrolactam

C

ca.-Wert approximate figure
Cadmium cadmium
Cadmiumabgabe cadmium emission
cadmiumfrei cadmium-free
Cadmiumgelb cadmium yellow
cadmiumhaltig containing cadmium
Cadmiumorange cadmium orange
Cadmiumpigment cadmium pigment
cadmiumpigmentiert cadmium pigmented
Cadmiumrot cadmium red
Cadmiumsulfid cadmium sulphide
calciniert calcined
Calcinierung calcination
Calcinierungsbedingungen calcination conditions
Calcit calcite
Calciumacrylat calcium acrylate
Calciumborat calcium borate
Calciumcarbonat calcium carbonate
Calciummontanat calcium montanate
Calciumhydroxid calcium hydroxide
Calciumoktoat calcium octoate
Calciumoxid calcium oxide
Calciumseife calcium soap
Calciumstearat calcium stearate
Calciumsulfat calcium sulphate
Campher camphor
Caprolactam caprolactam
Capronsäure caproic acid
Carbamat carbamate
Carbamidharz urea resin
Carbamidsäure carbamic acid
Carbamidsäurechlorid carbamic chloride
Carbaminsäure carbamic acid
Carbonat carbonate
Carbonatisierung carbonation
Carbonatisierungsbremse carbonation inhibitor
Carbonisierung carbonation
Carbonsäure carboxylic acid
Carbonsäureamid carboxylic acid amide

Carbonsäureanhydrid carboxylic anhydride
Carbonsäureester carboxylate, carboxylic acid ester
Carbonsäuregruppe carboxyl group
Carbonylgruppe carbonyl group
Carbonylverbindung carbonyl compound
Carboxyanilidthiazol carboxyanilide thiazole
Carboxylat carboxylate
Carboxylatgruppe carboxyl group
Carboxylendgruppe terminal carboxyl group
Carboxylgruppe carboxyl group
carboxylgruppenaufweisend containing carboxyl groups
carboxylgruppenhaltig containing carboxyl groups
carboxyliert carboxylated
Carboxylierungsgrad carboxyl (group) content
Carboxylsäure carboxylic acid
Carboxylsäuregruppe carboxyl group
Carboxymethylcellulose carboxymethyl cellulose
carboxymethyliert carboxymethylated
Carnaubawachs carnauba wax
Casein casein
Cd-frei cadmium-free
Cellulose cellulose
Celluloseacetat cellulose acetate
Celluloseacetobutyrat cellulose acetate butyrate
Celluloseacetopropionat cellulose acetate propionate
Cellulosederivat cellulose derivative
Celluloseester cellulose ester
Celluloseether cellulose ether
Cellulosefaser cellulose fibre
Cellulosehydrat cellulose hydrate
Cellulosekleber cellulose adhesive
Cellulosenitrat cellulose nitrate
Cellulosepropionat cellulose propionate
Cellulosetriacetat cellulose triacetate
Celluloseverdicker cellulose thickener
Cer cerium
Cernaphthenat cerium naphthenate
Ceroxid cerium oxide
Cetylalkohol cetyl alcohol
Charakterisierung characterisation, identification
Charge batch
Chargengröße batch size
Chargenmenge batch size
Chargennummer batch number
Chargenproduktionsverfahren batch production (process)
Chargenprotokoll batch record
Chargenprozeß batch/discontinuous process
Chargenprüfung batch control
Chargenqualität batch quality
Chargenschwankungen batch variations
Chargenunterschiede batch variations/differences
chargenweise batch-wise, discontinuous(ly)
Chelat chelate
Chelatbildner chelator
Chelatisierung chelatisation
Chelatkomplexbildner chelator, chelating agent
Chelator chelator
chelierend chelating
Chemieanlage chemical plant
Chemieanlagenbau chemical plant construction
Chemieapparatebau chemical equipment construction
Chemiereaktor chemical reactor
Chemiewerkstoff polymer material, synthetic resin, plastic
Chemikalien chemicals
Chemikalienangriff chemical attack
chemikalienbeansprucht exposed/subjected to chemical attack
Chemikalienbeanspruchung exposure to chemicals
Chemikalienbelastung exposure to chemicals
chemikalienbeständig chemical resistant
Chemikalienbeständigkeit chemical resistance
Chemikaliendämpfe chemical fumes/vapours
Chemikalieneinfluß effect of chemicals
Chemikalieneinwirkung chemical attack
chemikalienfest chemical resistant
Chemikalienfestigkeit chemical resistance
Chemikalienlagerung immersion in chemicals
chemikalienresistent chemical resistant
Chemikalienresistenz chemical resistance
Chemikalientauglichkeit chemical resistance
Chemikalienversprödung chemical embrittlement
Chemikalienzeitstandverhalten long-term chemical resistance
Chemiker chemist
Chemilumineszenz chemiluminescence
chemisch chemical
chemisch neutral chemically inert
chemische Beanspruchung chemical attack
chemische Beständigkeit chemical resistance
chemische Reinigung dry cleaning
chemische Umsetzung chemical reaction
chemische Verfahrenstechnik chemical process engineering
chemische Verwertung chemical recycling
chemischer Aufbau chemical constitution
chemischer Umsatz chemical reaction
chemisches Recycling chemical recycling
Chemismus chemical action
Chemisorption chemisorption
Chinacridon quinacridone
Chinacridonrot quinacridone red
Chinolin quinoline
Chlor chlorine
Chlor-Fluorkohlenwasserstoff chlorofluorocarbon, CFC
Chloracetamid chloroacetamide
Chloratom chlorine atom
chlorbehandelt chlorinated
Chlorbenzol chlorobenzene
Chlorderivat chlorine derivative
Chlorethan chloroethane

chlorfrei chlorine-free
Chlorgehalt chlorine content
chlorhaltig containing chlorine
Chlorid chloride
Chloridgruppe chloride group
Chloridverfahren chloride process
chloriert chlorinated
Chlorierung chlorination
Chlorierungsgrad degree of chlorination
Chlorion chlorine ion
Chlorit chlorite
Chlorkautschuk chlorinated rubber
Chlorkautschukanstrich chlorinated rubber coating/finish **Chlorkautschukanstrichstoff** chlorinated rubber paint
Chlorkautschukfarbe chlorinated rubber paint
Chlorkautschuklack chlorinated rubber paint
Chlorkautschuklösung chlorinated rubber solution
Chlorkohlenwasserstoff chlorinated hydrocarbon
Chlormethylsilan chloromethyl silane
Chlorolefinharz chloroolefin resin
Chloropren chloroprene
Chloroprenkautschuk chloroprene rubber
Chloroprenpolymerisat chloroprene copolymer
Chlorparaffin chlorinated paraffin
Chlorphenol chlorophenol
Chlorphenyltriazin chlorophenyltriazine
Chlorphthalsäure chlorophthalic acid
Chlorphthalsäureanhydrid chlorophthalic anhydride
Chlorpolyethylen chlorinated polyethylene
Chlorpolyolefin chlorinated polyolefin
Chlorpolypropylen chlorinated polypropylene
Chlorpolysilan chloropolysilane
Chlorsilan chlorosilane
chlorsulfoniert chlorosulphonated
Chlortrifluorethen chlorotrifluoroethene
Chlorverbindung chlorine compound
Chlorwasserstoff hydrogen chloride
Chlorwasserstoffgas hydrogen chloride
Chlorwasserstoffsäure hydrochloric acid
Cholesterin cholesterol
cholesterisch cholesteric
Chrom chromium
Chrom-III-nitrat chromic nitrate
Chromanteil chromium content
Chromat chromate
chromatfrei chromate/chrome-free
chromatiert chromated
Chromatierung chromating
Chromation chromate ion
Chromatogramm chromatogram
Chromatographie chromatography
chromatographisch chromatographic
Chromatpigment chrome pigment
Chromchlorid chromium chloride
Chromgelb chrome yellow
chromhaltig chromium-containing, containing chromium
chromophor chromophoric
Chromophor chromophore
Chromoxid chromium/chromic oxide
Chromoxidgrün green chromium oxide
Chromoxidschicht chromium oxide layer/film
Chromsäure chromic acid
Chromschwefelsäure chromic acid
Chromverbindung chromium compound
chronisch chronic
Citrat citrate
Citronensäure citric acid
CKW *abbr. of* **Chlorkohlenwasserstoff,** chlorinated hydrocarbon
CKW-Dampfentfettung chlorinated hydrocarbon vapour degreasing
Co-Bindemittel co-binder, joint binder
Co-Lösemittel co-solvent
Co-Löser co-solvent
Co-Octoat cobalt octoate
Co-Salz cobalt salt
Co-Sikkativ cobalt drier
Cobalt- *see* **Kobalt-**
Codeziffer code number
Coflockulation co-flocculation
Coil-Coating-Decklack coil coating paint/enamel
Coil-Coating-Grundierung coil coating primer
Coil-Coating-Industrielack industrial coil coating paint/enamel/lacquer
Coil-Coating-Lack coil coating paint/enamel
Cokatalysator co-catalyst
Cokondensat co-condensate
cokondensiert co-condensed
Collodiumwolle collodion cotton, nitrocellulose
colorimetrisch colorimetric
Colösemittel co-solvent
Comonomer co-monomer
computeransteuerbar computer controlled
Computerausdruck computer printout
computergeführt computer controlled
computergeregelt computer controlled
computergesteuert computer controlled
computergestützt computer aided
computerintegriert computer integrated
Computerprogramm computer program
computerprogrammiert computer programmed
Computersteuerung computer control
computerüberwacht computer controlled/supervised
computerunterstützt comouter aided
COOH-Gehalt carboxyl (group) content
Copolyacrylat copolyacrylate
Copolyamid co-polyamide
Copolyester copolyester
copolymer copolymeric
Copolymer copolymer
Copolymerbaustein comonomer
Copolymerharz copolymer resin
Copolymerisat copolymer
Copolymerisatharz copolymer resin
Copolymerisation copolymerisation
copolymerisierbar copolymerisable

copolymerisieren to copolymerise
Coronabehandlung corona treatment
Coronabeständigkeit corona resistance
Coronaentladung corona discharge
Coronafestigkeit corona resistance
Coronagenerator corona generator
Coronavorbehandlung corona pretreatment
CPE *abbr. of* **chloriertes Polyethylen**, chlorinated polyethylene
Cu-Katalysator copper catalyst
Cu-Legierung copper alloy
Cu-Phthalocyanin copper phthalocyanine
Cumaronharz coumarone resin
Cumol cumol
Cumolhydroperoxid cumol hydroperoxide
Cyanacrylat cyanoacrylate
Cyanacrylatklebstoff cyanoacrylate adhesive
Cyanacrylsäure cyanoacrylic acid
Cyanacrylsäurebutylester butyl cyanoacrylate
Cyanacrylsäureester cyanoacrylate
Cyanacrylsäureethylester ethyl cyanoacrylate
Cyanacrylsäuremethylester methyl cyanoacrylate
Cyanamid cyanamide
Cyanatgruppe cyanate group
Cyangruppe cyanide group
Cyanharnstoff cyanourea
Cyanursäure cyanuric acid
Cyanursäurechlorid cyanuric chloride
cyclisch cyclic
cyclisiert cyclised
Cyclisierung cyclisation
cycloaliphatisch cycloaliphatic
Cyclodextrin cyclodextrin
Cyclohexan cyclohexane
Cyclohexanol cyclohexanol
Cyclohexanon cyclohexanone
Cyclohexanonharz cyclohexanone resin
Cyclohexanonperoxid cyclohexanone peroxide
Cyclohexen cyclohexene
Cyclohexylamin cyclohexylamine
Cyclohexyldicarbonsäure cyclohexyl dicarboxylic acid
Cyclohexylester cyclohexyl ester
Cyclohexylrest cyclohexyl group/radical
Cycloimidgruppe cycloimide group
Cycloiminoether cycloimino ether
Cyclokautschuk cyclised rubber, cyclorubber
Cyclokautschukfarbe cyclised rubber paint
Cycloolefin cycloolefin
Cycloolefinpolymer cycloolefin polymer
Cyclopentadien cyclopentadiene
Cyclopentadienharz cyclopentadiene resin
Cyclopentanol cyclopentanol
cyklisch cyclic

D

Dachbahn roofing sheet/membrane
Dachbeschichtung roof covering
Dachbeschichtungsmasse roof coating compound
Dachdämmung roof insulation
Dachdichtungsbahn roofing sheet/membrane
Dachfenster dormer window, skylight
Dachhimmel (car) roof liner, headlining
Dachisolierung roof insulation
Dachmasse roofing compound
Dachpappe roofing felt
Dachrinne roof gutter
Dammar dammar
Dämmasse insulating material
Dämmaterial insulating material
Dämmdicke insulation thickness
Dämmeffekt insulating effect
Dämmeigenschaften insulating properties
Dämmschicht insulating layer
dämmschichtbildend intumescent
Dämmschichtdicke insulation thickness
dämmschichterzeugend intumescent
Dämmstoff insulating material
Dämmstoffmaterial insulating material
Dämmung insulation
Dämmvermögen insulating properties
Dämmwirkung insulating effect
Dampf 1. steam. 2. vapour
Dampfalterung steam ageing
Dampfautoklave autoclave
Dampfbehandlung steam treatment
dampfbeheizt steam heated
dampfbremsende Schicht vapour barrier
dampfdicht vapour proof/impermeable
Dampfdichtheit vapour impermeability
Dampfdiffusionswiderstand vapour impermeability
Dampfdruck vapour pressure
dampfdurchlässig 1. vapour permeable. 2. water vapour permeable
Dampfdurchlässigkeit vapour permeability
Dampfentfettungsanlage solvent vapour degreasing plant
Dampferzeuger steam generating unit
dampfförmig in vapour form: **dampfförmige Phase** vapour phase
dampfgeheizt steam heated
Dampfkammer steam chamber
Dampfpermeation vapour permeation/diffusion
Dampfphase vapour phase
Dampfsperrbahn vapour barrier (sheeting)
Dampfsperre vapour barrier
Dampfsperrmasse vapour barrier composition
Dampfteildruck partial vapour pressure
Dampfundurchlässigkeit vapour impermeability
Dämpfung damping
Dämpfungseigenschaften damping properties/characteristics

Dämpfungseinrichtung damping mechanism
Dämpfungselement damping element
Dämpfungsfaktor damping factor
Dämpfungsmedium damping medium/fluid
Dämpfungsmodul damping modulus
Dämpfungsverhalten damping behaviour
Dämpfungsvermögen damping characteristics
Daten data
Datenabfrage data retrieval
Datenablage data filing system, data store
Datenabnahme data collection/acquisition/gathering
Datenabspeicherung data storage
Datenanalysator data analyser
Datenanzeige 1. data display. 2. data display unit
Datenarchivierung data filing
Datenaufbereitung data preparation
Datenaufschreibung data recording
Datenausgabe data output
Datenauswertung data evaluation
Datenbank data bank
Datenblatt data sheet
Datenblattangaben data sheet information, information given in data sheets
Dateneingabe data input
Datenerfassung data collection/acquisition/gathering
Datensammlung data collection/acquisition/gathering
Datentechnik data processing
Datenübertragung data transfer/transmission
Datenverarbeitung data processing
Datenverwaltung data management
dauerantistatisch permanently antistatic
Dauerbeanspruchbarkeit resistance to long-term stress, resistance to sustained loads
Dauerbeanspruchung long-term stress
Dauerbefeuchtung exposure to permanently damp/humid conditions
Dauerbelastbarkeit resistance to long-term stress
Dauerbelastung 1. long-term stress. 2. long-term exposure *(e.g. of painted surfaces to chemicals, weather etc.)*
Dauerbeständigkeit long-term stability
Dauerbetrieb, im in continuous operation
Dauerbetriebsgrenztemperatur maximum continuous service/operating temperature
Dauerbetriebstemperatur continuous service/operating temperature
Dauerbiegefestigkeit long-term flexural strength
Dauerbiegewechselfestigkeit flexural fatigue strength
Dauerbruch fatigue failure
Dauerdruckbelastung continuous compressive stress
Dauerdurchschlagfestigkeit long-term dielectric strength
Dauereinsatz continuous operation/use
dauerelastisch permanently flexible
Dauerelastizität permanent flexibility
Dauerfestigkeit 1. fatigue strength. 2. durability
Dauerfestigkeitsversuch fatigue test
Dauerfeuchtebeanspruchung exposure to permanently damp/humid conditions
dauerflexibel permanently flexible
Dauerflexibilität long-term flexibility
Dauergebrauch continuous use
Dauergebrauchseigenschaften long-term performance
Dauergebrauchstemperatur continuous working/operating temperature, long- term service temperature
Dauergebrauchstüchtigkeit long-term serviceability
dauerhaft durable, permanent, lasting, long-lasting
Dauerhaftigkeit durability, permanence
dauerklebrig permanently tacky
Dauerkontakt permanent/continuous contact
Dauerkühlung continuous cooling
Dauerlagerung 1. long-term ageing. 2. prolonged immersion
Dauerlast continuous/sustained load
Dauernaßbelastung exposure to permanently wet conditions
dauerplastisch permanently plastic
Dauerproduktion continuous production
Dauerschlagzähigkeit long-term impact strength
Dauerschwellfestigkeit fatigue strength
Dauerschwingfestigkeit fatigue strength
Dauerschwingversuch fatigue test
Dauerstandfestigkeit creep strength
Dauerstandversuch creep test
Dauertauchversuch long-term immersion test
Dauertemperatur sustained temperature
Dauertemperaturbelastung long-term exposure to high temperatures
Dauertemperaturbeständigkeit long-term heat resistance
Dauertemperaturgrenze long-term temperature limit
Dauertest 1. long-term test. 2. creep test
Dauertorsionsbiegebeanspruchung long-term torsional bending stress
Dauerverhalten long-term performance/behaviour
Dauerversuch long-term test
Dauervibration continuous vibration
Dauerwärmebelastbarkeit long-term heat resistance
Dauerwärmebeständigkeit long-term heat resistance, long-term thermal stability
Dauerwärmelagerung long-term heat ageing
Dauerwärmestabilität long-term heat resistance, long-term thermal stability
dauerwärmestandfest having long-term heat resistance
Dauerwechselfestigkeit resistance to long-term alternating stress
Dauerwitterungsstabilität long-term weathering resistance

DD-Lack polyurethane paint
DDK *abbr. of* **dynamische Differenzkalorimetrie**, differential scanning calorimetry, DSC
deblockiert unblocked
Decarboxylierung decarboxylation
Deckanstrich top coat, final coat, finish
Deckbeschichtung top coat, final coat, finish
Deckel lid
Deckenbelag ceiling tile
deckend opaque
Deckfähigkeit hiding power
Deckfilm top coat
Deckkraft hiding power
Decklack paint
Decklackbeschichtung top coat
Decklackenthaftung paint (film) detachment
Decklackierung top coat
Decklackschicht 1. top coat. 2. opaque finish: **Klar- und Decklackschichten** clear and opaque finishes
Decklasur glaze
Deckpulverlack powder coating
Deckschicht top coat
Deckstrich top coat
Deckstrichpaste top coating paste
Deckung opacity *(of a pigment or paint)*
Deckungsgrad hiding power, opacity *(of a pigment orpaint)*
Deckvermögen hiding power
defekt defective, faulty
Defekt defect, blemish
definiert clearly defined, definite
Deflockulation deflocculation
deflockulieren deflocculate
Deformation deformation
Deformationsamplitude deformation amplitude
Deformationsenergie deformation energy
Deformationsfestigkeit deformation resistance
Deformationsgeschwindigkeit 1. rate of deformation. 2. shear rate
Deformationsgrenze deformation limit
Deformationsmechanismus deformation mechanism
Deformationsspannung deformation stress
Deformationsverhalten deformation behaviour
Deformationswiderstand deformation resistance
deformierbar deformable
Deformierbarkeit deformability
deformieren to deform
Degradation degradation
degradiert degraded
dehnbar extensible
Dehnbarkeit 1.extensibilty, elasticity. 2. elongation at break
Dehnbeanspruchung 1. strain. 2. elongation, extension
Dehnbewegung movement due to expansion
dehnen to stretch
dehnfähig elastic, extensible
Dehnfähigkeit 1. extensibility, elasticity. 2. elongation at break
Dehnfuge expansion joint
Dehngeschwindigkeit rate of elongation/straining, straining rate
Dehngrenze yield point
Dehngrenzenspannung creep stress
Dehngrenzlinie creep curve
Dehnmeßstreifen strain gauge
Dehnspannung offset yield stress
Dehnspannungswert offset yield stress
Dehnspannwert offset yield stress
Dehnung 1. strain. 2. elongation, extension
Dehnung bei Streckgrenze elongation at yield
Dehnung bei Streckspannung elongation at yield stress, yield strain, tensile strain at yield
Dehnung beim Bruch elongation at break
Dehnung, bleibende residual/irreversible elongation
Dehnungsaufnehmer strain transducer
Dehnungsausschlag strain amplitude
Dehnungsfuge expansion joint
Dehnungsgeschwindigkeit rate of elongation/straining
Dehnungsgrad amount of strain/elongation
Dehnungsgrenze elastic limit
dehnungsinduziert strain-induced
Dehnungsinkrement strain increment
Dehnungsmeßstreifen strain gauge
Dehnungsspannung offset yield stress
Dehnungsverformung strain deformation
Dehnungsverhalten extensibility
Dehnungsvermögen extensibility
Dehnungswert strain, elongation
Dehnvermögen extensibility
Dekahydronaphthalin decahydronaphthalene
dekorativ decorative
Dekorierung decoration
Dekormaterial decorative material
Delamination delamination
Delaminationsrisiko risk of delamination
Delaminierung delamination
demineralisiert demineralised
Denkmal monument
Dentallabor dental laboratory
Depolymerisation depolymerisation
Deponie waste/rubbish tip/dump
Deprotonierung deprotonisation
Derivat derivative
Desagglomerieren deagglomeration
Desinfektionsmittel disinfectant
Desorption desorption
Desorptionsbedingungen desorption conditions
Desorptionsenergie desorption energy
Desorptionskinetik desorption kinetics
Desorptionskurve desorption curve
Desorptionsprozeß desorption process
Desorptionstemperatur desorption temperature
Desorptionszeit desorption time
Destillat distillate
Destillation distillation
Destillation, fraktionierte fractional distillation
Destillationsbedingungen distillation conditions

Destillationsgerät distillation apparatus
Destillationskolonne distillation column
Destillationsrückstände distillation residues
destillativ by distillation
destilliert distilled
Detektor detector
Detergentien detergents
detergentienbeständig detergent resistant
Detergentienfestigkeit detergent resistance
Detergentienlösung detergent solution
Dextrin dextrin
Di-2-ethylhexylphthalat diethylhexyl/dioctyl phthalate
Diacrylat diacrylate
Diacrylsäureester diacrylate
Dialdehyd dialdehyde
Dialkylaminoacrylat dialkylamino acrylate
Dialkylester dialkyl ester
Dialkylethanolamin dialkylethanolamine
Dialkylperoxid dialkyl peroxide
Dialkylperoxidgruppe dialkyl peroxide group
Dialkylphosphat dialkyl phosphate
Dialkylpolysiloxan dialkyl polysiloxane
Dialkylsulfid dialkyl sulphide
Dialkylverbindung dialkyl compound
Dialkylzinnlaurat dialkyltin laurate
Dialkylzinnmaleat dialkyltin maleate
Dialkylzinnmercaptid dialkyltin mercaptide
Dialkylzinnverbindung dialkyltin compound
Diallylammoniumverbindung diallyl ammonium compound
Diamin diamine
Diaminocarbonsäure diaminocarboxylic acid
Diaminodiphenylmethan diaminodiphenyl methane
Diaminopolydimethylsiloxan diamine polydimethyl siloxane
Diaminopolyether diaminopolyether
Diaminopropanol diaminopropanol
Dianhydrid dianhydride
Dianilinterephthalsäure dianiline terephthalic acid
Diarylgelbpigment diaryl yellow pigment
Diarylidgelb diarylide yellow
Diaryljodoniumgruppe diaryl iodonium group
Diarylperoxid diaryl peroxide
Diarylphosphat diaryl phosphate
Diatomenerde diatomaceous earth
Diazochinon diazoquinone
Diazoniumsalz diazonium salt
Diazoniumverbindung diazonium compound
Diazotierung diazotisation
Diazotierungsreaktion diazotisation reaction
Diazoverbindung diazo compound
dibasisch dibasic
Dibenzoylperoxid dibenzoyl peroxide
Dibutyladipat dibutyl adipate, DBA
Dibutylmaleat dibtyl maleate, DBM
Dibutylphthalat dibutyl phthalate, DBP
Dibutylsebacat dibutyl sebacate, DBS
Dibutylzinndiacetat dibutyltin diacetate
Dibutylzinndilaurat dibutyl tin dilaurate
Dicarbonsäure dicarboxylic acid
Dicarbonsäurechlorid dicarboxylic acid chloride
Dicarbonsäureester dicarboxylate
Dichlorchinacridon dichloroquinacridone
Dichlormethan dichloromethane
Dichroismus dichroism
dicht 1. dense. 2. tight, watertight, leakproof
Dichte density
Dichte, relative relative density, specific gravity
Dichtegradient density gradient
Dichten sealing
Dichtfaden sealant bead
Dichtfläche sealing face
dichtgepackt tightly packed
Dichtigkeit tightness, imperviousness
Dichtmaterial sealant
Dichtmittel sealant
Dichtring gasket
Dichtschnur sealant bead
Dichtstoff sealant
Dichtstoffindustrie sealant industry
Dichtung 1. joint, gasket. 2. seal
Dichtungsbahn waterproof(ing) membrane/sheet
Dichtungsband sealing tape
Dichtungsmasse sealant (compound)
Dichtungsmaterial sealant
Dichtungsmittel sealant
Dichtungsprofil sealing profile
Dichtungsring gasket
Dichtungssatz seal assembly
Dichtwirkung sealant effect
dick thick
Dickbettfliesenkleber thick-bed tile adhesive
Dickbettkleber thick-bed adhesive
Dickbettverfahren thick-bed method
Dicke thickness
Dickegradient thickness difference
Dickenabweichungen thickness variations
Dickenänderung change in thickness
Dickenschwankungen thickness variations
Dickentoleranz thickness tolerance
Dickenverlust reduction in thicknesss
Dickenverteilung thickness distribution
dickflüssig high-viscosity, viscous
Dickflüssigkeit high viscosity
dickpastös thick and paste-like
Dickschichtanstrich high-build finish
Dickschichtfarbe high-build paint
dickschichtig high-build
Dickschichtlack high-build paint
Dickschichtlasur high-build glaze
Dickschichtprimer high-build primer
Dickschichtsystem high-build system/formulation
Dickschichtüberzug high-build coating
dickviskos high-viscosity
Dicumylperoxid dicumyl peroxide
Dicyandiamid dicyanodiamide
Dicyclohexylether dicyclohexyl ether
Dicyclohexylphthalat dicyclohexyl phthalate

Dicyclopentadien dicyclopentadiene
Dicyclopentylacrylat dicyclopentyl acrylate
Dielektrikum dielectric
dielektrisch dielectric
dielektrische Verlustzahl loss index
dielektrischer Verlustfaktor dissipation factor
Dielektrizitätskonstante dielectric constant
Dielektrizitätskonstante, relative relative permittivity, dielectric constant
Dielektrizitätszahl dielectric constant, relative permittivity
Dien diene
Dienbasis, auf diene-based
Dienharz diene resin
Dienkautschuk diene rubber
Dienmonomer diene monomer
Dienöl diene oil
Dienpfropfkautschuk grafted diene rubber
Dienstleistung service
Diepoxid diepoxide
Diethanolamin diethanolamine
Diethylentriamin diethylene triamine
Diethylethanolamin diethyl ethanolamine
Diethylether (diethyl) ether
Diethylphthalat diethyl phthalate, DEP
Diethylsebacat diethyl sebacate, DES
Differential-Pulspolarographie differential pulse polarography
Differentialgleichung differential equation
Differentialkalorimetrie differential calorimetry
differentialkalorimetrisch differential calorimetric
Differentialphotodiode differential photodiode
Differentialthermoanalyse differential thermoanalysis
Differentialthermoanalysengerät differential thermoanalyser
differentialthermoanalytisch differential-thermoanalytical
Differentialthermometrie differential thermometry
Differentialwaage differential scales
Differenzdruck differential pressure, pressure difference
Differenzdrucksensor differential pressure transducer
Differenzdruckumformer differential pressure transducer
Differenzierung differentiation
Differenzkalorimetrie differential calorimetry
Differenzspektrometrie differential spectrometry
Differenztemperatur temperature differential
Differenzwärmestromkalorimetrie differential heat flow calorimetry
Diffraktometrie diffractometry
diffundieren to diffuse
diffus diffuse
Diffusion diffusion
Diffusionsbarriere diffusion barrier
diffusionsdicht diffusion resistant
Diffusionsdurchlaßwiderstand diffusion resistance
Diffusionsfähigkeit permeability
diffusionsfest diffusion resistant
Diffusionsfestigkeit diffusion resistance
Diffusionsgeschwindigkeit diffusion rate
Diffusionsgesetz diffusion law
diffusionshemmend diffusion-inhibiting
Diffusionsklebung diffusion bonding
Diffusionskoeffizient diffusion coefficient, diffusivity
Diffusionskonstante diffusion constant
Diffusionsrate diffusion rate
diffusionsregulierend diffusion-controlling
Diffusionsverhalten diffusion characteristics
Diffusionsvorgang diffusion process
Diffusionswiderstand diffusion resistance
Diffusionswiderstandszahl diffusion resistance coefficient
Diffusionszeit diffusion time
Diffusivität diffusivity, diffusion coefficient
Difluordichlormethan difluorodichloromethane
difunktionell bifunctional
digital digital
Diglycidylalkan diglycidyl alkane
Diglycidylether diglycidyl ether
Diglycidylverbindung diglycidyl compound
Diglykoleinheit diglycol group
Dihydrazid dihydrazide
Dihydroxyverbindung dihydroxy compound
Diisobutylen diisobutylene
Diisobutylketon diisobutyl ketone
Diisocyanat diisocyanate
Diisopropylketoxim diisopropyl ketoxime
Dilatometer dilatometer
dilatometrisch dilatometric
Dimensionieren dimensioning
Dimensionsänderungen dimensional changes
dimensionsbeständig dimensionally stable
Dimensionsbeständigkeit dimensional stability
Dimensionsgenauigkeit dimensional accuracy
dimensionsstabil dimensionally stable
Dimensionsstabilität dimensional stability
dimer dimeric
Dimer dimer
Dimeralkohol dimerised alcohol
Dimerisat dimer
Dimerisation dimerisation
Dimerisationsprodukt dimer
dimerisiert dimerised
Dimerisierung dimerisation
Dimerisierungsvorgang dimerisation reaction
Dimersäure dimeric acid
Dimethacrylatklebstoff dimethacrylate adhesive
Dimethanolpropionsäure dimethanol propionic acid
Dimethylacrylat dimethyl acrylate
Dimethyladipat dimethyl adipate
Dimethylalkylamin dimethyl alkylamine
Dimethylamid dimethylamide
Dimethylaminoethanol dimethylamino ethanol
Dimethylbenzol dimethyl benzene
Dimethylbenzylamin dimethyl benzylamine
Dimethylcarbonat dimethyl carbonate

Dimethylchinacridon dimethyl quinacridone
Dimethylethanolamin dimethyl ethanolamine
Dimethylformamid dimethyl formamide
Dimethylglutarat dimethyl glutarate
Dimethylolcarbonsäure dimethylolcarboxylic acid
Dimethylolharnstoff dimethylol urea
Dimethylolpropionsäure dimethylolpropionic acid
Dimethylphenol dimethyl phenol
Dimethylphthalat dimethyl phthalate, DMP
Dimethylpolysiloxan dimethyl polysiloxane
Dimethylsulfat dimethyl sulphate
Dimethylsulfoxid dimethyl sulphoxide
Dimethylterephthalat dimethyl terephthalate
Dimethylvinylsilanol dimethylvinyl silanol
Dinatriumsalz disodium salt
Dioctyladipat dioctyl adipate, DOA
Dioctylazelat dioctyl azelate, DOZ
Dioctylfumarat dioctyl fumarate, DOF
Dioctylmaleinat dioctyl maleinate, DOM
Dioctylphthalat dioctyl phthalate, DOP
Dioctylsebacat dioctyl sebacate, DOS
Diol diol
Diolacrylat diol acrylate
Diolefinkautschuk diolefin rubber
Diolefinpolymerisat diolefin polymer
Diolmethacrylat diol methacrylate
Diolpolyether diol polyether
Diolverbindung diol compound
Diorganopolysiloxan diorganopolysiloxane
Diorganosilan diorganosilane
Dioxan dioxan
Dioxin dioxin
Dioxinausstoß dioxin emission
Dioxolanon dioxolanone
Dipentaerythrit dipentaerythritol
Dipeptid dipeptide
Diphenylamin diphenylamine
Diphenylester diphenyl ester
Diphenylether diphenyl ether
Diphenylharnstoff diphenyl urea
Diphenyliodoniumsalz diphenyl iodonium salt
Diphenylmethan diphenylmethane
Diphenylmethandiisocyanat diphenylmethane diisocyanate, MDI
Diphenylnitrosamin diphenyl nitrosamine
Diphenylphthalat diphenyl phthalate
Diphenylsulfid diphenyl sulphide
Diphenylsulfongruppe diphenylsulphone group
Diphenylthioharnstoff diphenyl thiourea
Dipol dipole
Dipol-Dipol-Anziehung dipole-dipole attraction
Dipolbindung dipole bond
Dipolcharacter dipole character
Dipolkräfte dipole forces
Dipolmolekül dipole molecule
Dipolmoment dipole moment
Dipolorientierung dipole orientation
Dipolwechselwirkung dipole interaction
Direktverglasung direct glazing
Disilan disilane
Disilylsiloxanharz disilylsiloxane resin
diskontinuierlich discontinuous
Dispergieradditiv dispersing agent
Dispergieraggregat dispersing unit
Dispergieranlage stirrer, mixer, dispersing unit
dispergierbar dispersible
Dispergierbarkeit dispersibility
Dispergierbedingungen dispersion conditions
dispergieren to disperse
Dispergierenergie dispersion energy
Dispergierergebnis dispersing effect
Dispergierfähigkeit dispersibility
Dispergiergrad degree of dispersion
Dispergiergut millbase
dispergierhart difficult to disperse
Dispergierhilfe dispersion aid
Dispergierhilfsmittel dispersing agent
Dispergierleistung dispersing efficiency
Dispergiermedium dispersion medium
Dispergiermittel dispersing agent
Dispergierprobleme dispersing problems
Dispergierprozeß dispersing process/operation
dispergiert dispersed
dispergierte Phase disperse phase
Dispergierung dispersion
Dispergierwirksamkeit dispersing efficiency
Dispergierwirkung dispersing effect
Dispergierzeit dispersing time
Dispergierzusatz dispersing agent
disperse Phase disperse phase
Dispersion dispersion, emulsion
Dispersionsacrylat acrylic dispersion
Dispersionsanstrichmittel emulsion paint
Dispersionsanstrichstoff emulsion paint
Dispersionsbasislack emulsion undercoat
Dispersionsbindemittel dispersion binder
Dispersionsdichtstoff dispersion (based) sealant
Dispersionsfarbe emulsion paint
Dispersionsfassadenfarbe exterior/masonry emulsion paint
dispersionsgebunden dispersion-based *(paint)*
dispersionsgestrichen emulsion painted
Dispersionsgrundierung 1. emulsion primer. 2. emulsion primer coat
Dispersionsharz redispersible resin
Dispersionshilfsstoff dispersing agent
Dispersionskleber emulsion/dispersion/latex/water-based adhesive
Dispersionsklebmittel emulsion/dispersion/latex/water-based adhesive
Dispersionsklebstoff emulsion/dispersion/latex/water-based adhesive
Dispersionskraft dispersive force
Dispersionslack emulsion paint
Dispersionslackfarbe emulsion paint
Dispersionsmedium dispersion medium
Dispersionsmittel dispersing agent

Dispersionspartikel dispersion particle
Dispersionspolymer emulsion polymer
Dispersionspolymerisation emulsion polymerisation
Dispersionspulver redispersible powder
Dispersionssilikatfarbe silicate emulsion paint
Dispersionsstabilisator dispersion stabiliser
dispersionsstabilisierendes Additiv dispersing agent
Dispersionsstabilität dispersion stability
Dispersionsteilchen dispersion particle
Dispersionsträger dispersion medium
Dissolver high-speed stirrer/mixer, agitator. *Since a* **Dissolver** *is mainly used to mix and disperse, rather than to dissolve, it is best to use one of the terms given here, rather than the literal* dissolver.
Dissoziation dissociation
Dissoziationsbedingungen dissociation conditions
Dissoziationsenergie dissociation energy
Dissoziationsgleichgewicht dissociation equilibrium
Dissoziationsgrad degree of dissociation
Dissoziationskonstante dissociation constant
dissoziiert dissociated
Dissoziierung dissociation
Distanzhalter spacer, distance piece
Distanzring spacer ring
Distanzscheibe spacer disc
Distanzstück spacer
Disulfid disulphide
Dithiocarbamat dithiocarbamate
Diurethan diurethane
Divinylbenzol divinyl benzene
DMA dynamic-mechanical analysis
DMTA dynamic-mechanical thermoanalysis
Dochteffekt wick effect
Dodecylamin dodecylamine
Dokumentation documentation, documents
Dolomit dolomite
Domäne domain
Doppelbindung double bond
Doppelbindungsanteil double bond content
doppelbindungsfrei free from double bonds
Doppelbindungsgehalt double bond content
doppelbrechend double refractory, birefringent
Doppelbrechung double refraction, birefringence
Doppelschicht double layer
doppelseitig double-sided
doppelseitig verstärkter T-Stoß strap-supported joint
doppelwandig double-walled
Dornbiegeprüfung mandrel flex test
Dornbiegeversuch mandrel flex test
Dose can
Doseninnenschutzlack can lacquer
Dosenlack can lacquer
Dosier- und Mischmaschine metering and mixing unit
Dosieraggregat metering/feed/dispensing unit
Dosieranlage metering/feed/dispensing equipment
Dosierautomat automatic metering unit, automatic feeder/dispenser
Dosierbehälter feed tank
Dosiereinheit metering/feed/dispensing unit
Dosiereinrichtung metering/feed/dispensing equipment
dosieren to meter, to feed, to dispense, to add, to incorporate
Dosiergenauigkeit metering accuracy
Dosiergerät metering/feed unit
Dosiermenge metered/required amount
Dosierpumpe metering pump
Dosiersystem metering/feed/dispensing system
Dosierung 1. metering, feeding, adding, amount (added). 2. metering/feed unit
Dosierungenauigkeit inaccurate metering
Dosiervorrichtung metering/feed/dispensing unit
Dosierwaage weigh feeder
Dosierwerk metering/feed/dispensing unit
Dosierzylinder metering/feed cylinder
Dotierkomponente doping agent
dotiert doped
Dotierung doping
Dotierungsmittel doping agent
Drahtbürste wire brush
Drahtgewebe wire gauze/cloth
Drahtgitter wire netting
Drahtlack wire enamel
Drahtlackierung wire coating
Drahtnetz wire mesh/gauze
Drahtsieb wire mesh/gauze
drehbar rotatable
Drehbewegung rotary movement
Drehen turning
drehend rotating
Drehfilter rotary filter
Drehmasse torsional mass
Drehmoment torque
Drehmomentabfall torque decrease
Drehmomentanstieg torque increase
Drehmomentaufnehmer torque sensor/transducer
Drehmomentdefizit insufficient torque
Drehmomenteinstellung 1. torque adjustment. 2. torque adjusting mechanism
Drehmomenteröhung torque increase
Drehmomentgrenze maximum torque
drehmomentkonstant with constant torque
Drehmomentkorrektur torque correction
Drehmomentmaximum maximum torque
Drehmomentrheometer torque rheometer
Drehmomentsensor torque sensor/transducer
Drehmomentstab torque wrench
Drehmomentveränderung change in torque
Drehmomentverstärker torque amplifier
Drehmomentwaage torque meter
Drehrichtung direction of rotation
Drehrohrofen rotary tubular furnace
Drehschwingung torsional vibration

Drehsinn direction of rotation
Drehstab torsion bar
Drehstrom three-phase current
Drehstromgetriebemotor three-phase geared motor
Drehstrommotor three-phase motor
Drehteller rotary table/platform
Drehtisch rotary table/platform
Drehung rotation
Drehzahl 1. number of revolutions. 2. speed *(of a rotating element)*
Drehzahl, regelbare variable speed
drehzahlabhängig depending on speed
Drehzahländerung change in speed
Drehzahlanzeige speed indicator
Drehzahlbereich speed range
Drehzahleinstellung speed setting
Drehzahlerhöhung speed increase
Drehzahlerniedrigung speed reduction/decrease
Drehzahlgeber speed transducer/sensor
Drehzahlgrenze maximum speed
Drehzahlistwert actual speed
Drehzahlkonstanz constant speed
Drehzahlmesser speed counter
Drehzahlprogramm speed program
Drehzahlregelbereich speed range
Drehzahlregelung speed control
Drehzahlregler speed regulator
Drehzahlregulierung speed control
Drehzahlschwankungen speed variations
Drehzahlsollwert required speed
Drehzahlsteigerung speed increase
Drehzahlsteigerungsrate rate of speed increase
Drehzahlsteuerung 1. speed control. 2. speed control mechanism/system/device
drehzahlunabhängig independent of speed
drehzahlvariabel (of) variable speed
drehzahlveränderlich (of) variable speed
dreibasisch tribasic
dreibindig trivalent
dreidimensional three-dimensional
dreieckig triangular
dreifach triple, treble, threefold
Dreifachbindung triple bond
Dreihalskolben three-neck flask
dreiphasig three-phase
Dreipunktbiegebelastung three-point bending stress
Dreischichtsystem three-coat system/formulation
Dreiwalze triple roller, triple roll mill
dreiwertig trivalent
Druck 1. pressure, compression. 2. printing. 3. print, printed image
Druck-E-Modul modulus of elasticity in compression
Druck-Elastizitätsmodul modulus of elasticity in compression
Druckabbau pressure decrease/reduction
Druckabfall pressure decrease
druckabhängig pressure-dependent, depending on the pressure
Druckänderung change in pressure
Druckanstieg pressure increase, increase in pressure
Druckanwendung application of pressure
Druckanzeige pressure gauge
Druckanzeigegerät pressure gauge
Druckaufbau pressure build-up
Druckaufbringung application of pressure
Druckaufgabe application of pressure
Druckaufnehmer pressure transducer/sensor
Druckausgleich pressure compensation/equalisation
Druckbeanspruchbarkeit compression resistance
druckbeansprucht under pressure, under compressive stress
Druckbeanspruchung 1. compressive stress. 2. (applied) pressure
Druckbedarf pressure requirement(s)
Druckbehälter pressurised tank/vessel
druckbelastbar resistant to compression, resistant to compressive stress
Druckbelastbarkeit compressive strength
druckbelastet under pressure, under compressive stress
Druckbelastung 1. compressive stress. 2. (applied) pressure
Druckbereich pressure range
Druckbild printed image
Druckbindemittel printing ink binder
druckdicht resistant to pressure
Druckdifferenz pressure difference, difference in pressure
Druckdose pressure transducer
Druckeigenspannungen internal compressive stresses
Druckeinbruch drop in pressure
Druckeinwirkung action/effect/application of pressure
druckempfindlich pressure sensitive
Drucken printing
Druckentlastung pressure release/relief
Druckentspannung pressure release/relief
Drucker printer
Druckerschwärze printer's black
Druckertinte printing ink
Druckfarbe printing ink
Druckfarbenbindemittel printing ink binder
Druckfarbenfilm printing ink film
Druckfarbenharz printing ink resin
Druckfarbenlösemittel printing ink solvent
Druckfarbensektor printing ink industry/sector
druckfest resistant to pressure/compression
Druckfestigkeit compressive/crushing strength
Druckfilter pressure filter
Druckfirnis printing varnish
Druckfühler pressure transducer/sensor
Druckgeber pressure transducer/sensor
Druckgefälle pressure gradient
Druckgefäß pressurised tank/vessel
Druckgleichgewicht equilibrium pressure

Druckgradient pressure gradient
Druckharz printing ink resin
Druckindustrie printing industry
druckinduziert pressure-induced
Druckistwert actual pressure
Druckkammer compression chamber
Druckkessel pressure vessel
Druckkleber pressure sensitive adhesive
Druckknopf push button
Druckkontrollsystem pressure control system
Druckkraft compressive force
Drucklast pressure, compressive stress
drucklos pressure-less, at normal/atmospheric pressure
Druckluft compressed air
druckluftbetrieben pneumatically operated
Druckluftpistole compressed air spraygun
Druckluftspritzpistole compressed air spraygun
Druckluftzylinder compressed air cylinder
Druckmanometer manometer
Druckmaximum peak/maximum pressure
Druckmedium hydraulic fluid
Druckmeßaufnehmer pressure transducer/sensor
Druckmeßdose pressure transducer/sensor
Druckmesser pressure transducer/sensor
Druckmeßgeber pressure gauge
Druckmeßgerät pressure gauge
Druckmesßumformer pressure transducer/sensor
Druckminderung pressure reduction, reduction in pressure
Druckmodul compressive modulus
Druckobergrenze maximum pressure
Drucköl hydraulic oil
Druckölspeicher hydraulic accumulator
Druckplatte 1. pressure plate. 2. printing plate
druckpolymerisiert high-pressure polymerised
Druckprofil pressure profile
Druckprüfung compression test
Druckreaktor pressure reactor
Druckreduzierung pressure reduction, reduction of pressure
Druckreduzierventil pressure relief/release valve
Druckschalter push-button switch
Druckschärfe crispness of the printed image
Druckscherfestigkeit compressive shear strength
Druckscherrestfestigkeit residual compressive shear strength
Druckscherversuch compressive shear test
Druckschrift leaflet, publication
Druckschwankungen pressure variations/fluctuations
Druckschwellwert threshold pressure
Druckschwingungen pressure variations/fluctuations
Drucksensor pressure transducer/sensor
Drucksollwert required/set pressure
Drucksonde pressure transducer/sensor
Druckspannung compressive stres
Drucksteifigkeit compressive strength
Drucksubstrat printing surface
Drucktaste push button
drucktastengesteuert push-button controlled
Drucktastenschalter push-button switch
Drucktastensteuerung 1. push-button control. 2. push-button control unit
Drucktinte printing ink
Druckübertragung pressure transfer
Druckumformer pressure transducer
druckunabhängig independent of pressure
Druckunabhängigkeit independence of pressure
Druckunterschied pressure difference, difference in pressure
Druckverfahren printing process
Druckverlauf 1. pressure profile. 2. changes in pressure
Druckverlaufkurve pressure curve
Druckverlust pressure loss/drop, loss of pressure
Druckversuch compression test
Druckwalze printing roller/cylinder
DSC differential scanning calorimetry
DSC-Gerät differential scanning calorimeter
DSC-Messung diffferential scanning calorimetry
DSC-Wert DSC figure
DSD *abbr. of* Duales System Deutschland, Dual System *(see explanatory note under* Duales System*)*
DTMA dynamic thermomechanical analysis
Duales Abfallsystem Dual System *(see explanatory note under* Duales System*)*
Duales System Dual System *(this forms part of the German Packaging Directive which obliges packaging producers and distributors to set up their own system for collecting and recycling used transit and post-consumer packaging materials. The full German name is* **Duales System Deutschland***)*
Duktilität ductility
dunkel dark
dunkelbraun dark brown
dunkelfarbig dark coloured
dunkelgrau dark grey
dunkelpigmentiert dark coloured
Dunkelverfärbung darkening
Dunkelwerden darkening
dünn thin
Dünnbettfliesenkleber thin-bed tile adhesive
Dünnbettkleber thin-bed adhesive
Dünnbettverfahren thin-bed method
dünnflüssig low-viscosity, runny
Dünnflüssigkeit low viscosity
Dünnsäure spent acid
Dünnschichtbeschichtung thin film/coating
Dünnschichtchromatographie thin-layer chromatography
dünnschichtchromatographisch thin-layer chromatographic
dünnschichtig thin
Dünnschichtlasur thin-/low-build glaze
Dünnschichtsystem low-build

system/formulation/paint
Dünnschliff microsection
Dünnschnitt microtome section
Durchbiegung deflection
Durchbiegung beim Bruch deflection at break
durchfeuchtet wet through
Durchfluß 1. flow rate. 2. throughput
Durchflußfühler flow sensor, flow sensing element
Durchflußgeschwindigkeit flow rate
Durchflußleistung 1. flow rate. 2. throughput
Durchflußmenge 1. flow rate. 2. throughput
Durchflußmengenmesser flowmeter
Durchflußmeßgerät flowmeter
Durchflußmessung measurement/determination of flow rate
Durchflußschwankungen flow/throughput fluctuations
Durchflußstrom 1. flow. 2. flow rate
Durchflußwiderstand flow resistance
Durchflußzeit flow time *(in viscosity determinations)*
Durchführbarkeit feasibility, practicability
Durchgangswiderstand volume resistance
Durchgangswiderstand, spezifischer volume resistivity
durchgefärbt self-coloured
durchgehärtet fully cured
durchgetrocknet completely dry
Durchhärtung ful/complete cure
Durchhärtungszeit time to achieve complete cure
durchlässig permeable
Durchlässigkeit permeability
Durchlässigkeitseigenschaften permeability characteristics
Durchlässigheitswert permeability coefficient
Durchlaufanlage continuous plant, continuously operating plant
Durchlaufbetrieb continuous operation
Durchlaufgeschwindigkeit throughput speed/rate
Durchlaufofen tunnel oven
Durchlicht transmitted light
Durchlichtelektronenmikrosokop transmission electron microscope, TEM
Durchlichtmikroskopie optical microscopy
Durchlüftung ventilation
Durchmesser diameter
Durchmessererhöhung increase in diameter
Durchmesservergrößerung increase in diameter
durchnäßt soaked, wet through
Durchrostung rust penetration
Durchsatz throughput
Durchsatz, volumetrischer volume throughput, volumetric flow rate
Durchsatzeinbuße reduction in throughput
Durchsatzleistung throughput (rate)
Durchsatzmenge throughput (rate)
Durchsatzrate throughput (rate)
Durchsatzreduktion reduction in throughput
Durchsatzregulierung 1. flow control. 2. flow control device
Durchsatzsteigerung increase in throughput
durchsatzunabhängig independent of throughput
durchscheinend translucent
Durchschlagen strike-through
Durchschlagfeldstärke breakdown field strength
Durchschlagfestigkeit dielectric strength
Durchschlagspannung breakdown voltage
Durchschlagversuch breakdown test
Durchschnittskosten average costs
Durchschnittsleistung average output
Durchschnittsprobe representative sample
Durchschnittswert mean/average value
durchsichtig transparent
Durchstoßofen tunnel oven
Durchstoßversuch penetration test
Durchstoßweg deformation
durchtränkt impregnated
Durchtränkung impregnation
durchtrocknen to hard-dry, to become completely dry
durchvernetzt fully cured/crosslinked
Durchvulkanisation complete vulcanisation/cure
duromer thermosetting
Duromer thermoset (material)
Duroplast thermoset
duroplastisch thermoset(ting)
Düse nozzle
dynamisch dynamic
dynamisch-mechanische Analyse dynamic-mechanical analysis
dynamisch-thermomechanische Analyse dynamic-thermomechanical analysis
dynamische Differenzkalorimetrie differential scanning calorimetry
dynamische Viskosität dynamic viscosity

E

E-Modul modulus of elasticity, elastic modulus
E-Zugsmodul tensile modulus of elasticity
EB-Bandbeschichtung electron beam curing coil coating
EB-Lack electron beam curing paint
EB-Produktionslinie electron beam curing line
EB-Verfahren electron beam curing
EBC electron beam curing
eben plane
EBU-Harz stoving urethane resin
ECD electron capture detector
echt true *(e.g. solution)*
Echtzeitverfahren real time process
Ecke corner

Eckenfuge corner joint
Eckverbindung corner joint
edelmetalüberzogen rare metal-coated
Edelputz facing plaster
Edelstahl stainless steel
EDTA 1. ethylene diamine tetra-acetic acid. 2. ethylene dimanine tetra- acetate
EEL-Wert emergency exposure limit, EEL
Effektivvolumen effective volume
Effektpigment special effect pigment
Eiche oak
Eichenfurnier oak veneer
Eichenholz oak
Eichfaktor calibration factor
Eichkurve calibration curve
Eichmarke (calibration) mark
Eichmethode method of calibration
Eichsubstanz calibrating substance
Eichung calibration
Eichungsfaktor calibration factor
Eichvorrichtung calibrating device/unit
Eigenfarbe inherent colour
Eigenfestigkeit inherent strength
Eigenfrequenz natural frequency
Eigengeruch inherent smell/odour
Eigenklebrigkeit inherent tack
Eigenkontrolle in-house control
Eigenplastizität inherent plasticity
Eigenschaft property
Eigenschaften, kennzeichnende characteristic properties/features
Eigenschaften, mechanische mechanical properties
Eigenschaften, physikalische physical properties
Eigenschaften, schaumzerstörende antifoam properties
Eigenschaften, thermische thermal properties
Eigenschaften, toxikologische toxicological properties
Eigenschaftsänderung change in properties
Eigenschaftsanforderungen required properties
Eigenschaftsanisotropie anisotropic properties
Eigenschaftsbereich range of properties
eigenschaftsbezogen property-related
Eigenschaftsbild (general) properties
Eigenschaftseinbuße deterioration of properties
Eigenschaftskennwerte properties
Eigenschaftskombination combination of properties
Eigenschaftsmerkmale properties, characteristics
Eigenschaftsniveau properties: **hohes Eigenschaftsniveau** excellent properties
Eigenschaftsparameter property
Eigenschaftsprofil (general) properties
Eigenschaftsrichtswerte typical properties
Eigenschaftsschwankungen property variations
Eigenschaftsspektrum range of properties
Eigenschaftstabelle table of properties
Eigenschaftsveränderung change in properties
eigenschaftsverbessernd property-enhancing
Eigenschaftsvergleich comparison of properties
Eigenschaftsverhalten properties
Eigenschaftswert property
eigensicher intrinsically safe
Eigenspannung internal stress
Eigenspannungsfeld internal stresses
eigenspannungsfrei free from internal stresses
Eigenspannungszustand state of internal stress
Eigenstabilität inherent stability
eigenständig independent
Eigenviskosität inherent viscosity
Eignungsprüfung suitability test
EIM *abbr. of* **Elektroimpedanzmessung**, electroimpedance determination
Eimer bucket, tub
einachsial uniaxial
einachsig uniaxial
einarbeiten to incorporate, to mix in
einatmen to inhale
Einatmen inhalation
Einbetten embedding, potting
Einbettmaterial potting/embedding compound
Einbettungsmasse potting/embedding compound
Einbettungsmittel potting/embedding compound
Einbeulen denting, indentation
einbindig monovalent
Einbrennalkyd stoving alkyd (resin)
Einbrennalkydharz stoving alkyd (resin)
einbrennbar stovable
Einbrennbedingungen stoving/baking conditions
Einbrennbereich stoving/baking temperature range
Einbrenndauer stoving/baking time
Einbrenndecklack stoving/baking paint
einbrennen to stove/bake
einbrennfest resistant to stoving (temperatures)
Einbrennfüller stoving/baking filler
Einbrenngeschwindigkeit stoving rate
Einbrenngrundierung stoving/baking primer
Einbrenngrundlack stoving/baking primer
Einbrennhärter stoving hardener/catalyst
Einbrennharz stoving /baking resin
Einbrennklarlack clear stoving/baking varnish
Einbrennlack stoving paint/enamel
einbrennlackiert stove enamelled
Einbrennlackierung stoved/baked finish
Einbrennlacksystem stoving paint/enamel
Einbrennofen stoving oven
Einbrennprozeß stoving/baking process
Einbrennrückstand stoving residue
Einbrennstrukturlack textured stoving paint
Einbrennsystem stoving paint/enamel/lacquer
Einbrenntauchgrundierung stoving dip primer
Einbrenntemperatur stoving/baking temperature
Einbrenntemperaturbereich stoving/baking temperature range
Einbrenntyp stoving/baking resin
Einbrennurethanharz stoving urethane resin
Einbrennversuch stoving test/trial

Einbrennvorgang stoving/baking (process)
Einbrennzeit stoving/baking time
eindampfen to concentrate by evaporation
eindicken to concentrate
Eindickmittel thickener, thickening agent
Eindickung thickening
Eindickungsadditiv thickener, thickening agent
Eindickungsanfälligkeit tendency to thicken
eindiffundieren to diffuse into
eindimensional unidimensional
eindispergieren to mix/stir in
Eindringamplitude penetration amplitude
Eindringen penetration
Eindringgeschwindigkeit penetration rate, speed of penetration
Eindringgrad degree of penetration
Eindringhärte penetration hardness
Eindringkoeffizient penetration coefficient
Eindringtiefe 1. depth of penetration. 2. depth of indentation *(when measuring ball indentation hardness)*
Eindringvermögen penetrating power
Eindruckfestigkeit indentation resistance
Eindruckhärte indentation resistance
Eindruckkörper indenter
Eindrucktiefe depth of indentation
Eindruckversuch indentation test
Eindruckweg depth of indentation
Eindruckwiderstand indentation resistance
eindunsten to concentrate by evaporation
Einelektronenoxidation single-electron oxidation
Einfachbindung single bond
einfallend incident
Einfallswinkel angle of incidence
einfärbbar capable of being pigmented
Einfärbbarkeit pigmentability
einfärben to colour, to pigment
Einfärbung 1. colouring, pigmentation. 2. colour
Einflußfaktor influencing factor
Einflußgröße influencing factor
Einflußhöhe degree of influence
Einflußparameter influencing factor
Einfrierbereich glass transition range
Einfriertemperatur glass transition temperature
Einfriertemperaturbereich glass transition range
Einfrierungstemperatur glass transition temperature
Eingangskonzentration 1. original/initial concentration. 2. inlet concentration
Eingangspartikelgrößenverteilung original particle size distribution
Eingangstemperatur 1. original/initial temperature. 2.inlet temperature
eingebrannt stoved, baked
eingedickt thickened, concentrated
eingedrungen penetrated
eingefärbt coloured, pigmented
eingeschlossen entrapped
eingeschränkt limited
Einkapseln embedding, potting
Einkomponentendichtstoff one-pack sealant
Einkomponentendichtmasse one-pack sealant
Einkomponentenkleber one-pack adhesive
Einkomponentenklebstoff one-pack adhesive
Einkomponentenlack one-pack/-part paint
Einkomponentensilikonkautschuk one-pack silicone rubber
Einkomponentensystem one-pack/-component system/formulation
Einkomponentenurethansystem one-pack polyurethane system/paint/formulation
einkomponentig one-pack
einkondensiert condensed into the molecule
einkristallin mono-crystalline
Einlagerung immersion
Einlagerungsversuch immersion test
Einlagerungszeit time of immersion
Einlaßventil inlet valve
Einlauf inlet
einphasig single-phase
einpolymerisiert polymerised into the molecule
Einsatz use, usage, employment
Einsatzbedingungen conditions of use
einsatzbereit ready for use
Einsatzbreite range of uses/applications
Einsatzdeterminanten decisive factors for using...
einsatzfähig serviceable, fit for use, in working order
Einsatzgebiet field/area of use
Einsatzgrenztemperatur maximum working/operating temperature
Einsatzkonzentration concentration, amount used/added
Einsatzmenge amount (used) **nur mit wesentlich größeren Einsatzmengen...** only with much larger amounts...
Einsatzmöglichkeiten application possibilities, possible uses/applications
Einsatzschwerpunkt main application, most important application
Einsatzspektrum range of uses/applications
einsatzspezifisch application-oriented
Einsatztemperatur operating temperature
Einsatzzweck intended application
Einschichtbetrieb 1. one-shift operation. 2. one-shift plant
Einschichtdecklack one-coat finish
Einschichtengrundierung one-coat primer
einschichtig one-coat
Einschichtklarlack one-coat clear varnish/lacquer
Einschichtlack one-coat paint/varnish
Einschichtlackierung 1. one-coat paint application. 2. one-coat paint film
Einschichtsystem one-coat system/paint/formulation
einschnittig überlappt single-lap
einschnittige Laschung single strap joint, butt strap joint
einschnittige Überlappung simple lap joint
einseitig on one side, one-sided

Einspannhilfe clamping device
Einspannklemme clamp
Einspannlänge clamping distance
Einspannvorrichtung clamping device
Einsparung saving
Einstellung adjustment
Einstoffverpackung single-material pack
Einstrahlung irradiation
Eintauchen immersion
Eintauchzeit immersion time
Eintopfsystem one-pack system/paint/formulation
Einwirkungsdauer time of exposure
Einzelbestandteile individual components
Einzelbindemittel individual binders
Einzelhändler retailer
Einzelkomponenten individual components
Einzelmeßwerte individual results
Einzelmessungen individual determinations
Einzelmoleküle individual molecules
Einzelpartikel individual particle
Einzelprüfungen individual tests
Einzelresultate individual results
Einzelschicht individual layer
Einzelteichen individual particle
Eis ice
EIS electrochemical impedance spectroscopy
Eisen iron
Eisen(II)-Sulfat ferrous sulphate
Eisen-Konstantanthermoelement iron-constantan thermocouple
Eisenacetylacetonat iron acetyl acetonate
Eisenbahnkesselwagen rail tanker
Eisenblech 1. iron (test) panel. 2. iron/steel sheet
Eisenchlorid iron/ferric chloride
Eisenglimmer micaceous iron ore
eisenhaltig containing iron
Eisenhydrat iron/ferric hydroxide
Eisenhydroxid iron/ferric hydroxide
Eisenmetall ferrous metal
Eisenoxid iron/ferric oxide
Eisenoxidaufschlämmung iron oxide slurry
Eisenoxidbraun brown iron oxide
Eisenoxidgelb yellow iron oxide
Eisenoxidhydrat iron/ferric hydroxide
Eisenoxidpigment iron oxide pigment
Eisenoxidrot red iron oxide
Eisenoxidrotpigment red iron oxide pigment
Eisenoxidschwarz black iron oxide
Eisenoxidschwarzpigment black iron oxide pigment
Eisenrohr iron pipe
Eisensalz iron salt
Eisenspuren traces of iron
Eisenuntergrund iron substrate/surface
Eisessig glacial acetic acid
Eiweiß albumen, protein
elastifizierend elasticising
Elastifizierung elasticisation, plastication
elastisch elastic, flexible
elastische Rückdeformation elastic recovery
elastische Rückfederung elastic recovery
Elastizität flexibility, elasticity
Elastizitätseigenschaften flexibility, elasticity, elastic properties
Elastizitätsgrenze elastic limit
Elastizitätsmodul modulus of elasticity, elastic modulus
Elastizitätsverminderung reduction in elasticity/flexibility
Elastizitätsvermögen 1. flexibility. 2. elasticity
elastomer elastomeric
Elastomer elastomer
Elastomerblend elastomer blend
Elastomerfilm elastomer film
Elastomerlegierung elastomer blend/alloy
elastomermodifiziert elastomer modified
Elastomerphase elastomer phase
elektrisch electric(al)
elektrisch beheizt electrically heated
elektrisch isolierend electrically insulating
elektrisch leitend electrically conductive
elektrische Entladung electric discharge
elektrische Feldstärke electric field strength
elektrische Leistung power
elektrische Werte electrical properties
elektrischer Widerstand electrical resistance
elektrisches Treeing water treeing
elektroabscheidbar capable of being deposited cataphoretically/ electrophoretically/by electrodeposition
Elektroanwendungen electrical applications
Elektroausrüstung electrical equipment
Elektrobauteil electrical component
Elektrobeheizung electric heating
Elektrochemie electrochemistry
elektrochemisch electrochemical
elektrochemische Impedanzspektroskopie electrochemical impedance spectroscopy
elektrochemische Korrosion electrochemical corrosion
Elektrode electrode
Elektrodenanordnung electrode arrangement
Elektroenergie electrical energy
Elektrogerät electrical appliance
Elektroherd electric cooker/oven
elektrohydraulisch electrohydraulic
Elektroindustrie electrical industry
Elektroisolierband insulating tape
Elektroisolierharz electrical insulating resin
Elektroisolierlack electrical insulating varnish
Elektroisoliermasse electrical insulating compound
Elektroisolierverhalten electrical insulating properties
elektrokathodisch electrocathodic
Elektrokinetik electrokinetics
elektrokinetisch electrokinetic
Elektrokorrosion electrolytic corrosion
Elektrolack cataphoretic/electrophoretic paint
Elektrolaminat printed circuit board material
Elektrolumineszenz electroluminescence,

electrofluorescence
Elektrolyt electrolyte
Elektrolytempfindlichkeit sensitivity to electrolytes
elektrolythaltig containing electrolytes
elektrolytisch electrolytic
elektrolytische Korrosion electrolytic corrosion
elektrolytische Korrosionswirkung electrolytic corrosion
Elektrolytkonzentration electrolyte concentration
Elektrolytlösung electrolyte solution
elektrolytstabil resistant to (*or*: unaffected by) electrolytes **Elektrolytstabilität** resistance to electrolytes
Elektrolyttransport electrolyte transport
elektromagnetisch electromagnetic
elektromechanisch electromechanical
Elektromotor electric motor
elektromotorisch electric, electrical
Elektron electron
elektronegativ electronegative
Elektronegativität electronegativity
Elektronenakzeptor electron acceptor
Elektronenanordnung electron arrangement
Elektronenbestrahlung electron irradiation
Elektronendichte electron density
Elektronendonor electron donor
Elektroneneinfangdetektor electron capture detector
Elektronenemission electron emission
Elektronenfänger electron captor
Elektronenhülle electron sheath
Elektronenmikroskop electron microscope
Elektronenmikroskopie electron microscopy
elektronenmikroskopisch electron microscopic
elektronenmikroskopische Abbildung electron micrograph
Elektronenpaar electron pair
Elektronenrückstreuung electron back scattering
Elektronenschale electron shell
Elektronenspektrometer electron spectrometer
Elektronenspektroskopie electron spectroscopy
Elektronenspektrum electron spectrum
Elektronenstoß electron collision
Elektronenstrahl electron beam
Elektronenstrahler electron beam radiator
elektronenstrahlgehärtet electron beam cured
elektronenstrahlhärtbar electron beam curable
Elektronenstrahlhärtung electron beam curing
Elektronenstrahloszillograph electron beam oscillograph
Elektronenstrahlresonanzspektroskopie electron beam resonance spectroscopy
elektronenstreuend electron-scattering
Elektronenstreuung electron scattering
Elektronenstrom electron flow
Elektronentransfer electron transfer
Elektronenwolke electron cloud
Elektronik electronics
Elektronikbaustein electronic module
Elektronikindustrie electronics industry
Elektronikplatine printed circuit board
Elektronikschaltung electronic circut
Elektroniksteuerung 1. electronic control. 2. electronic control system
elektronisch electronic
elektroosmotisch electro-osmotic
elektrophil electrophilic
Elektrophorese electrophoresis, electrodeposition
Elektrophoresebeschichtung electrophoretic/electrodeposition coating
Elektrophoresegrundierung electrophoretic/electrodeposition primer
Elektrophoreselack electrophoretic/electrodeposition paint
elektrophoretisch electrophoretic, cataphoretic, electrodeposition
elektrophoretisch abscheidbar capable of being deposited electrophoretically/cataphoretically/by electrodeposition **elektrophoretisch abscheidbarer Lack** electrophoretic/electrodeposition paint
elektropositiv electropositive
Elektropotential electropotential
Elektrostatikspritzen electrostatic spraying
elektrostatisch electrostatic
elektrostatische Aufladbarkeit electrostatic charegability
elektrostatische Aufladung electrostatic charging
elektrosterisch electrosteric
Elektrotauchbad electrophoretic/electrodeposition bath
Elektrotauchgrundierung electrodeposition primer
Elektrotauchlack electrophoretic/electrodeposition paint
Elektrotauchlackierung 1. electrophoretic/electrodeposition painting. 2. electrophoretic finish
Elektrotauchüberzugsmittel electophoretic/electrodeposition paint
Elektrotechnik electrical engineering
elektrotechnisch electrical
Elektroverzinken galvanising
Elementarladung elementary charge
Elementarteilchen elementary particle
Eliminierung elimination
elliptisch elliptical
eloxiert anodised
Eloxierung anodisation
Email(le) enamel
Emaillackfarbe enamel
Emballage package, packaging container
Emballagenindustrie packaging industry
Emballageninnenschutzlack can coating lacquer
Emission emission
emissionsarm low-emission
Emissionsauflage emission directive
Emissionsbegrenzung limiting emissions

emissionsfrei emission-free
Emissionsgrenze emission limit
Emissionsgrenzwert emission limit
Emissionskoeffizient emission coefficient
Emissionslinien emission lines
Emissionsminderung reduction of emissions
Emissionsminderungsmaßnahmen measures to reduce emissions
Emissionsquelle source of emissions
Emissionsrate emission rate
Emissionsvorschriften emission regulations
Emissionswellenlänge emission wave length
Emissionswert emission value/figure
Emissivität emissivity
emittierend emitting
empfindlich delicate, sensitive
Empfindlichkeit sensitivity
Empfindlichkeitsfaktor sensitivity factor
Empfindlichkeitsschwelle sensitivity threshold
empirisch empirical
Emulgator emulsifier, emulsifying agent
emulgatorarm low-emulsifier, with a low emulsifier content
Emulgatorart type of emulsifier
emulgatorfrei free from emulsifier, emulsifier-free
Emulgatorgemisch emulsifier blend
emulgatorhaltig containing emulsifier
Emulgatormenge amount of emulsifier
emulgatorreich high-emulsifier, with a high emulsifier content
Emulgatorreste emulsifier residues
Emulgatorsystem emulsifier system
emulgierbar emulsifiable
Emulgierbarkeit emulsifiability
Emulgieren emulsification
emulgierend emulsifying
emulgierfähig emulsifiable
Emulgierhilfsmittel emulsifier, emulsifying agent
Emulgiermittel emulsifier, emulsifying agent
Emulgiersystem emulsifying system
emulgiert emulsified
Emulgierwirkung emulsifying effect
Emulsion emulsion
Emulsionscopolymer emulsion copolymer
Emulsionscopolymerisat emulsion copolymer
Emulsionsfarbe emulsion paint
Emulsionsharz emulsion resin
Emulsionshilfsmittel emulsifier, emulsifying agent
Emulsionskleber dispersion/emulsion/latex adhesive
Emulsionskraft emulsifying power
Emulsionslack emulsion paint
Emulsionspolymer emulsion polymer
Emulsionspolymerisation emulsion polymerisation
Emulsionspolymerisat emulsion polymer
emulsionspolymerisiert emulsion polymerised
Endaushärtung final curing
Endeigenschaften end properties
Enderzeugnis end product
Endfarbstärke final colour intensity
Endfestigkeit final strength
Endfeuchte final moisture content
Endgruppe terminal group
Endhärte final hardness
Endkonsistenz final consistency
Endnaßfestigkeit final wet strength
endotherm endothermic
endständig terminal
Endtemperatur final temperature
Endverbraucher end user
energetisch *adjective relating to energy*
energetische Wiederverwertung energy recovery/recycling
energetisches Recycling energy recovery/recycling
Energie energy, power
Energie, aufgenommene energy input, energy used
Energie, kinetische kinetic energy
Energieabsorber energy absorber
energieabsorbierend energy-absorbent
Energieabsorption energy absorption
energiearm low-energy
Energieaufnahme energy input/absorption, energy used
Energieaufnahmevermögen energy-absorbing capacity, capacity/ability to absorb energy
Energieaufwand amount of energy required
energieaufwendig energy-intensive, requiring a lot of energy
Energieausbeute energy yield
Energiebedarf energy requirements
Energiebilanz energy balance
Energiedichte energy density
Energiedifferenz energy difference
energiedispersiv energy-dispersive
Energiedissipation energy dissipation
energiedissipierend energy-dissipating
Energiedosis energy dose
Energieeinleitung energy input
Energieeinsatz energy usage, energy used
Energieeinsparung energy saving
energieelastisch energy-elastic
Energieelastizität energy elasticity
Energieemission energy emission
Energieersparnis energy saving
Energieerzeugung production of energy
Energiefluß energy flow
Energiefreisetzung energy release
Energiefreisetzungsrate energy release rate
Energiegehalt energy content
Energiegleichung energy equation
energiegünstig using less energy
Energiehaushalt energy balance
Energieinhalt energy content
energieintensiv energy-intensive
Energieknappheit energy shortage
Energiekosten energy costs
Energiemehraufwand increased energy requirement

Energieniveau energy level
Energienutzung energy recovery
Energiepotential energy potential
Energiequelle energy source, source of energy
Energierecycling energy recovery
energiereich energy-rich, high-energy
Energierückführung energy recovery
Energierückgewinnung energy recovery
energiesparend energy-saving
energiesparsam energy-saving
Energiesparung energy saving
Energiespeicherung energy storage, storage of energy
Energiespender energy source, source of energy
Energiestrom amount of energy
Energieträger, fossiler fossil fuel
Energieübertragung energy transfer
Energieumsatz energy conversion
Energieumsetzung energy conversion
Energieumwandlungsprozeß energy conversion process
Energieverbrauch energy consumption
Energieverbraucher energy consumer
Energieverknappung increasing energy shortage
Energieverlust energy loss
Energieversorgung 1. energy supply. 2. energy supply system
energieverzehrend energy-consuming
Energievorräte energy reserves
Energiezufuhr energy supply
engmaschig vernetzt closely crosslinked
engvernetzt closely crosslinked
Enolether enol ether
Enolform enol form
Entfärbung loss of colour
entfernen to remove
entfetten to degrease
Entfettung degreasing
Entfettungsanlage degreasing plant
Entfettungsbad degreasing bath
Entfettungslösung degreasing solution
Entfettungsmittel degreasing agent
entfeuchtet 1. dehumidified. 2. dried
Entfeuchtung dehumidification
entflammbar flammable
Entflammbarkeit flammability
Entflammbarkeitsklasse flammability rating
Entgasung degassing
Enthaftung detachment *(e.g. of paint film)*
Enthaftungsmittel release agent
Enthalpie enthalpy
Enthalpierelaxation enthalpy relaxation
enthalpisch enthalpic
entionisiert deionised
Entlackung paint stripping
Entlackungsanlage paint stripping plant
Entlackungslösung paint stripping solution
Entlackungsmittel paint stripping solution/agent
Entladung discharge
Entladung, elektrische electric discharge
Entladungsgeschwindigkeit discharge rate
Entlastung stress removal
Entlastungskurve stress removal curve
Entlastungszeit stress removal time
entleeren to empty
Entleerungsöffnung outlet, discharge opening
Entlüfter deaerating agent
Entlüftung 1. ventilation. 2. deaeration
Entlüftungsanlage ventilation plant
Entlüftungsmittel deaerating agent
Entmischung separation
Entmischungserscheinungen signs of separation
Entmischungstendenz tendency to separate
Entnetzungsvorgang de-wetting (process)
Entropie entropy
entropieelastisch entropy-elastic, rubbery, rubbery-elastic
Entropieelastizität entropy elasticity, rubber-like elasticity
Entropiezunahme entropy increase
entropisch entropic
Entrostung removal of rust
entsalzt desalinated
Entschalungsmittel mould oil
Entschäumer defoaming agent, defoamer, antifoam (agent)
Entschäumeradditiv defoaming agent
Entschäumerwirkung defoaming effect
entschäumungsaktiv defoaming
Entschäumungsmittel defoaming agent, defoamer, antifoam (agent)
Entschichten paint stripping
entschlichtet de-sized
Entsorgung 1. (waste) disposal. 2. discharge
entsorgungsfreundlich easily/readily disposed of
Entsorgungsleitung discharge line
Entsorgungsprobleme (waste) disposal problems
Entspannung 1. relief, relaxation *(of stress or pressure)*. 2. recovery *(after compression)*
Entspannungsversuch stress relaxation test
Entstaubungsanlage dust removal unit
Entwässerung 1. drainage. 2. dewatering, removal of water
Entwickler (photographic) developer
Entwicklung, verfahrenstechnische technical development
Entwicklungsaktivität development activity
Entwicklungsarbeit development work
Entwicklungsaufwand development costs
Entwicklungslabor development laboratory
Entwicklungsmöglichkeiten development potential/possibilities
Entwicklungsphase development phase
Entwicklungsprodukt development product
Entwicklungstendenz development trend
Entwicklungszeit development period
Entwurf design
Entwurfsdaten design data
Entwurfsphase design stage
Entwurfsstadium design stage

Entzunderung descaling
entzündlich flammable
Entzündlichkeit flammability
Entzündungspunkt ignition point
Entzündungsquelle source of ignition
Entzündungstemperatur ignition temperature
entzündungswidrig flame resistant
Entzündungswidrigkeit flame resistance
Enzym enzyme
enzymatisch enzymatisch
EP-Anstrichmittel epoxy paint
EP-Beschichtung 1. epoxy coating. 2. epoxy paint
EP-Harz epoxy resin
EP-Klebstoff epoxy adhesive
EP-Laminat epoxy laminate
EP-Mörtel epox mortar
EP-Reaktionsharz epoxy resin
EP-Reaktionsharzmasse catalysed epoxy resin
EP-Spachtel epoxy-based knifing filler, epoxy-based stopper
EP-Werkstoff epoxy material
EPDM-Gummi ethylene-propylene diene rubber
EPIC-Harz epoxy-isocyanate resin
Epichlorhydrin epichlorohydrin
Epichlorhydrinkautschuk epichlorohydrin rubber
Epoxi-, epoxi- *see* **Epoxid-, epoxid-** *or* **Epoxy-, epoxy-**
Epoxid epoxide
Epoxid-Aminaddukt epoxy-amine adduct
Epoxidacrylat epoxy acrylate
Epoxidalkohol epoxy alcohol
Epoxidäquivalent epoxy equivalent
Epoxidäquivalentgewicht epoxy equivalent
Epoxidation epoxidation
Epoxidbeschichtung 1. epoxy coating. 2. epoxy paint
Epoxidbindemittel epoxy binder
Epoxiddiacrylat epoxy diacrylate
Epoxiddichtmasse epoxy sealant (compound)
Epoxidester epoxy ester
Epoxidesterharz epoxy ester resin
Epoxidfunktion epoxy function
Epoxidfunktionalität epoxy functionality
epoxidfunktionell epoxy-functional
Epoxidgehalt epoxy/epoxide value
Epoxidgruppe epoxy/epoxide group
Epoxidhärter epoxy hardener
Epoxidhärter epoxy hardener
Epoxidharz epoxy resin
Epoxidharzanstrichsystem epoxy paint
Epoxidharzbasis, auf epoxy-based
Epoxidharzbeton epoxy concrete
Epoxidharzester epoxy ester
Epoxidharzestrich epoxy screed
Epoxidharzgrundierung 1. epoxy primer. 2. epoxy primer coat
Epoxidharzgruppe epoxy/epoxide group
Epoxidharzhartpapier epoxy paper laminate
epoxidharzimprägniert epoxy (resin) impregnated
Epoxidharzklarlack clear epoxy lacquer
Epoxidharzklebstoff epoxy adhesive
Epoxidharzlack epoxy paint
Epoxidharzlösung epoxy resin solution
epoxidharzmodifiziert epoxy modified
Epoxidharzsystem epoxy system/paint/formulation
epoxidharzvergütet epoxy resin modified
Epoxidharzversiegelung epoxy sealant
Epoxidharzzusammensetzung epoxy composition
epoxidiert epoxidised
Epoxidkleber epoxy adhesive
Epoxidklebstoff epoxy adhesive
Epoxidlack epoxy paint
Epoxidlackschicht epoxy film
Epoxidlaminat epoxy laminate
Epoxidmasse epoxy compound
epoxidmodifiziert epoxy modified
Epoxidreaktivmasse epoxy compound
Epoxidrest epoxide/epoxy group
Epoxidring epoxide/epoxy ring
epoxidtypisch typical of epoxy resins
Epoxidumhüllung epoxy encapsulation
Epoxidverbindung epoxy/epoxide compound
Epoxidweichmacher epoxy plasticiser
Epoxidwert epoxy value
Epoxidzweikomponentenkleber two-pack epoxy adhesive
Epoxyacrylat epoxy acrylate
Epoxyacrylatharz epoxy acrylic resin
Epoxyalkoxysilan epoxy alkoxysilane
Epoxybindemittel epoxy binder
Epoxyd-, epoxyd- *see* **Epoxid-, epoxid- or Epoxy-, epoxy-**
Epoxygruppe epoxy/epoxide group
Epoxyharzschichtstoff epoxy laminate
Epoxymethacrylat epoxy methacrylate
Epoxyphenolharz epoxy phenolic resin
Epoxysilan epoxysilane
Epoxysilikat epoxy silicate
Epoxysystem epoxy paint/system/formulation
Epoxyterpen epoxy terpene
Epoxyverschnitt epoxy blend
Epoxyzinkstaubfarbe epoxy-based zinc-rich paint
ERA evaporation rate analysis
Erdalkali alkaline earth
Erdalkalicarbonat alkaline earth carbonate
Erdalkalimetall alkaline earth metal
Erdalkalioxid alkaline earth oxide
Erdalkalisalz alkaline earth salt
Erdalkalisulfat alkaline earth sulphate
Erdfarbe earth pigment
Erdöl petroleum, mineral oil
Erdölförderung petroleum extraction
Erdölfraktion petroleum fraction
Erdölharz petroleum resin
Erdölindustrie petroleum industry
Erdölkrise oil crisis
Erdölprodukte petroleum products

Erdölreserven oil reserves
Erdölressourcen oil resources
Erfahrungsaustausch exchange of information
erfassen to determine, to register, to measure, to pick up, to include, to cover, to collect
Ergebnis result
Ergiebigkeit yield
Erholung recovery
Erichsendehnbarkeit Erichsen distensibility
Erichsentiefung Erichsen indentation
Erkenntnisse findings
Erkrankung infection
Erlenmeyerkolben Erlenmeyer flask
Ermüdung fatigue
Ermüdungsbeanspruchung fatigue stress
Ermüdungsbelastung fatigue stress
Ermüdungsbeständigkeit fatigue resistance
Ermüdungsbruch fatigue failure/fracture
Ermüdungseigenschaften fatigue properties
ermüdungsfest fatigue resistant
Ermüdungsfestigkeit fatigue strength
Ermüdungsprüfung fatigue test
Ermüdungsriß fatigue crack
Ermüdungsrißausbreitung fatigue crack propagation
Ermüdungsrißwachstum fatigue crack growth
Ermüdungsverhalten fatigue behaviour
Ermüdungsversagen fatigue failure
Ermüdungsversuch fatigue test
Ermüdungswiderstand fatigue resistance
Ermüdungswiderstandsfähigkeit fatigue resistance
Erniedrigung reduction, lowering, decrease
Erosion erosion
Ersatzprodukt alternative product, substitute
Ersatzstoff alternative material, substitute
Ersatzverfahren alternative process/method
Erscheinungsbild appearance
Erschütterung shock, vibration
erschütterungsempfindlich shock-sensitive
erschütterungsfest shock/vibration resistant
Erschütterungsfestigkeit shock/vibration resistance
erschütterungsfrei vibration-free
Erstarren solidification
erstarrt solidified, set
Erstarrungsgeschwindigkeit setting speed
Erstarrungspunkt freezing/crystallising/solidifying point
Erstarrungsverhalten setting characteristics
Erwärmen heating up
Erwärmungsgeschwindigkeit heating-up rate
Erwärmungsphase heating-up phase
Erwärmungszeit heating-up period
erwartungsgemäß as expected
erweichbar capable of being softened: **thermoplastisch erweichbar** capable of being softened by heat
erweichen to soften
Erweichung softening
Erweichungsbereich softening range
Erweichungscharakteristik softening characteristics
Erweichungsgebiet softening range
Erweichungspunkt softening point/temperature
Erweichungstemperatur softening point/temperature
Erweichungsverhalten softening characteristics
erweiterungsfähig expandable
Erythritol erythritol
ESCA-Methode electron spectroscopy for chemical analysis
ESH *abbr. of* **Elektronenstrahlhärtung**, electron beam curing
ESH-Technologie electron beam curing (technology)
ESR-Spektroskopie electron beam resonance spectroscopy
Essigester ethyl acetate
Essigsäure acetic acid
Essigsäureanhydrid acetic anhydride
Essigsäurebutylester butyl acetate
Essigsäureester ethyl acetate
Essigsäureethylester ethyl acetate
essigvernetzend acetic acid-curing
Ester ester
Esteramid ester amide
Esterdiol ester diol
Estergehalt ester content
Estergruppe ester group
Estergruppierung ester group
Esterimid ester imide
esterlöslich ester soluble
Esterrest ester group
Esterzahl ester value
Estrich screed
Estrich, schwimmender floating screed
Estrichmasse screed mortar
Estrichmörtel screed mortar
Ethan ethane
Ethanamin ethanamine
Ethanol ethanol, ethyl alcohol
Ethanolamid ethanolamide
Ethanolamin ethanolamine
ethanolisch ethanolic
ethanolische Lösung ethanol solution
ethanollöslich ethanol soluble
Ethanollöslichkeit solubility in ethanol
Ethanollösung ethanol solution
Ethanolverdünnbarkeit dilutability with ethanol
Ethen ethene
Ethenharz ethene resin
Ethenpfropfcopolymer ethene graft copolymer
Ether ether
Etheralkohol ether alcohol
Etheramin ether amine
Etherbindung ether linkage
Etherbrücke ether bridge
Etherdiol ether diol
Ethereinheit ether group
Etherester ether ester
Etherextrakt ether extract

Etherglied ether segment
Ethergruppe ether group
Etherrest ether group
Etherverknüpfung ether linkage
Ethoxygehalt ethoxyl content
Ethoxygruppe ethoxyl group/radical
ethoxyliert ethoxylated
Ethoxypropylacetat ethoxypropyl acetate
Ethylacetat ethyl acetate
Ethylacetoacetoxim ethyl acetoacetoxime
Ethylacrylat ethyl acrylate
Ethylalkohol ethyl alcohol
Ethylbenzol ethyl benzene
Ethylcellulose ethyl cellulose
Ethylchlorid ethyl chloride
Ethylcyanacrylat ethyl cyanoacrylate
Ethyldiglykol ethyl diglycol
Ethylen ethylene
Ethylen-Acrylatcopolymer ethylene-acrylate copolymer
Ethylen-Acrylatkautschuk ethylene-acrylate rubber
Ethylen-Acrylsäurecopolymer ethylene-acrylic acid copolymer
Ethylenanteil ethylene content
Ethylenchlorid ethylene chloride
Ethylenchlortrifluorethylen ethylene chlorotrifluoroethylene
Ethylendiamin ethylene diamine
Ethylendiamintetraessigsäure ethylene diamine tetra-acetic acid
Ethylendicarbonsäure ethylene dicarboxylic acid
Ethylengas ethylene
Ethylengehalt ethylene content
Ethylenglykol ethylene glycol
Ethylenglykoldimethacrylat ethylene glycol dimethacrylate
Ethylenglykolether ethylene glycol ether
Ethylenglykolmonoalkylether ethylene glycol monoalkyl ether
Ethylenglykolmonobutylether ethylene glycol monobutyl ether
Ethylenglykolmonoethylether ethylene glycol monoethyl ether
ethylenhaltig containing ethylene
ethylenisch ethylenic
Ethylenkette ethylene chain
Ethylenoxid ethylene oxide
Ethylenoxidgruppe ethylene oxide group
Ethylenvinylacetatcopolymer ethylene-vinyl acetate copolymer
Ethylester ethyl ester
Ethylglykol ethyl glycol
Ethylglykolacetat ethyl glycol acetate
Ethylgruppe ethyl group
Ethylhexanol ethyl hexanol
Ethylhexylacrylat ethyl hexyl acrylate
Ethylhexylglycidylether ethylhexyl glycidyl ether
Ethyloxazolin ethyl oxazoline
Ethylsilikat ethyl silicate
Ethylsilikatzinkstaubfarbe ethyl silicate zinc-rich paint
Etikett label
Etikettenkennzeichnung labelling
Etikettieren labelling
Etikettierklebstoff label adhesive
Etikettierung labelling
ETL *abbr. of* **Elektrotauchlack**, electrophoretic/electrodeposition paint
europäischer Binnenmarkt single European market
europaweit throughout Europe
eutektisch eutectic
EVAC ethylene-vinyl acetate, EVA
EWG *abbr. of* **Europäische Wirtschaftsgemeinschaft**, European Economic Community, EEC
EWG-Länder EEC countries, common market countries
ex-geschützt explosion-proof
Ex-Schutz protection against explosion
Ex-Schutzbereich explosion-proof area
Excimerlaser excimer laser
exotherm exothermic
Exothermie exothermic character
Expansionskoeffizient coefficient of expansion
experimentell experimental
explosibel explosive
explosionsartig explosive
explosionsfähig explosive
Explosionsgefahr risk of explosion
explosionsgefährlich potentially explosive, representing an explosive hazard
explosionsgeschützt explosion-proof
exponentiell exponential
exponiert exposed
Expositionsbedingungen exposure conditions
Expositionsdauer exposure time, time of exposure
Expositionszeit exposure time, time of exposure
Exsikkator dessiccator
Extender extender
Extenderfüllstoff filler
Extendermischung extender blend
extern external
Extinktion extinction
Extinktionsverhältnis extinction ratio
Extinktonskoeffizient extinction coefficient
extrahierbar extractable
Extrahierbarkeit extractability
extrahieren to extract
Extraktion extraction
extraktionsbeständig extraction resistant
Extraktionsbeständigkeit extraction resistance
Extraktionsflüssigkeit extractant
Extraktionsgeschwindigkeit extraction rate
Extraktionsmedium extractant
Extraktionsmethode method of extraction
Extraktionsmittel extractant
Extraktionsrückstand extraction residue
extraktionsstabil extraction resistant
Extraktionsverhalten extraction behaviour

Extraktionsversuch extraction test
Extrapolation extrapolation
extrapolieren to extrapolate
Extruder extruder
exzentrisch eccentric

F

F+E *abbr. of* **Forschung und Entwicklung**, research and development
Fabrikationshalle production shed
Fabrikationskontrolle production control
Fabrikationsprüfung production control
Fabrikationsstätte factory, plant, production facility
Fabrikationsüberwachung production control
Fabrikhalle factory shed
Fachausschuß technical committee
Fachberater technical adviser
Fachhochschule technical university
Fachleute experts
Fachliteratur technical literature
Fachmann expert
fadenförmig thread-like
Fadenmolekül thread-like molecule
Fahrrad bicycle
Fahrzeug vehicle
Fahrzeugbau vehicle construction
Fahrzeugdecklack automotive paint/enamel
Fahrzeuglack automotive paint/enamel
Fahrzeuglackierung vehicle painting
Fahrzeugwesen motor industry
Fallgewicht drop weight
Fallhöhe drop height
Fällung precipitation
Fällungsgeschwindigkeit precipitation rate
Fällungskieselsäure precipitated silica
Fällungsmittel precipitating agent
Fällungsparameter precipitation conditions
Fällungsverfahren precipitation (process)
Fallversuch drop impact test
Faltbiegeversuch folding endurance test
falzen to bend, to fold, to flex
Farbabweichungen colour deviations
Farbänderung colour change, change in colour
Farbanpassung colour matching
Farbanstrich paint film, coating
Farbanteil pigment content
farbarm pale, having little colour
Farbbatch pigment paste/concentrate
farbbeständig colourfast
Farbbeständigkeit colour stability, colourfastness
Farbbrillanz brilliant/sparkling colour
Farbdichte colour intensity
Farbdifferenz colour difference, difference in colour
Farbdruckvorrichtung colour printing unit
Farbe 1. paint. 2. (printing) ink. 3. colour
Farbechtheit colourfastness
Farben- und Lackindustrie coatings/paint industry
Farbenhersteller paint manufacturer
Farbenindustrie paint/coatings industry
Farbfilm paint film
Farbgebung colouring, imparting colour
Farbglanzpigment interference pigment
Farbglimmer coloured mica
Farbgranulat masterbatch
Farbhelligkeit brightness of colour
Farbhomogenität uniform colour
farbig coloured
Farbigkeit colour, colourfulness
Farbintensität colour intensity
farbintensiv strong in colour, intensely coloured
Farbkonstanz colour uniformity, uniform/consistent colour
Farbkonzentrat pigment concentrate
Farbkraft 1. colour intensity. 2. pigment strength
farbkräftig intensely coloured
Farblack 1. paint. 2. coloured varnish/lacquer
Farblasur coloured glaze
farblos colourless, neutral
Farbmessung colorimetry
Farbmetall non-ferrous metal
Farbmetrie colorimetry
Farbmetrik colorimetry
farbmetrisch colorimetric
Farbmittel 1. colorant, colouring agent. 2. pigment. 3. dye
Farbnachstellung colour matching
Farbnester pigment agglomerates
Farbnuance (colour) shade
Farbpalette colour range, range of colours
Farbpaste pigment paste
Farbpastenharz pigment paste resin
Farbpigment 1. pigment. 2. coloured pigment *(where a distinction is made between* **Farbpigmente** *and* **Weißpigmente**, *q.v.)*
Farbpigmentpaste pigment paste
Farbpulver powdered pigment
Farbreinheit colour purity
Farbreste paint residues
Farbrezeptformulierung paint formulation
Farbruß carbon black pigment
Farbsättigung colour saturation
Farbschicht paint film
Farbschlieren coloured streaks
farbschwach weak in colour
Farbschwankungen colour variations, variations in colour
Farbskala colour scale
Farbsortiment colour range
Farbspritzanlage paint spraying unit
Farbspritzer paint splashes
farbstabil colourfast
farbstabilisiert colour stabilised

Farbstabilität colourfastness
farbstark intensely coloured
Farbstärke colour intensity
Farbstärkeabfall decrease in colour intensity
Farbstärkeabnahme decrease in colour intensity
Farbstärkeanstieg increase in colour intensity
Farbstärkeentwicklung development of colour intensity
Farbstärkezunahme increase in colour intensity
Farbstoff dye(stuff). *Occasionally, the word is carelessly and wrongly used in place of* **Pigment**
Farbstoffadsorption dye adsorption
Farbstofflösung dye solution
Farbstoffnester local pigment concentrations
Farbstoffpaste pigment paste
Farbstreuung colour variation
Farbtafel colour chart
Farbteig pigment paste
farbtief deep in colour
Farbtiefe depth of colour
Farbton shade, hue, colour
Farbtonabweichung colour variation
Farbtonänderung change in colour
farbtonbeständig colourfast
Farbtonbeständigkeit colourfastness, colour stability
Farbtongenauigkeit exact colour
Farbtonkarte colour chart
Farbtonnuancierung tinting
Farbtonspektrum colour range, range of colours
Farbtontiefe colour intensity/saturation
Farbtonveränderung change in colour
Farbtonverschiebung colour shift
Farbtonvorlage specfied colour
Farbübergang colour change/transition
Farbumschlag colour change
Färbung coloration, colouring
Farbunterschied colour difference, difference in colour
Farbveränderung colour change, change in colour
Farbverstärkung colour enhancement
Farbverteilung pigment dispersion
Farbvertiefung deepening in colour
Farbwalze inking roller
Farbwechsel colour change, change in colour
Farbwert colour value
Farbwiedergabe colour rendition
Farbzahl colour value
Faser fibre
Faserleder reconstituted leather
Faserplatte fibreboard
faserverstärkt fibre reinforced
Faß drum
Fassade facade
Fassadenanstrich 1. exterior/masonry finish. 2. architectural paint
Fassadenbau curtain wall construction
Fassadenbeschichtung 1. exterior/masonry coating. 2. architectural paint
Fassadenfarbe exterior/architectural paint
FCKW *abbr. of* **Fluorchlorkohlenwasserstoff**, chlorofluorocarbon, CFC
Fe-Katalysator iron catalyst
Fehler 1. error, mistake, fault. 2. defect
Fehler, optischer surface defect
Fehleranalyse fault analysis
fehlerfrei free from defects, faultless, perfect
Fehlerfreiheit freedom from defects
fehlerhaft faulty, incorrect, wrong, defective
Fehlerquelle source of error
Fehlersuchliste trouble-shooting chart
Fehlfunktion malfunction
Fehlinformation wrong information
Fehlinterpretation misinterpretation, wrong interpretation
Fehlkonstruktion faulty design
Fehlleistung malfunction
Fehlmessung measuring error
Fehloperation malfunction
Fehlschlüsse wrong conclusions
Fehlstelle flaw, defect
Fehlverklebung poorly bonded joint
Fehlverleimung poorly bonded joint
Feindispergierung fine dispersion
feindispers fine-particle
feindispers verteilt finely dispersed
Feineinstellung fine adjustment
feingemahlen finely ground
Feinheit fineness
Feinjustierung fine adjustment
Feinkörnigkeit fine-particle character
Feinkreide finely powdered chalk
feinkristallin finely crystalline
Feinmahlaggregat fine grinding machine, pulveriser, pulverising unit
Feinmahlen fine grinding, pulverisation
feinmechanisch precision: **feinmechanische Teile** precision parts
feinpulvrig finely powdered
Feinruß finely divided carbon black
feinstgemahlen very finely ground
Feinstpartikel ultra-fine particle
feinstteilig with extremely fine particles
feinstverteilt extremely finely dispersed
feinteilig fine-particle
Feinvermahlung fine grinding
feinverteilt finely dispersed
Feinwerktechnik precision engineering
feinzellig fine-cell
Feldkonstante field constant
Feldstärke field strength
Feldstudie field study
Feldtest field trial
Feldversuch field trial
Fenster window
Fensterbau window construction
Fensterfuge window joint
Fensterprofil window profile
Fensterrahmen window frame
Ferromagnetismus ferromagnetism

Fertigprodukt finished product
Fertigteilbau industrialised building methods
Fertigung production, manufacture
Fertigungsanlage production plant
Fertigungsbedingungen production conditions
Fertigungsbetrieb factory, plant, works
Fertigungscharge production batch
Fertigungseinheit production unit
Fertigungseinrichtungen production equipment
Fertigungsfehler production fault
Fertigungsfluß production cycle
fertigungsgegeben production-related
Fertigungsgeschwindigkeit production rate/speed
Fertigungshalle production shed
Fertigungskontrolle production control
Fertigungslinie production line
Fertigungspanne production breakdown
Fertigungsparameter production/manufacturing conditions
Fertigungsstätte plant, factory, production facility
Fertigungssteuerung 1. production control. 2. production control system. 3. production control department
Fertigungsstraße production line
Fertigungsstrecke production line
fertigungstechnisch *adjective relating to manufacture or production*
Fertigungstechnologie production technology
Fertigungstoleranz production tolerance
Fertigungsüberwachung 1. production control. 2. production control system. 3. production control department
Fertigungsverfahren manufacturing/production process
Fertigungsverlauf manufacturing/production process
Fertigungsversuch production trial
Fertigungszeit production time
Fertigungszelle production unit
Fertigungszyklus production cycle
Fertigwaren finished products/goods
fest solid
Festbettverfahren solid bed process
Festgehalt solids content
festhaftend firmly adhering
Festharz solid resin
Festharzdispersion solid resin dispersion
Festharzgehalt solid resin content
Festharzkombination solid resin blend
Festigkeit strength
Festigkeit, mechanische mechanical strength
Festigkeitsabfall loss of strength
Festigkeitsabnahme decrease in strength
Festigkeitsanstieg increase in strength
Festigkeitsaufbau build-up of strength
Festigkeitsberechnung strength calculation
festigkeitsbestimmend *that which determines the strength, e.g. of a bonded joint:* **die festigkeitsbestimmenden Größen der Klebstoffe wie z.B...** adhesive properties which determine bond strength, e.g....
Festigkeitseigenschaften strength (characteristics)
Festigkeitseinbuße loss of strength
Festigkeitserhöhung increase in strength
Festigkeitskriterien strength criteria
festigkeitsmindernd reducing strength
Festigkeitsminderung reduction in strength
Festigkeitsmittelwert average/mean strength
Festigkeitssteigerung increase in strength
Festigkeitsverlust loss of strength
Festkörper solid
Festkörperanteil solids content
Festkörpergehalt solids content
Festkörperoberfläche solid surface
Festkörperreibung solid friction
festkörperreich high-solids
Festkörperzustand solid state
Feststoff solid matter/substance
Feststoffanteil solids content
Feststoffbereich solid phase
Feststoffdichtung solid seal
Feststoffgehalt solids content
Feststoffoberfläche solid surface
Feststoffpartikel solid particle
Feststoffteilchen solid particle
Festwerden solidification
fett long-oil
Fettalkohol fatty alcohol
Fettalkoholester fatty alcohol ester
Fettalkoholethoxylat fatty alcohol ethoxylate
Fettalkoholsulfat fatty alcohol sulphate
Fettalkyloxazolin fatty alcohol oxazoline
Fettamid fatty amide
fettbeständig grease resistant
Fettbeständigkeit grease resistance
fettdicht greaseproof
fettes Alkydharz long-oil alkyd (resin)
fettfrei free from grease
fettgedruckt printed in bold type
fettig greasy
fettlösend grease-dissolving
fettlöslich fat-soluble
Fettlösungsmittel grease solvent
Fettreste grease residues
Fettrückstände grease residues
Fettsäure fatty acid
Fettsäureabbau fatty acid degradation
Fettsäurealkyd fatty acid alkyd
Fettsäureamid fatty acid amide
Fettsäureanteil fatty acid content
Fettsäurederivat fatty acid derivative
Fettsäureester fatty acid ester
Fettsäuregehalt fatty acid content
Fettsäuremethylester methyl ester of a fatty acid
fettsäuremodifiziert fatty acid modified
Fettsäuremonoester fatty acid mono-ester
Fettsäurerest fatty acid radical
Fettsäuresalz fatty acid salt
Fettsäureseife fatty acid soap
Fettschicht grease film

Fettseife fatty acid soap
feucht damp, moist, humid
Feucht-Warmauslagerung ageing under warm, humid conditions
Feuchtbeanspruchung exposure to damp (conditions)
Feuchte 1. moisture, moisture content *(of a substance)*. 2. damp *(in the atmosphere)*. 3. humidity *(of air)*
Feuchteabsorption moisture absorption
Feuchteaufnahme moisture absorption
feuchtebedingt due to moisture
Feuchtebelastung exposure to damp (conditions)
Feuchtebeständigkeit moisture resistance
Feuchtebeständigkeitsklasse moisture resistance group
Feuchtebestandteil moisture content
Feuchtediffusion moisture diffusion
Feuchtediffusionskoeffizient moisture diffusion coefficient
Feuchteeinfluß effect of moisture
feuchteempfindlich moisture-sensitive
Feuchteempfindlichkeit sensitivity to moisture
Feuchtegehalt 1. moisture content. 2. humidity *(of air)*
feuchtehärtbar moisture curing/curable
feuchtehärtend moisture curing
Feuchtehaushalt moisture content
feuchteinduziert moisture induced
Feuchteklima damp/humid conditions
Feuchtesättigung moisture saturation
Feuchteschutz protection against damp
Feuchteschwankungen variations in moisture content
feuchteunempfindlich unaffected by moisture
Feuchteverhalten behaviour under damp/humid conditions
Feuchtewert 1. moisture content. 2. humidity *(of the air)*
Feuchtfestigkeit moisture resistance, resistance to damp (conditions)
Feuchthaltemittel humectant
Feuchthalter humectant
feuchthärtbar moisture curing
feuchthärtend moisture curing
feuchtheiß hot and humid
Feuchtigkeit 1. damp. 2. moisture. 3. moisture content
Feuchtigkeitsabnahme moisture loss
Feuchtigkeitsabsorption moisture absorption
Feuchtigkeitsalterung ageing under damp conditions
Feuchtigkeitsaufnahme moisture absorption
Feuchtigkeitsausschluß, unter under the exclusion of moisture
Feuchtigkeitsbeanspruchung exposure to damp (conditions)
Feuchtigkeitsbelastung exposure to moisture/damp
feuchtigkeitsbeständig moisture resistant, resistant to damp
Feuchtigkeitsbeständigkeit moisture resistance
feuchtigkeitsdicht moisture proof
Feuchtigkeitsdurchlässigkeit moisture permeability
Feuchtigkeitseinfluß, unter under the influence of moisture
feuchtigkeitsempfindlich moisture sensitive, affected by moisture
Feuchtigkeitsempfindlichkeit moisture sensitivity
feuchtigkeitsfest moisture resistant
Feuchtigkeitsgehalt 1. moisture content. 2. humidity *(of air)*
feuchtigkeitshaltend moisture retaining
feuchtigkeitshärtbar moisture curing
feuchtigkeitshärtend moisture curing
Feuchtigkeitshaushalt moisture content
Feuchtigkeitsisolierung insulation against moisture
Feuchtigkeitskonzentration moisture concentration
feuchtigkeitsleitend moisture-transmitting
Feuchtigkeitsnester damp patches
Feuchtigkeitsniveau 1. moisture content. 2. humidity *(of air)*
feuchtigkeitsreaktiv moisture reactive
Feuchtigkeitsresistenz moisture resistance
Feuchtigkeitsschutz protection against damp
Feuchtigkeitssperre moisture barrier
Feuchtigkeitsspuren traces of moisture
feuchtigkeitsundurchlässig impermeable to moisture
Feuchtigkeitsundurchlässigkeit moisture impermeability
feuchtigkeitsunempfindlich unaffected by moisture
Feuchtigkeitsverlust moisture loss
feuchtigkeitsvernetzend moisture curing
Feuchtigkeitsversuch moisture exposure test
Feuchtraum wet-process room
Feuchtschrankprüfung humidity cabinet test
Feuchtwarmlagerung ageing under warm, humid conditions
Feuer fire
feuerabweisend fire/flame resistant
feuerbeständig fire resistant
feuerfest 1. flame resistant. 2. refractory
feuergefährdet liable to catch fire
feuergefährlich flammable
Feuergefährlichkeit fire hazard
feuerhemmend flame/fire retardant
feuerlöschend fire extinguishing
Feuerrisiko fire risk
Feuerschutzmaterial flame retardant
Feuerschutzmittel flame retardant
Feuersgefahr risk of fire
feuersicher fire/flame resistant
Feuersicherheit fire safety
feuersicherheitlich *adjective relating to fire safety*
Feuersicherheitsempfehlungen fire safety recommendations

feuerverzinkt hot galvanised
Feuervorschriften fire regulations
Feuerweiterleitung flame spread
Feuerwiderstandsdauer fire endurance
feuerwiderstandsfähig fire/flame resistant
Feuerwiderstandsfähigkeit fire/flame resistance
Feuerwiderstandsverhalten fire/flame resistance
Fichte spruce
Fichtenholz spruce
FID flame ionisation detector
Filiformkorrosion filiform corrosion
Film film
Filmaussehen film appearance
Filmbildehilfsmittel film forming aid
filmbildend film forming
Filmbildetemperatur, minimale minimum film forming temperature
Filmbildner film forming agent
Filmbildnermaterial film forming substance
Filmbildung film formation
Filmbildungsbedingungen film forming conditions
Filmbildungsdauer time needed to form a film
Filmbildungseigenschaften film forming properties
Filmbildungshilfsmittel film forming aid
Filmbildungsmechanismus film forming mechanism
Filmbildungsprozeß film forming process, film formation
Filmbildungstemperatur film forming temperature
Filmdefekt film defect
Filmdehnung film extension/stretching
Filmdicke film thickness
Filmdickenmessung film thickness determinaton
Filmeigenschaften film properties
Filmflexibilität film flexibility
Filmgefüge film structure
Filmhärte film hardness
Filmmangel film defect
filmmechanische Eigenschaften mechanical film properties
Filmmerkmale film characteristics
Filmoberfläche film surface
Filmoberflächeneigenschaften film surface characteristics
Filmoptik film appearance
Filmqualität film quality
Filmschichtdicke film thickness
Filmstärke film thickness
Filmstörungen film defects
Filmstruktur film structure
Filmtransparenz film transparency
Filmziehgerät film casting instrument
Filterapparatur filtration unit
Filtergerät filtration unit
Filtergewebe filter fabric/cloth
Filterhilfsmittel filtration aid
Filterkerze cartridge/candle filter
Filterkerzenpaket candle filter assembly
Filterkuchen filter cake
Filterrückstand filtration residue
Filtertuch filter cloth
Filtration filtration
Filtrierbarkeit filtrability
filtrieren to filter
Filtrieren filtration
Filzschreiber felt-tip pen
Fingernageltest fingernail test
FIPG foamed-in-place gasket
Firnis varnish
Fischaugen fish eyes
Fixierstelle fixing point
Fixierungszeit fixing time
Flachbettkaschieranlage flat bed laminating plant
Flachdach flat roof
Fläche area, surface
flächendeckend 1. nationwide, covering the whole country, *or*: the whole area. 2. total, comprehensive
Flächeneinheit unit area
flächenförmig flat, two-dimensional
Flächengewicht weight per unit area
Flächenhaftung adhesion
Flächenwiderstand surface resistance
flächig flat, two-dimensional: **flächiger Auftrag** application over the entire surface *(e.g. of adhesive)*
Flammbehandlung flame treatment
Flammenausbreitung flame spread
Flammenausbreitungsgeschwindigkeit rate of flame spread
Flammenausbreitungsrichtung direction of flame spread
Flammenionisationsdetektor flame ionisation detector
flammfest flame/fire resistant
Flammfestausrüstung flame/fire proofing
Flammfestigkeit flame/fire resistance
flammgehemmt flame retardant
flammhemmend flame retardant
flammhemmend ausgerüstet flame retardant
flammhemmendes Additiv flame retardant
Flammhemmer flame retardant
Flammhemmung flame retardancy
Flammionisationsdetektor flame ionisation detector
Flammklasse flammability classification/group/rating
flammlöschend fire extinguishing
Flammpunkt flash point
Flammschutz flame/fire proofing
Flammschutzadditiv flame retardant
flammschutzausgerüstet flameproofed
Flammschutzausrüstung flameproofing, imparting flame retardant properties
Flammschutzeffekt flame retardant effect
Flammschutzeigenschaften flame retardant properties
Flammschutzfarbe flame retardant paint

Flammschutzkomponente flame retardant
Flammschutzmittel flame retardant
Flammschutzsystem flame retardant
Flammschutzwirkung flame retardant effect
flammsicher flameproof, flame resistant
Flammstrahlen flame treatment
flammstrahlvorbehandelt flame treated
Flammverzögerungseigenschaften flame retardant properties
Flammvorbehandlung flame pretreatment
flammwidrig flame resistant
Flammwidrigkeit flame resistance
Flansch flange
Fleck stain
Fleckenbildung formation of (unsightly) patches
Flexbilisator flexibiliser
flexibel flexible
flexibilisierend flexibilising
Flexibilisierung flexibilisation
Flexibilisierungseffekt flexibilising effect
Flexibilisierungsmittel plasticiser
Flexibilität flexibility
Flexodruck flexographic printing
Flexodruckfarbe flexographic printing ink
Flexofarbe flexographic printing ink
Flexographiedruckfarbe flexographic printing ink
Flickmörtel patching mortar
Fliese tile
Fliesenklebstoff tile adhesive
Fließband 1. conveyor belt. 2. assembly line
Fließbandfertigung assembly line production
Fließbandprozeß assembly line production
Fließbandverfahren assembly line production
Fließbereich flow region
Fließcharakteristik flow properties/characteristics
Fließeigenschaften flow properties/characteristics
Fließen flow
Fließen, laminares laminar flow
Fließen, Newtonsches Newtonian flow
Fließen, viskoses viscous flow
Fließexponent flow index
fließfähig free-flowing, pourable
Fließfähigkeit flow, flowability, ease of flow
Fließgeschwindigkeit flow rate
Fließgesetz flow law
Fließgesetzexponent flow law index
Fließgrenze yield point
Fließhilfe flow promoter
Fließindex melt flow index
Fließkontrollmittel flow control agent
Fließkorrektur flow correction
Fließkurve flow curve
Fließmechanismus flow mechanism
Fließmittel flow control agent, flow promoter
Fließprofil flow profile
Fließrichtung direction of flow
Fließschema flow diagram/sheet
Fließspannung tensile stress at yield, yield stress
Fließstörungen flow irregularities
Fließübergangsbereich flow transition region
Fließverbesserer flow promoter
Fließverhalten flow behaviour/properties/characteristics
Fließvorgang flow process
Fließwiderstand flow resistance
Flockenbildung flocculation
Flockulation flocculation
Flockulationsgrad degree of flocculation
flockulierend flocculating
flockuliert flocculated
Flockulierung flocculation
Flockung flocculation
flockungsfrei non-flocculating
Flockungsmittel flocculating agent
flockungsstabilisierend anti-flocculating
Flockungstendenz tendency to flocculate
Flotationsmittel flotation agent
Flüchte volatile matter, volatiles
Flüchtegehalt volatile content
flüchtig volatile
flüchtige Bestandteile volatile matter/content
Flüchtigkeit volatility
Flüchtigkeitsgrad (degree of) volatility
Flugrost flash rust, rust bloom
Flugzeugbau aircraft construction
Flugzeugbauindustrie aircraft/aviation industry
Flugzeugindustrie aircraft/aviation industry
Flugzeugtriebwerk aircraft engine
fluidisieren to fluidise
Fluidisierungstrockner fluidising dryer
Fluidität fluidity
Fluor fluorine
Fluoracrylat fluoroacrylate
Fluoracrylsäure fluoroacrylic acid
Fluoralkylchlorsilan fluoroalkylchlorosilane
fluoralkylmodifiziert fluoroalkyl-modified
Fluoralkylrest fluoroalkyl group/radical
Fluoranthren fluoroanthrene
Fluoratom fluorine atom
Fluorbasis, auf fluorine-based
Fluorchlorkohlenwasserstoff chlorofluorocarbon, CFC
Fluordecklack fluororesin topcoat
Fluorelastomer fluoroelastomer
Fluoreszenz fluorescence
Fluoreszenzfarbe fluorescent colour
Fluoreszenzfarbstoff fluorescent dye
Fluoreszenzpigment fluorescent pigment
Fluoreszenzstrahlung fluorescent radiation
fluoreszierend fluorescent
Fluorether fluoroether
fluorfrei fluorine-free
Fluorgehalt fluorine content
Fluorglimmer fluoromica
fluorhaltig containing fluorine
Fluorharz fluoropolymer, fluoroplastic
Fluorharzbindemittel fluoropolymer binder
Fluorid fluoride
fluoriert fluorinated
Fluorierung fluorination
fluorisiert fluorinated

Fluorkautschuk fluororubber
Fluorkohlenwasserstoff fluorohydrocarbon
Fluorkohlenwasserstoffpolymer fluoropolymer
Fluorkunstharz fluoropolymer, fluoroplastic, fluorocarbon polymer
Fluorkunststoff fluoropolymer, fluoroplastic, fluorocarbon polymer
Fluorlack fluoropolymer paint
Fluormonomer fluoromonomer
Fluoroborsäure fluoroboric acid
Fluoroglykol fluoroglycol
FLuorogruppe fluoro group
Fluorolefin fluoroolefin
Fluorosilylmonomer fluorosilyl monomer
Fluorpolyester fluoropolyester
Fluorpolyether fluoropolyether
Fluorpolymer fluoropolymer, fluorocarbon polymer
Fluorpolyol fluoropolyol
Fluorsilicon fluorosilicone
Fluorsiliconelastomer fluorosilicone elastomer
Fluorsiliconfett fluorosilicone grease
Fluorsiliconkautschuk fluorosilicone rubber
Fluorsiliconöl fluorosilicone fluid
Fluortensid fluorosurfactant
Fluorthermoplast fluorothermoplastic
Fluortrichlormethan fluorotrichloromethane
Fluorüberzug fluoropolymer coating/film
Fluorverbindung fluorine compound
flüssig liquid, fluid
Flüssigadditiv liquid additive
Flüssigbeton liquid concrete
Flüssigchromatographie liquid chromatography
Flüssigharz liquid resin
flüssigisotrop liquid-isotropic
Flüssigkeit liquid, fluid
Flüssigkeit, Newtonsche Newtonian liquid
Flüssigkeit, nicht-Newtonsche non-Newtonian liquid
Flüssigkeit, wärmeaustauschende heat exchanging fluid
Flüssigkeitsaufnahme absorption of liquid
Flüssigkeitschromatographie liquid chromatography
Flüssigkeitsdiffusion liquid diffusion
Flüssigkeitsfraktion liquid fraction
Flüssigkeitsoberfläche liquid surface
Flüssigkeitsoberflächenspannung surface tension
Flüssigkeitsphase liquid phase
Flüssigkeitsrandwinkel contact angle, angle of contact
Flüssigkeitsreibung liquid friction
Flüssigkeitssäule liquid column
Flüssigkeitsscherkräfte liquid shear forces
Flüssigkeitsschicht liquid film/layer
Flüssigkeitsspiegel liquid level
Flüssigkeitsströmung liquid flow
Flüssigkleber liquid adhesive
Flüssigklebstoff liquid adhesive
Flüssigkristall liquid crystal
Flüssigkristallfolie liquid crystal film
flüssigkristallin liquid crystalline
Flüssigkristallpigment liquid crystal pigment
Flüssiglackierung painting with conventional paints *(as opposed to* **Pulverlackierung**, *q.v.)*
Flüssigpolymer liquid polymer
Flußmittel flux
Flußsäure hydrofluoric acid
Flußschema flow diagram/sheet
Fluten flow coating
Fokussierung focussing
Folgeanstrich subsequent coat
Folgebeschichtungen subsequent coatings
Folie 1. film, sheeting *(plastic)*. 2. foil *(metal)*
Foliendickenwert film thickness
Folienkaschierklebstoff film laminating adhesive
Foliensubstrat film surface
forciert forced
forciert trocknend force drying
forcierte Trocknung force drying
Förderband conveyor belt
Fördergurt conveyor belt
Formaldehyd formaldehyde
Formaldehyd-Tanninharz formaldehyde-tannin resin
Formaldehydabgabe formaldehyde emission
formaldehydarm low-formaldehyde, with a low formaldehyde content
Formaldehydemission formaldehyde emission
Formaldehydfänger formaldehyde interceptor
formaldehydfrei free from formaldehyde
Formaldehydfreiheit freedom from formaldehyde
Formaldehydgehalt formaldehyde content
Formamid formamide
Formänderung change in shape, dimensional change
formbeständig dimensionally stable
Formbeständigkeit dimensional stability, shape retention
Formbeständigkeit in der Wärme 1. heat resistance. 2. deflection temperature, heat distortion temperature
Formbeständigkeit in der Wärme nach Martens Martens heat distortion temperature
Formbeständigkeit in der Wärme nach Vicat Vicat softening point
Formbeständigkeit nach Martens Martens heat distortion temperature
Formel formula
formschlüssig positive *(connection or joint)*
formsteif rigid, stiff
Formstoff moulded material/plastic
Formtrennmittel mould release agent
formuliert formulated
Formulierung formulation
Formulierungsbestandteil formulation constituent
Formveränderung change in shape
Forschung research
Forschung, angewandte applied research
Forschungsarbeit research work

Forschungsaufwand research costs/expenditure
Forschungseinrichtungen research facilities
Forschungsgruppe research group
Forschungsinstitut research institute
Forschungsprojekt research project
Fortentwicklung improvement, further development, improved version
Fortschrittsbericht progress report
fossiler Brennstoff fossil fuel
fossiler Energieträger fossil fuel
Foto-, foto- *see* **Photo-, photo-**
Fourier-Transform-Infrarotspektroskopie FTIR spectroscopy, Fourier transform infrared spectroscopy
Fraktalanalytik fractal analysis
fraktionierte Destillation fractional distillation
Fraktionierung fractionation
Fräsen milling
Freialterung outdoor weathering
freibewittert naturally weathered
Freibewitterung outdoor weathering
Freibewitterungsbedingungen outdoor weathering conditions
Freibewitterungsbeständigkeit outdoor weathering resistance
Freibewitterungsdauer outdoor weathering period/duration
Freibewitterungsergebnisse outdoor weathering results
Freibewitterungsprüfung outdoor weathering test
Freibewitterungsstand outdoor weathering station
Freibewitterungsstation outdoor weathering station
Freibewitterungsverhalten outdoor weathering performance
Freibewitterungsversuch outdoor weathering test
Freibewitterungszeit outdoor weathering period
freifließend free-flowing
Freilandbewitterung outdoor weathering
Freiluftbeständigkeit outdoor weathering resistance
Freiluftbewitterung outdoor weathering
Freiluftbewitterungsanlage outdoor weathering station
Freiluftklima outdoor conditions
Freiluftprüfstand outdoor weathering station
Freilufttauglichkeit suitability for outdoor use
Freiluftversuch outdoor weathering test
Freisetzung release, liberation
Freistation outdoor weathering station
Fremdharz extender/blending resin, extender polymer
Fremdstoff foreign substance, contaminant
fremdvernetzend requiring the addition of a crosslinking agent
Frequenz frequency
Frequenzbereich frequency range
Frequenzverdopplung doubling the frequency
Friktion friction
friktionsarm low-friction
friktionsbedingt due to friction
Friktionsenergie frictional energy
Friktionswärme frictional heat
Frischbeton fresh concrete
frischgefällt freshly precipitated
Frischluft fresh air
Frischluftgerät breathing apparatus
Frischlufthaube breathing apparatus
Frischluftmaske breathing apparatus
Frischluftzufuhr fresh air supply
Frontteil front end
Frost frost
Frost-Taubeständigkeit freeze-thaw resistance
Frost-Taustabilität freeze-thaw resistance
Frost-Tauwechsellagerung freeze-thaw cycle
frostbeständig frost resistant
Frostbeständigkeit frost resistance
frostempfindlich frost sensitive, affected by frost
Frostschaden frost damage
Frostschutzmittel antifreeze
Frostwechsel freeze-thaw cycle
FTIR-Spektroskopie FTIR spectroscopy, Fourier transform infrared spectroscopy
Fuchsin fuchsin
Fuge joint
Fügefläche adherend surface
Fugeisen putty knife, trowel
Fügematerial adherend material
Fugenabdichten joint sealing
Fugenabstände joint spacings
Fugenbewegung joint movement
Fugenbruch joint failure
Fugendichtstoff joint sealant
Fugendichtung 1. joint seal. 2. joint sealing
Fugendichtungsband joint sealing/sealant tape
Fugendichtungsmasse 1. jointing filler/compound. 2. (tile) grout
Fugenflanke joint side
fugenfüllend gap-filling
Fugenfüller 1. jointing filler/compound. 2. (tile) grout
fugenlos jointless
Fugenmasse 1. jointing filler/compound. 2. (tile) grout
Fugenspachtel 1. jointing filler/compound. 2. (tile) grout
Fugentiefe joint depth
Fugenvergußmasse joint filler
Fugenverschluß joint sealing
Fugenversiegelungsmasse joint sealant
Fügeoffenzeit open assembly time
Fügestelle joint
Fügestoff adherend material
Fügeteil adherend
Fügeteilbruch adherend fracture/failure
Fügeteilfestigkeit adherend strength
Fügeteilgeometrie adherend dimensions
Fügeteiloberfläche adherend surface
Fügeteilwerkstoff adherend material

Fügeverfahren joining/bonding process
Fülle 1. body *(of a paint)*. 2. build *(of a paint film)*
Füller 1. filler. 2. knifing filler, stopper, surfacer
Füllergemisch filler blend
Füllfaktor bulk factor
Füllgrad filler content/loading
Füllgut contents *(of a package, container, bottle etc.)* **Isolierschicht zwischen Verpackung und Füllgut** insulating film between the pack and its contents; **Verpackungen für technische Füllgüter** packs for industrial products
Füllkraft high-build characteristics
füllkräftig high-build
Füllmasse knifing filler, stopper, surfacer
Füllmittel filler
Füllpulver powdered filler
Füllstoff filler
Füllstoffaggregat filler agglomerate
Füllstoffanteil filler content
füllstoffarm with a low filler content
Füllstoffart type of filler
Füllstoffaufnahme filler tolerance
Füllstoffgehalt filler content
füllstoffhaltig filled
Füllstoffkonzentration filler concentration/content
Füllstoffkorn filler particle
Füllstoffmenge amount of filler, filler content
Füllstoffnester local filler concentrations
Füllstofforientierung filler particle orientation/alignment
Füllstoffpartikel filler particle
Füllstoffpulver powdered filler
füllstoffrei unfilled
füllstoffreich with a high filler content
Füllstoffsorte type of filler
Füllstoffteilchen filler particle
füllstoffverträglich filler-compatible, compatible with fillers
Füllstoffzugabe 1. addition of filler. 2. amount of filler (added)
Füllungsgrad filler content/loading
Fumarat fumarate
Fumarsäure fumaric acid
Fumarsäureester fumaric ester
fungizid fungicidal
Fungizid fungicide
funktionalisiert functionalised
Funktionalität functionality
funktionell functional
Furan furan
Furanharz furan resin
Furfurol furfurol
Furfurylalkohol furfuryl alcohol
Furnier veneer
Fußboden floor
Fußbodenausgleichmasse jointless flooring compound, floor screed
Fußbodenbelag floorcovering
Fußbodenegalisiermasse jointless flooring compound, floor screed
Fußbodenklebstoff flooring adhesive
Fußbodenleiste skirting board
Fußbodenmasse flooring compound
Fußbodenplatte floor tile
Fußbodenversiegelung floor sealant (composition)
Fußleiste skirting board
Futtermaterial lining material
Futterstoff lining material
Futterstoffe, aufbügelbare iron-on interlinings

Galvanik electroplating
Galvanikanlage electroplating plant/unit
Galvanikbad electroplating bath
galvanisch electrolytic
galvanisch hergestellt made by electrodeposition
galvanische Korrosion electrolytic corrosion
galvanisierbar electroplatable
Galvanisierflüssigkeit electroplating solution
Galvanisierschicht galvanised finish
galvanisiert galvanised
Galvanisierungslösung electroplating solution
Galvanobeschichtung electroplating
Gammabestrahlung gamma irradiation
Gammabestrahlungsanlage gamma irradiation plant
Gammaspektroskopie gamma spectroscopy
Gammastrahlen gamma rays
Gammastrahlendetektor gamma ray detector
Gammastrahlenresistenz gamma ray resistance
Gammastrahlung gamma radiation
Garage garage
Garagenboden garage floor
Gardinenbildung curtaining, sagging
Gardner Farbskala Gardner colour scale
Gardner Farbwert Gardner colour value/number
Gardner Farbzahl Gardner colour value/number
Gas gas
Gasabgabe gas release/emission
Gasabsorption gas absorption
gasabspaltend gas-producing
Gasabspaltung gas evolution
gasartig gas-like
Gasausbeute gas yield
gasbeheizt gas heated
Gasbeton cellular/aerated concrete
Gasbläschen gas bubbles
Gasblase gas bubble
gasbremsend gas/vapour impermeable
Gasbrenner gas burner
Gaschromatogramm gas chromatogram

Gaschromatographie gas chromatography
gaschromatographisch gas chromatographic
gasdicht gas-tight, vapour proof
Gasdichtheit gas impermeability
Gasdichtigkeit gas impermeability
Gasdiffusion gas diffusion
Gasdruck gas pressure
gasdurchlässig gas permeable
Gasdurchlässigkeit gas permeability
Gasentladung gas discharge
Gasflamme gas flame
gasförmig gaseous
Gasfreisetzung gas release/emission
Gasgemisch gas mixture
Gasinnendruck internal gas pressure
Gaskonstante gas constant
Gaskonstante gas constant
gasnitriert gas nitrided
Gasnitrierung gas nitriding
Gaspermeabilität gas permeability
Gasphase gas/vapour phase
Gasphasenfluorierung gas phase fluorination
gassensitiv sensitive to gases
Gassperreigenschaften gas impermeability (properties)
Gasspühlsystem gas purging system
gasundurchlässig gas-impermeable, impermeable to gases
Gasundurchlässigkeit gas impermeability
GC *abbr. of* **Gaschromatographie**, gas chromatography
gealtert aged
Gebäude building
gebeizt pickled
Gebinde pack(age), container
gebleicht bleached
geblockt blocked
gebondet bonded
Gebrauchsbedingungen conditions of use
Gebrauchsdauer pot life
Gebrauchseigenschaften functional properties, performance characteristics
Gebrauchsfähigkeit serviceability
gebrauchsfertig ready-to-use
Gebrauchsfestigkeit final/maximum strength
Gebrauchsgegenstände consumer goods/articles
Gebrauchsgüter consumer goods/articles
Gebrauchskonzentration working concentration
Gebrauchslebensdauer working/service life
gebrauchstauglich suitable for the job
Gebrauchstauglichkeit serviceability
Gebrauchstemperatur working/service/operating temperature
Gebrauchstemperaturbereich working/service/operating temperature range
gebrauchstüchtig serviceable
Gebrauchstüchtigkeit serviceability
gecoatet coated
gedämpft damped
Gedankenaustausch exchange of ideas
gedeckt opaque
gedeckt eingefärbt pigmented
gedruckte Schaltung printed circuit
geeicht calibrated
geeignet suitable
geerdet earthed
Gefahr danger, hazard
Gefahr, gesundheitliche health hazard
Gefährdung exposure to danger
Gefährdung, gesundheitliche health hazard
Gefährdungspotential potential hazard
Gefahrenklasse hazard classification
Gefahrenspotential potential hazard
Gefahrenssymbol hazard symbol
Gefahrenstoffverordnung dangerous substances directive
Gefahrgüter hazardous/dangerous goods
Gefahrguttransportvorschrift dangerous goods transport ordinance
Gefahrgutverordnung Eisenbahn German regulation covering the transport of dangerous goods by rail
Gefahrgutverordnung Straße German regulation covering the transport of dangerous goods by road
Gefahrkennzeichen hazard symbol/marking
gefährlich dangerous, hazardous
Gefährlichkeit hazard
Gefahrstoff dangerous/hazardous substance
Gefahrstoffverordnung dangerous/hazardous substances directive
gefällt precipitated
gefalzte Überlappung rebated lap joint
gefärbt pigmented, dyed
gefiltert filtered
Gefrierpunkt freezing point
Gefriertruhe freezer, deep-freeze
Gegenion counter-ion
Gegenmaßnahmen countermeasures
Gegenstand object
Gehalt content
gehärtet cured, hardened
Gehäuse housing
gehobelt planed
geklebt bonded
geknäuelt entangled
gekoppelt coupled, linked
gekratzt scratched
gekrümmt curved
Gel-Chromatogramm gel chromatogram
Gel-Chromatographie gel chromatography
geladen charged
geladen, negativ negatively charged
geladen, positiv positively charged
Geländer 1. banisters *(indoors)*. 2. parapet *(of balcony)*
gelartig gel-like
gelascht strapped
Gelatine gelatine
gelb yellow
Gelbfärbung yellow colouring

gelbgold gold(en) yellow
Gelbildung gel formation, gelling
gelblich yellowish
Gelbpigment yellow pigment
Gelbstich yellow tinge
gelbstichig with a yellow tinge
Gelbstichindex yellowness index
Gelbtönung yellowing
gelförmig gel-like
Gelfraktion gel fraction
Gelieren gelation, gelling
Gelierpunkt gel point
Gelierungsneigung tendency to gel
Gelierungstendenz tendency to gel
Gelierzeit gel time
gelöst dissolved
Gelöstes solute
Gelpartikel gel particle
gelpermeationschromatographisch gel permeation chromatographic
Gelpermeationschromatographie gel permeation chromatography
Gelstruktur gel structure
Gelteilchen gel particle
Gelzustand gel state
gemahlen ground
gemeinsamer Binnenmarkt single European market
Gemenge mixture
Gemisch mixture
gemischtdispers of mixed/non-uniform particle size
Gemischtfraktion mixed fraction
Genauigkeit accuracy, precision
Genauigkeitsgrad degree of accuracy
genehmigungsbedürftig requiring official approval/permission
Genehmigungsbedürftigkeit need for official approval/permission
genehmigungspflichtig requiring official approval/permission
Geometrie geometry, shape
geometrische Gestaltung design
gepfropft grafted
geprimert primed
geprüft tested
gequollen swelled
Gerade straight line
Geradengleichung straight line equation
Gerät instrument
geräuschdämpfend sound-deadening
Geräuschdämpfung noise reduction
Geräuschdämpfungsmaßnahmen noise reduction measures
Geräuschniveau noise level
Geräuschpegel noise level
Geräuschreduzierung noise reduction
geräuschstark noisy
Geräuschverminderung noise reduction
Gerbsäure tannic acid
geritzt scratched
Germanium germanium
Germaniumderivat germanium derivative
Germaniumtetrachlorid germanium tetrachloride
Geruch smell, odour
geruchlos odourless
Geruchlosigkeit freedom from odour
geruchsarm low-odour
Geruchsarmut low-odour properties
geruchsbelästigend evil-smelling, malodorous
Geruchsdichtigkeit odour impermeability
geruchsfrei free from odour
Geruchsfreiheit freedom from odour
Geruchsintensität odour intensity
geruchsintensiv strong smelling, smelly, malodorous
geruchsneutral odourless
Geruchsneutralität freedom from odour
Geruchsprobleme odour problems
geruchsschwach low-odour
Geruchsschwelle odour threshold
Geruchssperre odour barrier
Gesamtabmessungen overall dimensions
Gesamtbeurteilung overall assessment
Gesamtdeformation total deformation
Gesamtdehnung total elongation
Gesamteinsatzmenge total amount used
Gesamtenergieaufnahme total energy input, total energy used
Gesamtenthalpie total enthalpy
Gesamtfestigkeit total strength
Gesamtfeuchte total moisture content
Gesamtfilmdicke total film thickness
Gesamtheizleistung total heating capacity
Gesamtleistung 1. total power. 2. total output. 3. overall performance
Gesamtlösemittel total solvent (content)
Gesamtmenge total amount
Gesamtpigmentierung total pigment content
Gesamtproduktionskosten total production costs
Gesamtreaktionszeit total reaction time
Gesamtschalldruckpegel overall noise level
Gesamtschallpegel overall noise level
Gesamtschichtdicke total film/coating thickness
Gesamtschrumpfung total shrinkage
Gesamtschwindung total shrinkage
Gesamttemperaturniveau overall temperature
Gesamtverhalten general/overall performance
Gesamtvolumen total volume
Gesamtwechselwirkungsenergie total interactive energy
gesättigt saturated
geschäftete Rohrverbindung scarfed/tapered tubular joint
geschäumt foamed
geschliffen ground, sanded
geschlossen 1. closed, shut. 2. enclosed. 3. continuouous *(e.g. film)* 4. closed-loop
geschlossene Wartezeit closed assembly time
geschlossene Zeit closed assembly time
geschlossenporig closed cell
geschlossenzellig closed-cell

Geschlossenzelligkeit 1. closed-cell character. 2. closed-cell content
geschmacklos tasteless
geschmacksarm having little taste, with little taste, almost tasteless
Geschmacksfreiheit freedom from taste
geschmacksneutral tasteless
Geschmacksneutralität freedom from taste
geschmeidig soft, supple
Geschmeidigkeit softness, suppleness
geschmolzen melted, molten
geschult trained
geschweißt welded
Geschwindigkeit speed, velocity
Geschwindigkeitsaufnehmer speed transducer
Geschwindigkeitsdifferenz speed/velocity difference
Geschwindigkeitsfernsteuerung 1. remote speed control. 2. remote speed control mechanism
Geschwindigkeitsfühler speed transducer
Geschwindigkeitsgefälle 1. speed/velocity gradient. 2. shear rate
Geschwindigkeitsgradient 1. velocity gradient. 2. shear rate
Geschwindigkeitskennlinie velocity curve
Geschwindigkeitskonstante velocity constant
Geschwindigkeitskonstanz constant speed/velocity
Geschwindigkeitsprogramm speed programme
Geschwindigkeitsprofil speed/velocity profile
Geschwindigkeitsregler speed regulator
Geschwindigkeitssensor speed sensor/transducer
Geschwindigkeitsverteilung velocity distribution
Geschwindigkeitszunahme speed/velocity increase
Gesetzesauflage directive
Gesetzesregelungen rules and regulations
Gesetzgeber legislator
Gesetzgebung legislation
Gesetzgebungsmaßnahmen legislative measures
gesintert sintered
gespritzt sprayed
Gestalt shape
gestalterisch design: **gestalterische Möglichkeiten** design possibilities
Gestaltfaktor design factor
Gestaltfestigkeit dimensional strength
Gestaltung design
Gestaltung, geometrische design
Gestaltung, konstruktive structural design
Gestaltungsfreiheit design freedom
Gestaltungshinweise design guidelines
Gestaltungsmöglichkeiten design possibilities/options
Gestaltungsrichtlinien design guidelines
gestaltungstechnisch design: **gestaltungstechnische Richtlinien** design guidelines
Gestaltungsvorschlag suggested design
gestrahlt grit/sand blasted
gestreckt stretched, extended
gestreut scattered
gestrichelt dashed *(curve)*
gestrichen coated
gesundheitlich unbedenklich non-toxic
gesundheitliche Gefahr health hazard
gesundheitliche Gefährdung health hazard
gesundheitliche Probleme health problems
gesundheitliche Unbedenklichkeit non-toxicity
Gesundheitsbeeinträchtigung health impairment
gesundheitsgefährdend harmful to health, toxic
Gesundheitsgefährdung health hazard/risk
Gesundheitsgesetze health regulations
Gesundheitsrisiko health risk
gesundheitsschädigend hazardous to health
gesundheitsschädlich hazardous to health
Gesundheitsschädlichkeit a danger to health
Gesundheitsunschädlichkeit not a danger to health
getempert conditioned
getönt tinted
Getränkedose drinks can
Getränkeverpackung beverage/drinks pack
getränkt impregnated
Getriebegehäuse gear housing
getrocknet dried
Gew.Anteile parts by weight
Gewächshaus greenhouse
Gewebe (woven) fabric
Gewerbe trade, industry, business, craft, occupation *(or any combination of these terms, according to context)*
gewerbehygienisch *adjective relating to workshop or factory hygiene*
Gewerbetoxikologe industrial toxicologist
Gewerbetoxikologie industrial toxicology
gewerblich industrial
gewerbliche Arbeitnehmer blue-collar workers
gewerbliche Mitarbeiter blue-collar workers
Gewicht weight
Gewicht, spezifisches specific gravity
Gewichtsabnahme weight loss, loss of weight
Gewichtsänderung change in weight
Gewichtsanteile parts by weight
Gewichtsdosierung 1. weigh feeding. 2. weigh feeder
Gewichtseinsparungen weight savings
Gewichtsersparnis weight saving
gewichtsintensiv heavy
Gewichtskonstanz constant weight
Gewichtskonzentration concentration by weight
Gewichtsmittel weight average
Gewichtsprozent percent by weight
gewichtsprozentig percent by weight
Gewichtsreduzierung weight reduction
Gewichtsschwankungen weight fluctuations/variations
Gewichtsteile parts by weight
Gewichtsverhältnis ratio by weight

Gewichtsverlust weight loss, loss of weight
Gewichtszunahme weight increase, increase in weight
Gewindesicherung thread locking
GFL *abbr. of* **Grenzflächenladung**, interfacial charge
GGVE *abbr. of* **Gefahrgutverordnung Eisenbahn,** German regulation covering the transport of dangerous goods by rail
GGVS *abbr. of* **Gefahrgutverordnung Straße**, German regulation covering the transport of dangerous goods by road
gießfähig pourable
Gießharz casting resin
Gießlack paint for application by flooding
Gießmasse casting compound
Gießverhalten pouring characteristics
Gießviskosität pouring consistency
giftig poisonous, toxic
Giftigkeit toxicity
Giftmüll toxic waste
Giftstoff toxic substance
gilbend yellowing
Gilbung yellowing
gilbungsbeständig non-yellowing
Gilbungsbeständigkeit resistance to yellowing
Gilbungsfestigkeit resistance to yellowing
Gilbungsfreiheit freedom from yellowing
Gilbungsneigung tendency to (become) yellow
gilbungsresistent non-yellowing, resistant to yellowing
Gilbungsresistenz resistance to yellowing
gilbungsstabil non-yellowing, resistant to yellowing
Gips 1. plaster (of Paris). 2. gypsum
Gipsfaserplatte plasterboard
Gipsformling plaster moulding
Gipskarton plasterboard
Gipskartonplatte plasterboard
Gipsplatte plasterboard
Gipsputz gypsum plaster
Gitterschnitt cross-hatching
Gitterschnittgerät cross-hatch adhesion test apparatus/instrument
Gitterschnittkennwert cross-hatch adhesion
Gitterschnittmethode cross-hatch adhesion test
Gitterschnittprobe cross-hatch adhesion test specimen
Gitterschnittprüfung cross-hatch adhesion test
Gitterschnittversuch cross-hatch adhesion test
Gitterschnittwert cross-hatch adhesion
Gitterstabiliserung lattice stabilisation
Glanz gloss
Glanzabbau reduction in gloss
Glanzabfall reduction/decrease in gloss
Glanzbeeinträchtigung gloss deterioration
Glanzbeständigkeit gloss retention
glänzend 1. glossy. 2. gloss *(paint)*
Glanzerhalt gloss retention
Glanzfarbe gloss paint
Glanzgrad (degree/amount of) gloss, percentage gloss
Glanzgradverlust gloss reduction, reduction in gloss
Glanzhaltevermögen gloss retention
Glanzhaltung gloss retention
Glanzlack gloss paint/enamel
Glanzlackschicht gloss coat/finish
glanzlos matt
Glanzmaximum maximum gloss
glanzmindernd gloss-reducing
Glanzminderung gloss reduction, reduction in gloss
Glanzniveau degree of gloss
Glanzschleier haze
glanzsteigernd increasing/enhancing gloss
Glanzüberzug gloss coat
Glanzverbesserer gloss enhancing agent
Glanzverlust reduction in gloss, gloss reduction
glanzverstärkend gloss-enhancing
Glanzverstärker gloss enhancing agent
Glanzverstärkung gloss enhancement
Glanzwert gloss value
Glanzwinkel gloss angle
Glanzzunahme increase in gloss
Glas glass
glasartig glassy
Glasartigkeit glass-like character
gläsern glassy
Glasfarbe glass paint
Glasfaser 1. glass fibre. 2. optic fibre
Glasfaserkabel fibre-optic cable, optic fibre cable
Glasfaserkunststoff glass fibre reinforced plastic, GRP
Glasfaserlaminat glass fibre laminate
Glasfasermatte glass fibre mat
glashart glassy
glasiert glazed
glasig glassy
Glaskapillare glass capillary
glasklar crystal clear
Glasklarheit crystal clarity
Glasklebung 1. bonded glass joint. 2. bonding of glass
Glaspapier glass paper
Glasperlen glass beads
Glasplatte glass plate
Glaspunkt glass transition temperature
Glasscheibe glass pane
Glastemperatur glass transition temperature
Glasübergang glass transition
Glasübergangsbereich glass transition range
Glasübergangsgebiet glass transition zone
Glasübergangspunkt glass transition temperature
Glasübergangstemperatur glass transition temperature
Glasumwandlung glass transition
Glasumwandlungsbereich glass transition zone
Glasumwandlungspunkt glass transition temperature
Glasumwandlungstemperatur glass transition temperature

Glasumwandlungstemperaturbereich glass transition temperature zone
Glasur glaze
Glasverklebung 1. bonding of glass. 2. bonded glass joint
Glaszustand glassy state
glatt smooth
glattpoliert polished
gleichförmig uniform, even
Gleichgewicht equilibrium
Gleichgewicht, thermisches thermal equilibrium
Gleichgewichtsabsorption equilibrium absorption
Gleichgewichtsdampfdruck equilibrium vapour pressure
Gleichgewichtsdruck equilibrium pressure
Gleichgewichtskonzentration equilibrium concentration
Gleichgewichtskurve equilibrium curve
Gleichgewichtsrandwinkel equilibrium contact angle
Gleichgewichtsreaktion equilibrium reaction
Gleichgewichtssystem equlibrium system
Gleichgewichtswassergehalt equilibrium moisture content
Gleichgewichtszustand state of equilibrium
Gleichmäßigkeit uniformity, evenness
Gleichstrom direct current, d.c.
Gleichung equation
Gleiteigenschaften surface slip characteristics/properties
Gleitlack non-stick paint
Gleitmittel lubricant
Gleitüberzug non-stick coating
Gleitung sliding, slippage
Gleitwiderstand sliding resistance
Glimmer mica
glimmerartig mica-like
Glimmerfüllstoff mica filler
Glimmerlamelle mica platelet
Glimmerpigment mica pigment
Glimmerplättchen mica flakes
global 1. global, world-wide. 2. overall, total, general
Glucose glucose
Glucoselösung glucose solution
Glührückstand ash
Glühtemperatur incandescence temperature
Glühung incandenscence
Glühverlust loss on ignition
Glutarsäure glutaric acid
Glutarsäureanhydrid glutaric anhydride
Glutbeständigkeit incandescence resistance
Glycerid glyceride
Glycerin glycerin
Glycerinacrylester glycerin acrylate
Glycidoxypropylgruppe glycidoxypropyl group
Glycidylacrylat glycidyl acrylate
Glycidylamin glycidylamine
Glycidylanteil glycidyl content
Glycidylester glycidyl ester
Glycidylether glycidyl ether
glycidylfrei glycidyl-free
Glycidylgehalt glycidyl content
Glycidylgruppe glycidyl group
glycidylgruppenhaltig containing glycidyl groups
Glycidylmethacrylat glycidyl methacrylate
Glycidylrest glycidyl radical
Glycidylsilan glycidyl silane
Glycin glycine
Glykol glycol
Glykolallylether glycol allyl ether
Glykolderivat glycol derivative
Glykoldiacetat glycol diacetate
Glykoleinheit glycol group
Glykolether glycol ether
Glykoletheracetat glycol ether acetate
Glykolgruppe glycol group
Glykolmonovinylether glycol monovinyl ether
Glykololigomer glycol oligomer
Glykolsäurebutylester butyl glycolate
Glykolyse glycolysis
Glyoxal glyoxal
Glyptalharz glyptal resin, glycerin-phthalic acid resin
Glyzerin glycerin
Gold gold
goldbraun golden brown
goldfarben gold
goldgelb golden yellow
Goldpigment gold pigment
Goniospektralphotometer goniospectrophotometer
goniospektralphotometrisch goniospectrophotometric
GPC gel permeation chromatography
gradkettig straight-chain
Graffiti graffiti
graffitiabweisend anti-graffiti
Granulat granules, granulated material *(e.g. hot melt adhesive)*
Graphit graphite
gravimetrisch gravimetric, by weight
Gravitationskraft gravitational force
Grenzfläche interface
Grenzflächenadhäsion interfacial adhesion
grenzflächenaktiv surface active
Grenzflächenebene 1. interfacial zone. 2. glueline
Grenzflächeneigenschaften interfacial properties
Grenzflächenenergie interfacial energy, surface tension
Grenzflächenenergiewert interfacial energy, surface tension
Grenzflächenerscheinung interfacial surface phenomenon
Grenzflächenkraft interfacial force
Grenzflächenladung interfacial charge
grenzflächennah near the interface
Grenzflächenpolymerisation interfacial polymerisation
Grenzflächenschicht interfacial layer

Grenzflächenspannung interfacial surface tension
Grenzflächenviskosität interfacial viscosity
Grenzgeschwindigkeit maximum/limiting speed
Grenzkonzentration maximum/limiting concentration
Grenzschicht boundary layer
Grenzschichtfestigkeit boundary layer strength
Grenzspannung limiting/maximum stress
Grenztemperatur limiting/maximum temperature
Grenzviskosität limiting/maximum viscosity
Grenzwert limiting/maximum value
Grenzwert, oberer upper limit
Grenzwert, unterer lower limit
Grenzzone boundary zone
Griff, trockener dry handle
griffest touch dry
Griffigkeit handle, feel
grob coarse
Grobagglomerat coarse agglomerate
grobdispers coarse-particle
grobgemahlen coarsely ground
grobkoaguliert coarsely coagulated
Grobkornanteil coarse particle content
grobkörnig coarse-particle
grobkristallin coarsely crystalline
grobporig coarse-cell
Grobpulver coarse powder
grobteilig coarse-particle
Großbehälter large tank
Großcharge large batch
großdimensioniert large
Größe 1. size. 2. quantity
große Produktionsserien long runs
große Stückzahlen large numbers
Größenordnung order of magnitude
großflächig with a large surface area
Großgebinde large container
Großhändler wholesaler
Großmolekül macromolecule
Großprojekt major project
Großserieneinsatz large-scale use
Großserienfertigung large-scale production
großtechnisch industrial, large-scale
großvolumig large-volume
grün green
Grundanstrich primer coat
Grundbeschichtung primer coat(ing)
Grundeigenschaft basic property
Grundfarbe 1. primer. 2. basic/base paint
Grundforderung basic requirement
Grundformulierung basic formulation
Grundgerät basic instrument
Grundgerüst basic framework
Grundharz base resin
Grundieranstrichstoff primer
Grundierharz primer resin
Grundiermittel primer
Grundierschicht primer coat
Grundierung 1. primer. 2. priming. 3. primer coat
Grundierungsauftrag application of primer: **unmittelbar nach dem Grundierungsauftrag** immediately after application of the primer *or*: after the primer has been applied
Grundierungsfilm primer coat
Grundierungsmittel primer
Grundlack primer
Grundlackierung primer coating
Grundlackschicht primer coat
Grundlagenforschung basic/fundamental research
Grundmolekül base molecule
Grundmonomer base monomer
Grundpolymer base polymer
Grundprinzipien basic/fundamental principles
Grundproblem basic problem
Grundrezeptur basic formulation
Grundschicht primer/base coat
Grundschichtüberzug primer/base coat
Grundstoff raw material
Grundstruktur basic structure
Grundvoraussetzung basic condition
Grüner Punkt Green Dot *(this is a symbol printed on packs and containers considered recoverable and recyclable through the* **Duales System** *q.v.)*
Grünfestigkeit green strength
Gruppe group, radical
Gruppierung group
Gummi rubber *(see explanatory note under* **Kautschuk***)*
gummiähnlich rubbery, rubber-like
gummiartig rubbery, rubber-like
Gummidichtung rubber seal
gummielastisch rubbery, rubbery-elastic
Gummielastizität rubber elasticity
Gummierklebstoff remoistenable adhesive
Gummifußboden rubber floor(ing)
Gummiindustrie rubber industry
Gummikleber rubber adhesive
Gummilatex rubber latex
Gummilösung rubber solution
Gummimilch rubber latex
Gummiprofil rubber profile
Gußasphalt poured asphalt
Gußeisen cast iron
gut löslich readily soluble
Güte quality
Güteanforderungen quality requirements
Gütebestimmung quality specification
Gütefaktor quality factor
Gütegrad quality
Gütekriterien quality criteria
Gütenorm quality standard
Güterichtlinie quality guideline
Güterwaggon goods wagon
Gütesicherung quality assurance
Gütesteuerung quality control
Güteüberwachung quality control
guthaftend with good adhesion

H

Haarriß hairline crack
haftabweisend non-stick
Haftbrücke primer coat
Hafteigenschaften adhesive properties, adhesion
haftend adhering
Haftetikett self-adhesive label
Haftetikettierung presssure sensitive labelling, labelling using pressure sensitive adhesives
Haftfähigkeit adhesion
haftfest having good adhesion
Haftfestigkeit adhesive/bond strength, adhesion
Haftfestigkeitsmängel loss of adhesion
Haftfestigkeitsprüfung adhesion test
Haftfestigkeitsverminderung loss of adhesion
Haftfläche adherend surface
Haftgrund adherend surface
Haftgrundierung primer coat
Haftgruppe adhesive group/radical
Haftharz 1. primer resin. 2. adhesive resin
hafthindernd 1. non-stick, anti-adhesive. 2. interfering with adhesion
Haftklebeband adhesive tape
Haftklebedispersion pressure sensitive adhesive dispersion
Haftklebeeigenschaften pressure sensitive properties
Haftklebeetikett self-adhesive label
Haftklebemittel pressure sensitive adhesive
haftklebend self-adhesive
Haftkleber pressure sensitive adhesive
Haftklebeverfahren pressure sensitive bonding
Haftklebstoff pressure sensitive adhesive
Haftklebstoffdispersion pressure sensitive adhesive dispersion
Haftkraft adhesive force, adhesion
Haftmechanismus adhesive mechanism
haftmindernd reducing adhesion
Haftmittel 1. adhesive. 2. adhesion promoter
Haftprimer primer
Haftprüfung adhesion test
Haftreibung static friction
Haftreibungskoeffizient coefficient of static friction
Haftreibungszahl coefficient of static friction
Haftschicht primer coat
Haftschmelzklebstoff hot melt pressure sensitive adhesive
Haftschwierigkeiten adhesion problems
Haftsystem adhesive, adhesive system/formulation
Haftung 1. adhesion. 2. liability, responsibility
Haftung-bleibende permanent adhesion
Haftungsabweisung non-stick properties
Haftungseigenschaften adhesive properties, adhesion
Haftungseinbuße loss of adhesion
Haftungsfähigkeit adhesion, adhesive properties
Haftungsfehler poor/faulty adhesion
Haftungskraft adhesive force
Haftungsmängel poor/faulty adhesion
Haftungsmechanismus adhesion mechanism
haftungsmindernd adhesion-reducing
Haftungsprobleme adhesion problems
Haftungsschäden poor/faulty adhesion
Haftungsschwierigkeiten adhesion problems
Haftungsspektrum adhesive properties
Haftungssteigerung improved adhesion
Haftungsverbesserer coupling/bonding agent
haftungsverbessernd improving adhesion
Haftungsverlust loss of adhesion
Haftungsvermögen adhesion
haftungszerstörend destroying adhesion
Haftverbesserer coupling/bonding agent
haftverbessernd improving adhesion
Haftverbesserung improvement of adhesion
Haftverhalten adhesion
Haftverlust loss of adhesion
haftvermittelnd adhesion promoting
Haftvermittler adhesion promoter, coupling/bonding agent
Haftvermittlung adhesion promoting
Haftvermittlungsschicht primer coat
Haftvermögen adhesion
Haftwert adhesion
Haftwirkung adhesive effect
Haftzugfestigkeit tensile bond strength
Haftzugversuch tensile bond test
Haftzusatz adhesion promoter
halbautomatisch semi-automatic
halbblockiert semi-blocked
halbfest semi-solid
halbfett medium-oil, 50% oil length
halbflexibel semi-flexible
halbflüssig semi-liquid
halbhart semi-rigid, medium-hard
halbkristallin semi-crystalline
halbleitend semi-conducting
Halbleiter semi-conductor
Halbleiterbauelement semi-conductor module
Halbleiterbaustein semi-conductor module
Halbleiterlaser semi-conductor laser beam
Halbleitertechnik semi-conductor technology/engineering
halbquantitativ semi-quantitative
halbspröde semi-brittle
halbsteif semi-rigid
halbtechnisch semi-industrial
halbtransparent semi-transparent
halbtrocknend semi-drying
Halbwertszeit half-life
Halogen halogen
halogenfrei halogen-free
Halogengehalt halogen content
halogenhaltig containing halogens
Halogenhydrin halogen hydrin
Halogenid halide
halogeniert halogenated
Halogenion halogen ion
Halogenkohlenwasserstoff halogenated

hydrocarbon
Halogenlicht halogen light
Halogenradikal halogen radical
Halogensilan halogenated silane
Halogenverbindung halogen compound
Halogenwasserstoff hydrogen halide
HALS hindered amine light stabiliser
haltbar durable
Haltbarkeit durability
Hämatit hematite
Hammerschlageffekt hammer finish
Hammerschlageffektlackierung hammer finish
Hammerschlaglack hammer finish paint
Hand, aus einer from a single source, from one source
Handapplikation manual application, application by hand
Handelsbezeichung trade/brand name
Handelsname trade/brand name
Handelspartner trading partner
Handelsprodukt commercial product
handelsüblich commercial
Handhabung handling
Handhabungsautomat robot
handhabungsfreundlich easy to handle
Handhabungsgerät 1. handling device. 2. robot
Handhabungsrichtlinien handling instructions
handhabungssicher safe to handle
Handhabungssystem handling/robotic system
Handpistole manually operated spraygun
Handrührgerät hand mixer
Handschuhe (protective) gloves
Handspritzen hand/manual spraying
Handspritzpistole manually operated spraygun
Handspritzverfahren hand spraying (process)
handtrocken touch-dry
Handwerk craft, trade
Handwerker craftsman
Handwerksbetrieb workshop
Hantieren handling
Harnstoff urea
Harnstoff-Formaldehydharz urea/urea-formaldehyde/UF resin
Harnstoff-Formaldehydkondensat urea-formaldehyde condensate
Harnstoff-Formaldehydleim urea-formaldehyde adhesive
Harnstoffabkömmling urea derivative
Harnstoffbindung urea linkage
Harnstoffbindungsanteil urea linkage content
Harnstoffbrücke urea bridge
Harnstoffderivat urea derivative
Harnstoffgruppe urea group
Harnstoffharz urea/urea-formaldehyde/UF resin
Harnstoffharzschichtstoff urea laminate
Harnstoffleim urea adhesive
Harnstoffrest urea group
Harnstoffverbindung urea compound
hart hard
härtbar curable
Härte hardness
Härteabfall decrease in hardness
Härtegrad (degree of) hardness
hartelastisch energy-elastic
Hartelastizität energy elasticity
Härteprüfer hardness tester
Härteprüfgerät hardness tester
Härteprüfung hardness test
Härter hardener, catalyst
Härteranteil amount of hardener
Härterkomponente hardener, catalyst
Härterkonzentration hardener concentration
Härtermenge amount of hardener
Härterpaste catalyst paste
Härterpulver powdered hardener/catalyst
Härterradikal hardener radical
Härterüberschuß excess hardener/catalyst
Härterunterschuß less hardener/catalyst
Härterzugabe addition of hardener
Härteskala hardness scale
Härtetemperatur curing temperature
Härtezeit curing time
Härtezyklus curing cycle
Hartfaserplatte hardboard
Hartharz hard resin
Hartholz hard wood
Hartpapier paper-based laminate
Hartschaumkern rigid foam core
Härtung curing
Härtungsablauf curing reaction
härtungsauslösend initating curing
Härtungsbedingungen curing conditions
Härtungsbeschleuniger accelerator
Härtungscharakteristik curing characteristics
Härtungsdauer cure time
Härtungseigenschaften curing characteristics
Härtungsgeschwindigkeit curing/setting speed
Härtungsgrad degree of cure
Härtungskatalysator catalyst, hardener, curing agent
Härtungsmechanismus curing mechanism
Härtungsmittel catalyst, hardener, curing agent
Härtungsofen curing oven
Härtungsprogramm curing schedule
Härtungsprozeß curing process
Härtungsreaktion curing reaction
Härtungsschrumpf curing shrinkage
Härtungsschwund curing shrinkage
Härtungsstufe curing stage/phase
Härtungssystem catalyst
Härtungstemperatur curing temperature
Härtungsverhalten curing characteristics/properties
Härtungsverlauf curing pattern
Härtungszeit cure time
Harz resin
Harz-Härtermischung resin-catalyst mix
Harzaddukt resin adduct
Harzansatz resin-catalyst mix
Harzanteil resin content
harzarm low-resin, with a low resin content
harzartig resin-like

Harzaufnahme resin pick-up/take-up
Harzbasis, auf resin-based
Harzbestandteil resin constituent
harzbildend resin-forming
Harzdispersion resin dispersion
Harzdomäne resin domain
Harzemulsion resin emulsion
Harzester rosin ester
Harzfeststoff solid resin content
harzgebunden resin bonded
Harzgehalt resin content
Harzgemisch resin blend
harzgetränkt resin impregnated
Harzgitter polymer lattice
Harzgruppe group of resins
harzimprägniert resin impregnated
Harzkomponente resin (component)
Harzkonzentration resin concentration
Harzlösung resin solution
Harzmatrix resin matrix
Harzmenge amount of resin
Harzmerkmale resin characteristics
Harzmischung resin blend
harzmodifiziert resin modified
Harzmolekül resin molecule
Harzpartikel resin particle
Harzpulver powdered resin
harzreich resin-rich, with a high resin content
Harzsäure rosin acid
Harzschicht resin coating/film
Harzschmelze resin melt
Harzseife rosin soap
Harzteilchen resin particle
Harzüberschuß excess resin
harzüberzogen resin coated
harzverfestigt resin impregnated
Harzverschnitt resin blend
Harzviskosität resin viscosity
Harzzusammensetzung resin composition
Haube 1. hood. 2. bonnet
Hauptbestandteil main constituent/component
Hauptbindemittel main binder
Haupteinsatzgebiet main field of use
Haupterweichungsbereich glass transition range
Hauptkette main chain
Hauptketten-LCP linear liquid crystal polymer
Hauptkettenflüssigkeitskristall linear liquid crystal
Hauptkettenpolymer linear polymer
Hauptkriterium main criterion
Hauptlichtquelle main light source
Hauptlösemittel main solvent
Hauptvalenz primary valency
Hauptvalenzbindung primary valency bond
Hauptvalenzkräfte primary valency forces
hauseigen in-house
Haushalt household
Haushaltsgeräte household/domestic appliances
Haushaltsgerätelack household appliance paint
Haushaltsklebstoff household adhesive
Haushaltsreiniger household cleaner
Haushaltssektor household sector
Hautbildung skinning
Hautbildungsresistenz skinning resistance
Hautbildungszeit skinning time
Hauterkrankung skin infection
Hautkontakt skin contact
hautreizend skin-irritant
Hautreizmittel skin irritant
Hautreizung skin irritation
Hautschutzcreme barrier cream
Hautschutzsalbe barrier cream
Hautverhinderungsmittel anti-skinning agent
Hautverhütungsmittel anti-skinning agent
Hautverträglichkeit skin tolerability
Hazenfarbskala Hazen colour scale
HBB hydrogen bridge bond
HDI hexamethylene diisocyanate
Hebelsystem system of levers
Heckteil rear end
Hectorit hectorite
Hefe yeast
Heimwerker do-it-yourself worker
Heimwerkerklebstoff DIY adhesive
heiß hot
heißaktivierbar capable of being heat activated
heißhärtbar heat curable/curing
heißhärtend hot setting, heat curing
Heißhärtung heat curing
Heißklebfolie hot melt film adhesive
Heißluft hot air
Heißluftalterung hot air ageing
Heißluftbeständigkeit hot air resistance
Heißluftlagerung hot air ageing
Heißluftofen hot air oven
Heißluftschrank hot air oven
Heißlufttrocknung 1. hot air drying. 2. hot air drying unit
Heißpressen hot pressing
heißschmelzend hot melt
Heißschmelzfolie hot melt film adhesive
Heißschmelzkleber hot melt adhesive
Heißschmelzklebstoff hot melt adhesive
Heißschmelzmasse hot melt compound
Heißschmelzplastisol hot melt plastisol
Heißschmelzverklebung hot melt bonding
Heißsiegelautomat automatic heat sealing unit
heißsiegelbar heat sealable
Heißsiegelbarkeit heat sealability
Heißsiegelbeschichtung heat sealable coating
Heißsiegeletikettierung heat-seal labelling
heißsiegelfähig heat sealable
Heißsiegelgerät heat sealing instrument
Heißsiegelklebstoff heat sealing adhesive
Heißsiegellack heat sealing lacquer
Heißsiegelmasse heat sealing compound
Heißsiegelschicht heat sealable coating
Heißsiegelung heat sealing
Heißspritzverfahren hot spraying (process)
heißsprühbar hot sprayable
Heißverklebung heat bonding
Heißversiegelung heat sealing

Heißverträglichkeit high-temperature compatibility
Heißwasser hot water
Heißwasserbelastung 1. immersion in hot water. 2. exposure to hot water
heißwasserbeständig hot water resistant
heißwasserlöslich soluble in hot water, hot water soluble
Heizbad heating bath
heizbar heatable
Heizeinrichtung heating equipment
Heizelement heating element
Heizenergie heating energy
Heizfläche heating surface/area
Heizgeschwindigkeit heating-up rate
Heizkörperlack radiator paint
Heizkosten heating costs
Heizleistung heating capacity
Heizofen oven
Heizöl fuel oil
heizölbeständig fuel oil resistant
Heizöllagertank fuel oil (storage) tank
Heizöltank fuel oil (storage) tank
Heizölwanne fuel oil drip pan
Heizpatrone cartridge heater
Heizrate heating-up rate
Heizschlange heating coil
Heizstrahler heating tunnel/section
Heiztrommel heating drum
Heizwert calorific value
heizwertarm with/having a low calorific value
heizwertreich with/having a high calorific value
Heizwiderstand heating resistance
Heizzone heating section/zone
hell bright
hellblau light blue
hellfarbig light coloured
hellgrau light grey
Helligkeit brightness
Helligkeitsgrad brightness
Helligkeitsunterschied difference in brightness
Hemmung inhibition
Heptan heptane
Herbizid herbicide
hermetisch geschlossen hermetically sealed
Herstell- *see* **Herstellungs-**
Hersteller manufacturer, producer
Herstellerangaben information supplied by the manufacturer
Herstellung production, manufacture
Herstellungsabschnitt production stage
Herstellungsbedingungen production conditions
Herstellungsdatum manufacturing/production date
Herstellungshilfsmittel processing aid
Herstellungshinweise production guidelines
Herstellungskosten manufacturing/production costs
Herstellungspalette product range, range of products
Herstellungsparameter production conditions
Herstellungsprogramm 1. product range, range of products. 2. production schedule/plan/programme
Herstellungsprozeß production/manufacturing process
herstellungstechnisch production, manufacture: **herstellungstechnische Vorteile** production advantages
Herstellungstechnologie production technology
Herstellungstoleranz production tolerance
Herstellungsverfahren manufacturing/production process
Heteroatom heteroatom
heterocyklisch heterocyclic
Heteroflockulation heteroflocculation
heterogen heterogeneous
Heterogenität heterogeneity
heteropolar heteropolar
Heteropolykondensat heteropolycondensate
Heteropolymer heteropolymer
Hexa hexamethylene tetramine
Hexaalkoxymethylmelamin hexaalkoxymethylmelamine
Hexadekanol hexadecanol
Hexafluorantimonsäure hexafluoroantimonic acid
Hexafluorarsensäure hexafluoroarsenic acid
Hexafluorphosphorsäure hexafluorophosphoric acid
Hexafluorpropen hexafluoropropene
Hexafluorpropylen hexafluoropropylene
Hexafluorsilikat hexafluorosilicate
hexafunktionell hexafunctional
Hexaglykolether hexaglycol ether
Hexahydrophthalsäure hexahydrophthalic acid
Hexahydrophthalsäurediglycidylester diglycidyl hexahydrophthalate
Hexahydrophthalsäureanhydrid hexahydrophthalic anhydride
Hexamethoxymethylamin hexamethoxymethylamine
Hexamethoxymethylmelaminharz hexamethoxymethyl melamine resin
Hexamethoxymethylmelamin hexamethoxymethyl melamine
Hexamethyldisilazan hexamethyl disilazane
Hexamethylenbrücke hexamethylene bridge
Hexamethylendiamin hexamethylene diamine
Hexamethylendiisocyanat hexamethylene diisocyanate, HDI
Hexamethylentetramin hexamethylene tetramine
Hexamethylolmelamin hexamethylolmelamine
Hexan hexane
Hexandiolacrylat hexanediol acrylate
Hexandioldiglycidylether hexanediol diglycidyl ether
Hexanol hexanol
Hexaphenylether hexaphenyl ether
Hilfslösemittel co-solvent
Hilfsmittel additive
Hilfsstoff additive

Hinderung, sterische steric hindrance
Hinterfüllmaterial back filling material
Hinterfüllung back filling
Hinterfüttern back filling
Hinterfütterungsmasse backing mix
Hinterschneidung undercut
Hitze heat
Hitzeaktivierung heat activation
Hitzealterung heat ageing
hitzebeansprucht exposed/subjected to high temperatures
hitzebelastbar heat resistant
Hitzebelastbarkeit heat resistance
Hitzebelastung exposure to heat, exposure to high temperatures
hitzebeständig heat resistant
Hitzebeständigkeit heat resistance
Hitzeeigenschaften heat resistance
Hitzeeinwirkung, unter when heated, when exposed to high temperatures
hitzeempfindlich heat sensitive, affected by heat
hitzefest heat resistant
hitzehärtbar heat/hot curing, heat setting
hitzehärtend heat setting/curing
Hitzehärtung heat/hot curing, heat setting
Hitzelagerung heat ageing
Hitzereaktivierung heat activation
hitzeresistent heat resistant
Hitzeresistenz heat resistance
Hitzeschild heat shield
Hitzeschockfestigkeit heat shock resistance
Hitzeschutzmittel heat stabiliser
hitzestabil heat resistant
Hitzestabilisator heat stabiliser
hitzestabilisiert heat stabilised
Hitzestabilisierung heat stabilisation
Hitzestabilität thermal stability
Hitzestau heat accumulation
Hitzesterilisation heat sterilisation
hitzesterilisierbar heat sterilisable
hitzesterilisiert heat sterilised
hitzeunempfindlich unaffected by heat
hitzevernetzbar heat/hot curing
HKW *abbr. of* **Halogenkohlenwasserstoff**, halogenated hydrocarbon
HL-Dispergiermittel high-performance dispersing agent
HLB-Wert hydrophilic-lipophilic balance
HMMM-Harz hexamethoxymethyl melamine resin
Hobbock drum
Hobbybereich hobby sector
hochabriebfest highly abrasion resistant
hochabsorbierend highly absorbent
hochadhäsiv highly adhesive
hochaggresiv highly corrosive/aggressive
hochaktiv highly reactive, high-reactivity
hocharomatisch highly aromatic
hochauflösend high-resolution
Hochbau building *(as opposed to* **Tiefbau**, *q.v.)*
Hochbauten high-rise buildings
hochbeanspruchbar heavy-duty
hochbeansprucht highly stressed
hochbelastbar heavy-duty
hochbelastet highly stressed
Hochbelastung high stress
hochbeständig highly resistant
hochcarboxyliert highly carboxylated
hochchemikalienbeständig having excellent chemical resistance
Hochchleistungsanlage high-capacity/-performance plant
hochdehnfähig highly elastic
hochdicht 1. high-density. 2. very dense. 3. very tight/watertight
hochdispers fine-particle
Hochdruck high pressure
Hochdruckdosieranlage high-pressure feed/metering unit
Hochdruckdüse high-pressure nozzle
Hochdruckflüssigkeitschromatographie high-pressure liquid chromatography
Hochdrucklaminat high-pressure laminate
Hochdruckpolyethylen low-density polyethylene, LDPE
Hochdruckpolymerisation high-pressure polymerisation
Hochdruckpumpe high-pressure pump
Hochdruckreaktor high-pressure reactor
Hochdruckspritzen high-pressure spraying
Hochdruckventil high-pressure valve
hochelastisch highly flexible
hochempfindlich highly sensitive
hochenergetisch high-energy
hochentzündbar highly flammable
hochentzündlich highly flammable
hochfest high-strength
hochflexibel highly flexible
hochfrequent high-frequency, HF
Hochfrequenzabschirmung high-frequency screening
Hochfrequenzbeaufschlagung exposure to a high-frequency field
Hochfrequenzfeld high-frequency field
Hochfrequenzgenerator high-frequency generator
Hochfrequenzpotential high-frequency potential
hochgebrannt high-fired
hochgefüllt highly filled
Hochglanz high-gloss
Hochglanzdeckschicht high-gloss finish
Hochglanzdispersionsfarbe high-gloss emulsion paint
Hochglanzdruckfarbe high-gloss printing ink
hochglänzend high-gloss
Hochglanzfarbe high-gloss paint
Hochglanzklarlack clear high-gloss lacquer/varnish
Hochglanzlackierung high-gloss finish
Hochglanzlackschicht high-gloss finish
hochglanzpoliert highly polished
Hochglanzsystem high gloss paint/system/formulation

hochglanzverchromt high-polish chromium plated
hochhitzebeständig high-temperature resistant
hochhitzestabil high-temperature resistant
hochkohäsiv highly cohesive
hochkondensiert highly condensed
hochkonzentriert highly concentrated
hochkristallin highly crystalline
hochlegiert high-alloy
Hochleistungs- 1. high-capacity/-performance. 2. high-speed. 3. heavy-duty
Hochleistungsdichtstoff high-performance sealant
Hochleistungsdispergiermttel high-performance dispersing agent
hochleistungsfähig heavy-duty, high-performance
Hochleistungskatalysator high-performance catalyst
Hochleistungsklebstoff high-performance adhesive
Hochleistungsmischer heavy-duty mixer
hochleitfähig highly conductive
hochmethylverethert highly methyl etherified
hochmolekular high-molecular weight
hochpigmentiert highly pigmented
hochpoliert highly polished
hochpolymer high-polymer
Hochpolymer high polymer
Hochpolymerisat high polymer
hochpräzis high-precision
Hochpräzisionswaage high-precision balance
hochprozentig highly concentrated
hochqualitativ high-quality
hochreaktiv highly reactive
hochreflektierend highly reflective
hochrein high-purity
hochschlagbeansprucht subjected to high impact (stress)
hochschlagfest high-impact
hochschlagzäh high-impact
hochschmelzend high-melting
hochsensibel highly sensitive
hochshorig with a high Shore hardness
hochsiedend high-boiling
Hochsieder high-boiling solvent
Hochspannung high voltage/tension
Hochspannungsbelastung exposure to high voltage
Hochspannungsbereich high-voltage range
Hochspannungsfeld high-voltage/tension field
Hochspannungskriechstromfestigkeit high-voltage tracking resistance
Hochspannungslichtbogenbeständigkeit high-voltage arc resistance
Hochspannungsprüffeld high-voltage test bed
Hochspannungsprüfung high-voltage test
höchstdruck maximum pressure
Höchstkraft maximum force
höchstmolekular ultra-high molecular weight
hochstoßfest high-impact
höchstschlagzäh ultra-high impact
Höchsttemperatur maximum temperature
Höchstwert maximum value
höchstzulässig maximum permissible
hochtechnologisch high-tech
Hochtemperaturanwendung high-temperature application
hochtemperaturbelastbar high-temperature resistant
Hochtemperaturbelastbarkeit high-temperature resistance
Hochtemperaturbereich, im at high temperatures
hochtemperaturbeständig high-temperature resistant
Hochtemperaturbeständigkeit high-temperature resistance
Hochtemperatureigenschaften high-temperature properties
Hochtemperatureinsatz, für for high-temperature use
Hochtemperaturen high temperatures
hochtemperaturfest resistant to high temperatures
Hochtemperaturpolymer high-temperature polymer
Hochtemperaturthermostat high-temperature thermostat
Hochtemperaturverbrennungsanlage high-temperature incinerator
Hochtemperaturverhalten high-temperature behaviour/performance
Hochtemperaturversuch high-temperature test
hochtourig high-speed
hochtransparent highly transparent
Hochvakuum high-vacuum
Hochvakuumbedampfen high-vacuum metallisation
Hochvakuummetallisierung high-vacuum metallisation
hochverdichtet tightly compressed
hochverdünnt highly diluted
hochverethert highly etherified
hochvernetzt highly crosslinked
hochverschleißfest extremely hard-wearing, extremely wear resistant
hochverschleißwiderstandsfähig extremely hard-wearing, extremely wear resistant
hochviskos high-viscosity
hochwärmebelastet subjected to high temperatures
hochwärmebeständig high-temperature resistant
hochwärmestabilisiert high-temperature stabilised
hochwertig 1. high-quality. 2. high-valency
hochwetterecht highly weather resistant
hochwirksam highly effective
hochzäh very tough
Hochziehen lifting *(of paint film)*
hochzugfest high-tensile
höhenverstellbar vertically adjustable, adjustable in height

Höhenverstellung vertical adjustment
höherenergetisch high-energy
höherfest high-strength
höhermolekular high-molecular weight
höhersiedend high-boiling
höherviskos high-viscosity
Hohlkügelchen hollow microsphere
Hohlkugeln hollow microspheres
Hohlpartikel hollow particle
Hohlraum void, space
Hologramm hologram
Holographie holography
holographisch holographic
Holz wood
Holzanstrich 1. wood finish. 2. coating/painting of wood. 3. wood lacquer/varnish
Holzaußenanstrich exterior wood finish
Holzbeize wood stain
Holzbrett wooden board
Holzbruch failure in the wood *(i.e. bonded joint)*
Holzfarbstoff wood dye
Holzfaser wood fibre
Holzfaserplatte (wood) chipboard
Holzfeuchte moisture content of wood
holzfressend wood-eating
Holzfurnier wood veneer
Holzfußboden wooden floor
Holzgrundierung wood primer
Holzimprägnierung 1. wood impregnation. 2. wood impregnating agent
Holzindustrie timber industry
Holzklebstoff wood adhesive
Holzkonservierung wood preservation
Holzlack wood lacquer/paint
Holzlackierung wood finish
Holzlasur wood glaze/varnish/lacquer
Holzleim wood glue/adhesive
Holzleimbau engineering timber construction
Holzmasse wood pulp
Holzmöbel wooden furniture
Holzoberfläche wood surface
Holzöl wood oil
Holzölalkydharz wood oil alkyd (resin)
Holzplatte wooden board
Holzprobekörper wooden test piece
Holzschutzanstrichmittel wood preservative
Holzschutzemulsion wood preservative emulsion
Holzschutzlack wood lacquer
Holzschutzlasur wood glaze
Holzschutzlasurfarbe wood glaze
Holzschutzmittel woodcare product, wood preservative
Holzschutzzusatzmittel wood preservative
Holzspanplatte (wood) chipboard, particleboard
Holztiefenimprägnierung pressure impregnation (of wood)
Holzverleimung glued wooden joint
Holzversiegelung wood sealant
Holzwerkstoff wood-based material
Holzzellulose wood cellulose
holzzerstörend wood-destroying
homogen homogeneous
homogenisieren to homogenise
Homogenisierhilfe homogenising aid
Homogenisierleistung homogenising efficiency
Homogenisierung homogenisation
Homogenisierwirkung homogenising effect
Homogenität homogeneity
Homolog homologue
homolytisch homolytic
homopolar homopolar
Homopolykondensat homopolycondensate
homopolymer homopolymeric
Homopolymer homopolymer
Homopolymerisat homopolymer
Homopolymerisation (homo)polymerisation
Honigwaben honeycomb
Honigwabenstruktur honeycomb structure
Hookesch Hookean
Hookesches Gesetz Hooke's Law
horizontal horizontal
Horizontalausführung, in horizontal
Horizontalbauweise, in horizontally constructed
horizontales Ausschwimmen floating
horizontales Pigmentausschwimmen floating
Horizontallage, in horizontal
HPL-chromatographisch HPL chromatographic
HPLC high-pressure liquid chromatography
HPLC-Untersuchung HPLC test, high-pressure liquid chromatography
HS-Lack high-solids paint
HS-Lösung high-solids solution
Hüllschicht enveloping/outer layer
Huminsäure humic acid
Hüttentechnik metallurgical engineering
hybrid hybrid
Hybridharz hybrid/composite resin
Hybridpolymer hybrid polymer
Hybridsystem hybrid/composite system
Hydantoinesterimid hydantoin ester imide
Hydophobie water repellency, hydrophobic properties
hydophobierend making water repellent
Hydratation hydration
Hydrationswasser water of crystallisation
Hydratisation hydration
hydratisieren to hydrate
Hydratwasser water of crystallisation
Hydraulik hydraulics, hydraulic system
Hydraulikaggregat hydraulic unit
Hydraulikanlage hydraulics, hydraulic system
Hydraulikbehälter hydraulic accumulator
Hydraulikblock hydraulic unit
Hydraulikdruck hydraulic pressure
hydraulikdruckabhängig depending on the hydraulic pressure
hydraulikdruckunabhängig irrespective of the hydraulic pressure
Hydraulikflüssigkeit hydraulic fluid
Hydraulikgerät hydraulic unit
Hydraulikkreis hydraulic circuit

Hydrauliköl hydraulic oil
Hydrauliköltank hydraulic oil reservoir
Hydraulikpumpe hydraulic pump
Hydrauliksystem hydraulic system, hydraulics
Hydraulikventil hydraulic valve
Hydraulikzylinder hydraulic cylinder
hydraulisch hydraulic
hydraulisch abbindend hydraulic
hydraulisch erhärtend hydraulic
hydraulisches Bindemittel hydraulic adhesive/cement
Hydrazid hydrazide
Hydrazin hydrazine
Hydrazinderivat hydrazine derivative
Hydrazoverbindung hydrazo compound
Hydrazylradikal hydrazyl radical
hydrierbar capable of hydrogenation
hydriert hydrogenated
Hydrierung hydrogenation
Hydrobasislack water-based/waterborne paint
Hydrochinon hydroquinone
Hydrodecklack waterborne/water-based topcoat
hydrodynamisch hydrodynamic
Hydroeinschichtlack waterborne/water-based one-coat paint
Hydrofluoralkan hydrofluoroalkane
Hydrofluorchlorkohlenwasserstoff hydrochlorofluorocarbon, HCFC
Hydrofluorkohlenwasserstoff hydrofluorocarbon, HFC
Hydrogrundierung waterborne/water-based primer
Hydrogrundlack waterborne/water-based primer
Hydroklarlack waterborne/water-based clear lacquer
Hydrokorrosionsschutzfarbe waterborne anti-corrosive paint
Hydrolack waterborne/water-based paint
hydrolisieren to hydrolyse
Hydrolyse hydrolysis
hydrolyseanfällig hydrolysable
Hydrolysebeständigkeit resistance to hydrolysis
hydrolyseempfindlich subject to hydrolysis
Hydrolysegrad degree of hydrolysis
hydrolyseinstabil not resistant to hydrolysis
Hydrolyseresistenz hydrolysis resistance
hydrolysierbar hydrolysable
hydrolysiert hydrolysed
hydrolytisch hydrolytic
Hydrometallicgrundlack waterborne/water-based metallic primer
Hydroperoxid hydroperoxide
Hydroperoxidgruppe hydroperoxide group
Hydroperoxyradikal hydroperoxy radical
hydrophil hydrophilic
hydrophil eingestellt made hydrophilic
Hydrophilie hydrophilic properties/character
hydrophiliert made water miscible
Hydrophilität hydrophilic properties/character
hydrophob hydrophobic, water repellent
Hydrophobie water repellency, hydrophobic properties
hydrophobierend making water repellent
hydrophobiert water repellent, made water repellent
Hydrophobierung making water repellent, imparting water repellent properties, waterproofing
Hydrophobierungsmittel water repellent
Hydrophobierungszusatz hydrophobic agent
Hydrophobierwirkung water repellent effect
Hydrophobizität hydrophobicity, water repellency, hydrophobic properties
hydrophobmodifiziert made water repellent
Hydrophobzusatz water repellent
Hydropumpe hydraulic pump
Hydrosilanharz hydrosilane resin
Hydrospeicher hydraulic accumulator
hydrostatisch hydrostatic
Hydrosystem water-based system
Hydroterephthalsäure hydroterephthalic acid
Hydroxid hydroxide
hydroxidfunktionell hydroxy-functional
Hydroxidgruppe hydroxyl group
Hydroxidradikal hydroxyl group
Hydroxidrest hydroxyl group
Hydroxyacetal hydroxyacetal
Hydroxyacrylester hydroxy acrylate
Hydroxyalkylacrylat hydroxyalkyl acrylate
Hydroxyalkylacrylester hydroxyalkylacrylic ester
Hydroxyalkylamid hydroxyalkylamide
Hydroxyalkylamin hydroxyalkylamine
Hydroxyalkylester hydroxyalkyl ester
Hydroxyalkylmethacrylat hydroxyalkyl methacrylate
Hydroxyalkylpyridin hydroxyalkyl pyridine
Hydroxyalkylvinylether hydroxyalkyl vinyl ether
Hydroxyallylether hydroxyallyl ether
Hydroxylharz hydroxy resin
Hydroxyäquivalent hydroxy equivalent
Hydroxybenzylalkohol hydroxybenzyl alcohol
Hydroxybenzylamin hydroxybenzylamine
Hydroxycapronsäure hydroxycaproic acid
Hydroxycarbonsäure hydroxycarboxylic acid
Hydroxychinolin hydroxyquinoline
Hydroxycyclohexylphenylketon hydroxycyclohexylphenyl ketone
Hydroxyester hydroxyester
Hydroxyethylacrylat hydroxyethyl acrylate
Hydroxyethylcellulose hydroxyethyl cellulose
Hydroxyethylmethacrylat hydroxyethyl methacrylate
Hydroxyethylpropylester hydroxyethylpropyl ester
Hydroxyfunktionalität hydroxy functionality
hydroxyfunktionell hydroxyfunctional
Hydroxygruppe hydroxyl group
Hydroxylacrylat hydroxyacrylate
Hydroxylaminchlorhydrat hydroxylamine chlorohydrate
Hydroxylaminhydrochlorid hydroxylamine hydrochloride

Hydroxylaminoamid hydroxylalkylaminoamide
Hydroxylaminsulfat hydroxylamine sulphate
hydroxylarm with a low hydroxyl content
Hydroxylendgruppe terminal hydroxyl group
Hydroxylfunktionalität hydroxyl functionality
hydroxylfunktionell hydroxy-functional
Hydroxylgehalt hydroxyl content
Hydroxylgruppe hydroxyl group
hydroxylgruppenenthaltend containing hydroxyl groups
hydroxylgruppenfrei free from hydroxyl groups
Hydroxylgruppengehalt hydroxyl group content
hydroxylgruppenhaltig containing hydroxyl groups
hydroxyliert hydroxylated
Hydroxylradikal hydroxyl radical/group
hydroxylreich with a high hydroxyl group content
Hydroxylverbindung hydroxyl compound
Hydroxylwert hydroxyl value
Hydroxylzahl hydroxyl value
Hydroxymethylgruppe hydroxymethyl group
Hydroxymethylmethacrylat hydroxymethyl methacrylate
Hydroxynaphthaldehyd hydroxynaphthaldehyde
Hydroxypolyether hydroxypolyether
Hydroxypropylacrylat hydroxypropyl acrylate
Hydroxypropylester hydroxypropyl ester
Hydroxypropylmethacrylat hydroxypropyl methacrylate
Hydroxyradikal hydroxyl radical/group
Hydroxysiloxan hydroxysiloxane
Hydroxystearinsäure hydroxystearic acid
Hydroxyurethan hydroxyurethane
Hydroxyvaleriansäure hydroxyvaleric acid
Hydroxyvinylverbindung hydroxyvinyl compound
Hygiene hygiene
Hygroskopie hygroscopicity
hygroskopisch hygroscopic
Hygroskopizität hygroscopicity, hygroscopic properties
Hypophosphit hypophosphite
Hysterese hysteresis
Hysteresekurve hysteresis curve
Hystereseschleife hysteresis loop

I

IC-Harz coumarone-indene resin
idealelastisch ideal-elastic
idealplastisch ideal-plastic
Identifizierung identification
Ilmenit ilmenite
Imid imide
Imidazol imidazole
Imidglied imide segment
Imidgruppe imide group
Imidharz imide resin
imidisiert imidised
Imidpigment imide pigment
Iminogruppe imino group
Iminoverbindung imino compound
Immersionsversuch immersion test
Immission pollution
Immissionsschutz pollution control
Immissionsschutzgebung anti-pollution legislation, pollution control legislation
immissionsschutzrechtliche Regelung anti-pollution regulations, pollution control regulations
Immissionsschutzrechtsetzung anti-pollution legislation, pollution control legislation
Immobilisierung immobilisation
Impägniermaschine impregnating machine
Impedanz impedance
Impedanzmessung impedance measurement/determination
Impedanzspektroskopie impedance spectroscopy
Impedanzspektroskopie, elektrochemische electrochemical impedance spectroscopy
impedanzspektroskopisch impedance spectroscopic
imprägnieren to impregnate
Imprägnierharz impregnating resin
Imprägnierlack impregnating varnish
Imprägnierlösung impregnating solution
Imprägniermittel impregnating agent
Imprägnierparameter impregnating conditions
Imprägniersystem impregnating solution
Imprägniertiefe depth of impregnation
Imprägnierung impregnation
inaktiv inert
Inbetriebnahme putting into operation *(e.g. plant)*
Inbetriebsetzung starting-up
Indan indane
Inden indene
Inden-Cumaronharz coumarone-indene resin
Indencarbonsäure indene carboxylic acid
Indenharz indene resin
indifferent inert
Indifferenz inertness
Indikator indicator
induktionshärtend induction-curing
Induktionsheizgerät induction heater, induction heating unit
Induktionsheizung induction heating
Induktionskraft inductive force
Induktionsperiode induction period
Induktionsphase induction phase
Induktionsspule induction coil
Induktionszeit induction period
induktiv inductive
Induktivheizung inductive heating
Induktivität inductivity
Industrie, papiererzeugende paper-producing

industry
Industrieabgase industrial waste gases
Industriealkohol industrial alcohol
Industrieanlage industrial plant
Industrieatmosphäre industrial atmosphere
Industriebekleidung industrial clothing
Industrieboden industrial floor
Industriebodenbelag industrial floor screed
Industriedecklack industrial paint
Industriedichtstoff industrial sealant
Industrieeinbrennlack industrial stoving paint/enamel
Industriefarbe industrial paint
Industriegebinde industrial container
Industriegrundierung industrial primer
Industriehandschuhe industrial gloves
Industriehygiene industrial hygiene
Industriekleber industrial adhesive
Industrieklebstoff industrial adhesive
Industrieklima industrial atmosphere
Industrielack industrial paint
Industrielacksektor industrial paints sector
Industrieländer industrialised countries
industriell industrial
Industriemaßstab industrial/production scale
Industriemüll industrial waste
Industrienationen industrialised countries
Industrienorm industrial standard
Industrieregale industrial shelving
Industriereinigungsmittel industrial cleaner
Industrieschutzbekleidung industrial protective clothing
Industriesilicon industrial silicone
Industriesparte branch of industry
Industriestaaten industrialised countries
Industrieverbrennungsanlage industrial incineration plant
Industrieverpackung industrial packaging
Industriewaage industrial scales
Industriezweig (branch of) industry
inert inert
Inertgas inert gas
Inertisierung neutralisation
Informationen information *(this word does **not** have a plural)*
Informationsaustausch exchange of information
Informationsquelle source of information
Informationsschrift information sheet
infrarot infrared, IR
Infrarotbereich infrared range
Infrarotbestrahlungslampe infrared lamp
Infrarotdifferenzspektrometrie infrared differential spectrometry
Infrarotlicht infrared light
Infrarotmeßfühler infrared sensor
Infrarotofen infrared oven
infrarotspektrographisch infrared spectrographic
infrarotspektroskopisch infrared spectroscopic
Infrarotspektroskopie infrared spectroscopy
Infrarotspektrum infrared spectrum
Infrarotstrahlen infrared rays
Infrarotstrahler infrared heater
Infrarotstrahlung infrared radiation
Infrarotstrecke infrared heating section
Infrarottemperaturmeßgerät infrared thermocouple
Infrarottunnel infrared heating tunnel
Infrarotuntersuchung infrared test
Infraschallbereich infrasonic range
Infrastruktur infrastructure
Ingenieurbau constructional/civil engineering
Ingenieurholzbau engineering timber construction
Inhaltsstoff contents
inhärent inherent
inhibierend inhibiting
Inhibierung inhibition
Inhibitionseffeklt inhibiting effect
Inhibitor inhibitor
Inhibitorwirkung inhibiting effect
inhomogen inhomogeneous
Inhomogenitäten inhomogeneities
Initiator initiator
Initiatorsuspension initiator suspension
Initiatorsystem initiator
Initiatorzerfall initiator breakdown
initiieren to initiate
Initiierung initiation
Initiierungsmittel initiator
Injektionsdruck injection pressure
injizieren to inject
inkompatibel incompatible
Inkompatibilität incompatibility
Inkrafttreten coming into force
Inkrement increment
inkrementell incremental
inländisch home, domestic
Inlandsgeschäft home/domestic business
innenbeheizt internally heated
Innenbeschichtung 1. interior/inside coating. 2. interior finish. 3. painting the inside
Innendispersionsfarbe interior emulsion paint
Innendurchmesser inside diameter
Inneneinsatz interior/inside use
Innenemulsgiersystem internal emulsifier
Innenfarbe interior paint
Innenfuge inside joint
Innenrohrauskleidung pipe lining
Innenschutzlack can coating lacquer
Innenschutzlackierung can coating
Innentemperatur inside temperature
Innenwandfarbe interior paint
innerbetrieblich in-plant, in-house
innere Gleitwirkung internal lubrication
innere Reibung internal friction
innere Weichmachung internal plasticisation
inneres Gleitmittel internal lubricant
innermolekular intramolecular
inokuliert inoculated
Insektizid insecticide
instabil unstable

Instabilität instability
Installateur plumber, fitter
installierte Leistung installed/connected load
Instandhaltung maintenance
Instandhaltungsmaßnahmen maintenance measures
Instandsetzung repair
Integration integration
integriert integrated
Intensität intensity
Intensitätsabfall decrease in intensity
Intensitätsabnahme decrease in intensity
Intensitätsänderung change in intensity
Intensitätsanstieg increase in intensity
Intensitätsveränderung change in intensity
Intensitätsverringerung decrease in intensity
Intensitätszunahme increase in intensity
Intensitätszuwachs increase in intensity
Interferenz interference
Interferenzeffekt interference effect
Interferenzfarbe interference colour
Interferenzfeld interference field
Interferenzgrundmuster interference pattern
Interferenzlinien interference lines
Interferenzmikroskop interference microscope
Interferenzmuster interference pattern
Interferenzpigment interference pigment
Interferenzstreifen interference strip
Interferogramm interferogram
Interferometrie interferometry
interkristallin intercrystalline
interlaminar interlaminar
interlaminare Scherfestigkeit interlaminar shear strength
intermittierend intermittent
intermolekular intermolecular
intern plastifiziert internally plasticised
Interpenetration interpenetration
interpenetrierend interpenetrating
Interpolation interpolation
interpoliert interpolated
intiieren to initiate
intramolekular intramolecular
intrinsisch intrinsic
intrinsisch leitfähig intrinsically conductive
intumeszent intumescent
Intumeszenzmaterial intumescent compound
invers inverse
Iodoniumsalz iodonium salt
Ion ion
Ionenabgabe ion release
Ionenaustausch ion exchange
Ionenaustauscheffekt ion exchange effect
Ionenaustauschpigment ion exchange pigment
Ionenbeweglichkeit ion mobility
ionenbildend ion-forming
Ionenbindung ionic bond
Ionenchromatographie ion chromatography
Ionenfänger ion interceptor
Ionengitter ionic lattice structure
Ionenkettenpolymerisation ion chain polymerisation
Ionenkomplex ion complex
Ionenkonzentration ion concentration
Ionenleitfähigkeit ionic conductivity
Ionenquelle ion source
ionenreaktiv ionically reactive
Ionenstärke ionic strength
Ionenstoß ion collision
Ionenstrahl ion beam
Ionenverbindung ionic bond
Ionisationsbeständigkeit ionisation resistance
Ionisationsenergie ionisation energy
ionisch ionic
ionisierbar ionisable
ionisierend ionising
ionisierfähig ionisable
ionisiert ionised
Ionisierung ionisation
ionitrierbehandelt ionitrided
Ionitrieren ionitriding
ionitriert ionitrided
ionogen ionic
Ionogenität ionic character
ionomer ionomeric
Ionomerbeschichtung ionomer coating
Ionomerharz ionomer resin
IPDI isophorone diisocyanate
IPN interpenetrating network
IPN-Polymer interpenetrating network polymer, IPN polymer
IR-Bestrahlung infra-red irradiation
IR-Härtung infrared curing
IR-Laserstrahl infrared laser beam
IR-Licht infrared light
IR-Spektralbereich infrared spectral range
IR-Spektroskopie IR/infrared spectroscopy
IR-spektroskopisch IR/infrared-spectroscopic
IR-Strahlung IR/infrared radiation
IR-Strahlungspyrometer infrared radiation pyrometer
IR-Temperaturmeßgerät infrared thermocouple
IR-Trocknungsapparatur infrared drying equipment
IR-Trocknungskammer infrared drying oven
IR-Wärmequelle infrared heat source
irreversibel irreversible
Isobutanol isobutanol, isobutyl alcohol
Isobuten isobutene
Isobutylacrylat isobutyl acrylate
Isobutylmethylacrylat isobutyl methacrylate
Isobutylstearat isobutyl stearate
Isochinacridon isoquinacridone
Isochinolin isoquinoline
isochor isochoric, at constant volume
Isochore isochore
isochron isochronous
Isocyanat isocyanate
Isocyanatanteil isocyanate content
isocyanatfrei isocyanate-free
isocyanatfunktionell isocyanate-functional
isocyanatgehärtet isocyanate-cured

Isocyanatgruppe isocyanate group
isocyanatgruppenhaltig containing isocyanate groups
isocyanathärtbar isocyanate-curing
Isocyanathärter isocyanate catalyst/hardener
Isocyanatharz isocyanate resin
Isocyanatkleber isocyanate adhesive
Isocyanatkomponente isocyanate component
Isocyanatlack isocyanate paint
Isocyanatpräpolymer isocyanate prepolymer
Isocyanatrest isocyanate group
Isocyanatverbindung isocyanate compound
isocyanatvernetzend isocyanate-curing
Isocyanatvernetzer isocyanate curing agent
Isocyansäure isocyanic acid
Isocyanurat isocyanurate
Isocyanuratring isocyanurate ring
Isododecan isododecane
isoelektrisch isoelectric
Isolation insulation
Isolationseigenschaften insulating properties
Isolationslack insulating varnish
Isolationsschicht insulating layer
Isolationsvermögen insulating properties
Isolationswerte insulating properties
Isolationswiderstand insulation resistance
Isolator insulator
Isoliereigenschaften insulating properties
isolierend insulating
Isolierglasscheibe sealed unit
Isoliergrund barrier coat
Isolierlack insulating varnish
Isoliermasse insulating compound
Isoliermedium insulating medium
Isolierschicht insulating layer
Isolierstoff insulating material
isoliert insulated
Isolierung insulation
Isolierverglasung double glazing
Isoliervermögen insulating properties
Isolierwerkstoff insulating material
Isolierwerte insulating properties
Isolierwirkung insulating effect
isomer isomeric
Isomer isomer
Isomerie isomerism
isomerisiert isomerised
Isomerisierung isomerisation
isometrisch isometric
Isooctan isooctane
Isooctanol isooctanol
Isoparaffin isoparaffin
Isophoron isophorone
Isophorondiamin isophorone diamine
Isophorondiisocyanat isophorone diisocyanate, IPDI
Isophthalat isophthalate
Isophthalsäure isophthalic acid
Isopren isoprene
Isoprenkautschuk isoprene rubber
Isopropanol isopropanol
Isopropanolamin isopropanolamine
isopropanolisch isopropanolic
Isopropenylnaphthalin isopropenyl naphthalene
Isopropylacetat isopropyl acetate
Isostearinsäure isostearic acid
isosterisch isosteric
isotaktisch isotactic
Isotaxieindex isotactic index
isotherm isothermal
Isotherme isotherm
isothermisch isothermal
Isothiazolinon isothiazolinone
isotrop isotropic
Isotropie isotropy
Isoverbindung iso-compound
Isoverglasung 1. double glazing. 2. sealed unit
Istgewicht actual weight
Istgröße actual value
Istkurve actual value curve
Istwert actual value

J

Jahreskapazität annual capacity
Jahresproduktion annual production
Jahreszeit, kalte winter months
jato *abbr. of* **Jahrestonnen**, tonnes p.a.
JFZ *abbr. of* **Jodfarbzahl**, iodine colour value
JIT just-in-time
Jod iodine
Jodfarbskala iodine colour scale
Jodfarbzahl iodine colour value
jodhaltig containing iodine
jodometrisch iodometric
Jodoniumverbindung iodonium compound
Just-in-time Produktion just-in-time production/manufacturing
Just-in-time Lieferung just-in-time delivery
justierbar adjustable
Justierbarkeit adjustability
justieren to adjust
Justierwert adjusted figure/value

K

K-Wert 1. K-value. 2. heat transfer coefficient, coefficient of heat transmission
Kaffee coffee

kalalysatorhaltig catalysed
Kalibrierprozedur calibrating method/procedure, method of calibration
Kalibrierung calibration
Kalibrierungskurve calibration curve
Kalibriervorrichtung calibrating device
Kaliglimmer potash mica, muscovite
Kalilauge caustic potash solution
Kaliumcarbonat potassium carbonate
Kaliumchlorid potassium chloride
Kaliumdichromat potassium dichromate
Kaliumferritpigment potassium ferrite pigment
Kaliumhydroxid potassium hydroxide
Kaliumpersulfat potassium persulphate
Kaliumtetrafluorborat potassium tetrafluoroborate
Kaliumverbindung potassium compound
Kaliwasserglas potash waterglass
Kalk lime
Kalkanstrich limewash, whitewash
Kalkfarbe limewash, whitewash
Kalkmilch milk of lime
Kalkputz lime plaster
Kalksandstein sand-lime brick
Kalkschlämme limewash, whitewash
Kalorimeter calorimeter
kalorimetrisch calorimetric
kalorisch caloric
kalt cold
kaltabbindend cold setting/curing
Kaltbiegeeigenschaften cold bending properties
Kälte cold
kalte, Jahreszeit winter months
Kältebeanspruchung low-temperature exposure
Kältebelastung low-temperature exposure
kältebeständig low-temperature resistant
Kältebeständigkeit low-temperature resistance
Kältebiegebeständigkeit cold bend resistance
Kältebiegeversuch cold bend test
Kältebruch low-temperature fracture
Kältedämmstoff low-temperature insulating material
Kälteeigenschaften low-temperature properties
kälteelastisch flexible at low temperatures
Kälteelastizität low-temperature flexibility
kältefest low-temperature resistant
kältefestigkeit low-temperature resistance
kälteflexibel flexible at low temperatures
Kälteflexibilität low-temperature flexibility
Kälteisolierung low-temperature insulation
Kältekammer low-temperature conditioning cabinet
Kältekontraktion low-temperature shrinkage
Kältekreis cooling circuit
Kältekreislauf cooling circuit
kältelagerstabil resistant to cold storage conditions
Kältelagerung low-temperature ageing
Kälteleistung cooling capacity/efficiency
Kältemittel coolant, refrigerant
Kaltentfetter cold degreasing agent
Kaltentfettung cold degreasing
Kaltentfettungsanlage cold degreasing plant
kalter Fluß cold flow
Kälteschälfestigkeit low-temperature peel strength
Kälteschlagbeständigkeit low-temperature impact strength
kälteschlagfest low-temperature impact resistant
Kälteschlagfestigkeit low-temperature impact strength
Kälteschlagwert low-temperature impact strength
kälteschlagzäh low-temperature impact resistant
Kälteschlagzähigkeit low-temperature impact strength
Kälteschockfestigkeit low-temperature shock resistance
Kälteschrank low-temperature test chamber
Kältesprödigkeitspunkt low-temperature brittleness point, brittleness temperature
Kältesprödigkeitsversuch low-temperature brittleness test
Kältestandfestigkeit low-temperature resistance
Kältestauchfestigkeit low-temperature compressive strength
Kältethermostat low-temperature thermostat
Kälteverhalten low-temperature properties/performance/behaviour
Kälteversprödung low-temperature embrittlement
Kälteversprödungsneigung tendency to become brittle at low temperatures
Kälteversprödungstemperatur low-temperature brittleness point, brittleness temperature
kältezäh low-temperature resistant
Kältezähigkeit 1. low-temperature resistance. 2. low-temperature toughness, low-temperature impact strength
Kaltfluß cold flow
kalthärtend room temperature curing, cold setting/curing
Kalthärter cold-curing catalyst
Kalthärtung cold-curing/-setting
Kalthärtungssystem cold-curing/-setting system/formulation
Kaltkaschieranlage cold laminating plant
Kaltkaschieren cold laminating
Kaltkleben cold bonding
Kaltreiniger cold cleaner
kaltschlagzäh low-temperature impact resistant
Kaltsiegelkleber contat adhesive
Kaltverträglichkeit low-temperature compatibility
Kaltwasser cold water
kaltwasserlöslich soluble in cold water, cold water soluble
Kalzium- *see* Calcium-
Kammer chamber
Kampher camphor
kampherähnlich camphor-like
Kanister canister, jerrycan
Kante edge
Kantenbedeckung edge coverage

Kantendeckung edge coverage
Kantenklebstoff edge glueing adhesive
Kantenklebung edge glueing
Kantenkorrosion edge corrosion, corrosion near the edges
kanzerogen carcinogenic, cancer-causing
Kanzerogenität carcinogenity
Kaolin kaolin
Kapazität 1. capacity. 2. capacitance *(electrical)*
kapillar capillary
Kapillarachse capillary axis
kapillarähnlich capillary-like
Kapillardurchmesser capillary diameter
Kapillare capillary
Kapillareffekt capillary effect
Kapillarenquerschnitt capillary cross-section
Kapillarenwand capillary wall
Kapillargaschromatographie capillary gas chromatography
Kapillarkräfte capillary forces
Kapillarlänge capillary length
Kapillarrheometer capillary rheometer
Kapillarrheometrie capillary rheometry
kapillarrheometrisch capillary-rheometric
Kapillarröhre capillary
Kapillarrohrthermostat capillary thermostat
Kapillarviskosimeter capillary viscometer
Kapillarviskosimetrie capillary viscometry
Kapillarwand capillary wall
Kapillarwirkung capillary action/effect
Kapillarzentrum capillary centre
kapseln to encapsulate, to embed
Karbon- *see* **Carbon-**
Karosserie car body
Karosseriebau car body construction
Karton card
Kartusche cartridge
Kartuschenpistole cartridge gun
Kaschieranlage laminating plant
kaschieren 1. to laminate. 2. to conceal, to hide
Kaschierkleber laminating adhesive
Kaschierkleberprogramm range of laminating adhesives
Kaschierklebstoff laminating adhesive
Kaschiermaschine laminator, laminating machine
Kaschierung lamination
Kaschierverfahren lamination, laminating process
Kaschiervorrichtung laminating unit
Kaschierwalze laminating roll
Kaschierwerk laminating unit
Kasein casein
kaseinbasierend casein-based
Kaseinklebstoff casein adhesive
Katalysator catalyst
Katalysatoraktivität catalyst activity
Katalysatorauswahl choice of catalyst
Katalysatorgehalt catalyst content
Katalysatorgift catalyst poison
Katalysatorkonzentration catalyst concentration
Katalysatormenge amount of catalyst
Katalysatorpaste catalyst paste
Katalysatorreste catalyst residues
Katalysatorrückstände catalyst residues
Katalysatorvergiftung catalyst poisoning
Katalyse catalysis
katalysieren to catalyse
katalysiert catalysed
Katalysierung catalysis, mixing with catalyst
katalytisch catalytic
Kataphorese cataphoresis, electrophoresis
Kataphoresebeschichtung cataphoretic/electrophoretic coating
Kataphoresebindemittel binder for cataphoretic/electrophoretic/ electrodeposition paints
Kataphoresegrund cataphoretic/electrophoretic primer coat
Kataphoresegrundierung cataphoretic/electrophoretic/electrodeposition primer
Kataphoreseharz resin for making cataphoretic/electrophoretic paints
Kataphoreselack cataphoretic/electrophoretic paint
Kataphoreseschicht cataphoretic/electrophoretic coating
Kataphoreseüberzug cataphoretic/electrophoretic coating
kataphoretisch cataphoretic, electrophoretic
kataphoretisch abscheidbar capable of being deposited cataphoretically/ electrophoretically/by electrodeposition
katastrophaler Bruch catastrophic failure
katastrophisches Versagen catastrophic failure
Kathode cathode
Kathodenstrahloszillograph cathode ray oscillograph
kathodisch cathodic
kathodisch abscheidbar capable of being deposited cataphoretically/ electrophoretically/by electrodeposition
kathodische Tauchlackierung cathodic electrodeposition painting
Kation cation
kationaktiv cationic
kationisch cationic
kationogen cationic
Kautschuk rubber: **Kautschuk** *is crude, i.e. unvulcanised rubber as opposed to* **Gummi** *(which is the word used to denote rubber in its vulcanised form). In English, both words are called* rubber *and no qualification (i.e. the words* vulcanised *and* unvulcanised*) is normally necessary. Where a distinction is made between* **Gummi** *and* **Kautschuk**, *the translation should be along the following lines:* **Durch Vulkanisation wird der plastische Silikonkautschuk in den elastischen Silikongummi übergeführt** vulcanisation converts silicone rubber from the plastic into the elastic state

Kautschukanteil rubber content
kautschukbasierend rubber-based
Kautschukbasis, auf rubber-based
kautschukelastisch rubber-elastic
Kautschukelastizität rubber elasticity, rubber-like elasticity
Kautschukhaftklebstoff rubber-based pressure sensitive adhesive
kautschukhaltig containing rubber, rubber-modified
Kautschukhydrochlorid rubber hydrochloride
Kautschuklatex rubber latex
Kautschukleim rubber(-based) adhesive
Kautschukmasse rubber mix/compound
Kautschukmischung rubber mix
kautschukmodifiziert rubber modified
Kautschukphase rubber phase
kautschuküberzogen rubber covered
Kegel-Platteviskosimeter cone-plate viscometer
Kehrwert reciprocal (value)
Keimbildner nucleating agent
Keimbildung nucleation
Keimbildungsgeschwindigkeit nucleation rate
Keller cellar
Kennbuchstabe code/identifying letter
Kenndaten characteristics, constants, properties
Kenngröße characteristic, constant, property
Kennwert characteristic (value), constant, property
Kennzeichnungspflicht (legal) obligation to provide special markings/labelling
kennzeichnungspflichtig requiring special markings/labelling
Keramik ceramics, ceramic material
Keramikfliese ceramic tile
Keramikmahlkörper porcelain balls
keramisch ceramic
Kerbe notch
Kerbeffekt notch effect
kerbempfindlich notch sensitive
Kerbempfindlichkeit notch sensitivity
kerbfest notch resistant
Kerbfestigkeit notch resistance
Kerbschlagbiegeversuch notched flexural impact test
Kerbschlagbiegezähigkeit notched flexural impact strength
Kerbschlagfestigkeit notched impact strength
Kerbschlagversuch notched impact test
Kerbschlagwert notched impact strength
kerbschlagzäh notch impact resistant
Kerbschlagzähigkeit notched impact resistance
Kerbschlagzugzähigkeit notched tensile impact strength
Kerbspannung notch stress
Kerbstellenwirkung notch effect
kerbunempfindlich notch resistant
Kerbunempfindlichkeit notch resistance
Kerbwirkung notch effect
kerbzäh notch resistant
Kern core
Kerneigenschaften most important properties
Kernlage core layer
Kerntechnik nuclear engineering
Kesselwagen road tanker
Ketimin ketimine
Ketocarbonylgruppe keto-carbonyl group
Ketogruppe keto group
Keton ketone
ketonartig ketone-like
Ketonformaldehydharz ketone formaldehyde resin
Ketongruppe ketone group
Ketonharz ketone resin
Ketonhydroperoxid ketone hydroperoxide
Ketonperoxid ketone peroxide
Ketosäure keto acid
Ketoximogruppe ketoximo group/radical
Kette chain
Kettenabbau chain degradation
kettenabbauend chain-degrading
Kettenabbruch chain termination
Kettenabschnitt chain segment
Kettenaktivierung chain activation
kettenartig chain-like
Kettenaufbau chain structure
Kettenausrichtung chain alignment/orientation
Kettenbaustein chain segment
Kettenbeweglichkeit chain mobility
Kettenbruch chain fission/scission
Kettenbruchstück chain segment
Kettenende chain end
Kettenformation chain formation
kettenförmig chain-like
Kettengebilde chain structure
Kettenlänge chain length
Kettenmolekül chain molecule
Kettenorientierung chain orientation
Kettenradikal chain radical
Kettenreaktion chain reaction
Kettensegment chain segment
Kettenspaltung chain fission/scission
Kettenstart chain initiation
Kettenstarter chain initiator
Kettenstartmolekül chain initiating molecule
kettensteif with rigid chains
Kettensteifigkeit chain rigidity
Kettenstruktur chain structure
Kettenstück chain segment
Kettenteil chain segment
Kettenübertragung chain transfer
Kettenübertragungsreaktion chain transfer reaction
Kettenübertragungsreagenz chain transfer reagent
Kettenverkürzung chain shortening
Kettenverlängerer chain extender, curing/crosslinking agent
Kettenverlängerung chain extension
Kettenverlängerungsreaktion chain extension reaction
Kettenverschlingung chain entanglement

Kettenverzweigung chain branching
Kettenwachstum chain growth
Kiefer pine
Kiefernharz pine oleoresin
Kieselerde silica
Kieselgel silica gel
Kieselgur kieselguhr
Kieselsäure 1. silicic acid. 2. silicon dioxide, silica
Kieselsäuregel silica gel
Kieselsäuregerüst silica skeleton
kindersicher child-proof, child-resistant
kinematische Viskosität kinematic viscosity
Kinetik kinetics
kinetisch kinetic
kinetische Energie kinetic energy
Kitt mastic, putty
klar clear, transparent
Kläranlage sewage treatment/purification plant
Klarlack clear varnish/lacquer
Klarlackschicht clear varnish/lacquer film
klarlöslich completely soluble *(i.e. to form a clear solution)*
Klassiereinheit grading unit
Klassierung grading
Klassifizierung classification
Klebanwendungen adhesive applications
Klebband adhesive tape
Klebbandrolle roll of adhesive tape
klebbar bondable
Klebbarkeit bondability
Klebbedingungen bonding conditions
Klebbschichtung adhesive coating
Klebdichtungsmasse adhesive sealant
Klebe-, klebe- *see* **Kleb-, kleb-**
Klebeffekt tack
Klebeigenschaften adhesive properties
Kleben bonding
klebend 1. tacky, sticky. 2. adhesive
Kleber adhesive
Kleber- *see* **Klebstoff-**
Klebfaden adhesive bead
Klebfähigkeit adhesive properties
Klebfestigkeit bond/adhesive strength
Klebfilm film adhesive
Klebfläche adherend surface
Klebflächenabmessungen adherend surface dimensions
Klebflächenvorbehandlung preparation of adherend surfaces
Klebfolie 1. film adhesive. 2. self-adhesive film
Klebfolienbuchstabe self-adhesive letter
klebfrei tack-free
Klebfreiheit freedom from tack
Klebfreitrocknung drying to a tack-free finish
klebfreundlich easy to bond/stick
Klebfuge glueline
Klebfugenbruch glueline failure
Klebfugendicke glueline thickness
Klebfugengeometrie glueline form
Klebfugenrand glueline edge
klebgerecht suitable for bonding: **klebgerechte Konstruktion** correct joint design
Klebgrundstoff adhesive resin
Klebhaftung adhesion
Klebharz adhesive resin
Klebkraft adhesive power/strength
Kleblinie glueline
Kleblösung solvent-based adhesive, adhesive solution
Klebmaterial adhesive
Klebmittel adhesive
Klebmörtel adhesive mortar
Klebnaht glueline
Klebneigung tendency to stick
Klebparameter bonding conditions
Klebpartner adherend
Klebpistole glue gun
Klebproblem bonding problem
Klebprozeß bonding process/operation
klebrig tacky, sticky
Klebrigkeit tack
Klebrigkeit, oberflächliche surface tack
Klebrigkeitsdauer open time
Klebrigkeitsverlust loss of tack
klebrigmachend tackifying
Klebrigmacher tackifier, tackifying agent
Klebrigmacherharz tackifying resin
klebrigweich soft and tacky
Klebrohstoff adhesive resin
Klebschicht adhesive film/layer
Klebschichtbreite adhesive film width
Klebschichtdicke adhesive film thickness
Klebschichteigenschaften adhesive film properties
Klebschichtelastizität adhesive film flexibility
Klebschichtfestigkeit adhesive film strength
Klebschichtschwindung adhesive film shrinkage
Klebschichtstärke adhesive film thickness
Klebspalte glueline
Klebstelle bond area
Klebstoff adhesive
klebstoffabweisend non-stick, anti-adhesive
Klebstoffallergie adhesive allergy, allergy to adhesives
Klebstoffansatz adhesive mix
Klebstoffanwender adhesive user
Klebstoffanwendung adhesive application
Klebstoffart type of adhesive
Klebstoffauftrag application of adhesive: **bei dickerem Klebstoffauftrag** if the adhesive is applied more thickly
Klebstoffauftragmaschine glue applicator
Klebstoffaushärtung curing of the adhesive
Klebstoffauswahl choice of adhesive
Klebstoffauswahlsystem adhesive selection system
klebstoffbeschichtet adhesive coated
Klebstoffbestandteil adhesive constituent
Klebstoffbett adhesive layer
Klebstoffbranche adhesives industry
Klebstoffdämpfe adhesive fumes
Klebstoffdatenbank adhesive data bank

Klebstoffeigenschaften adhesive properties
Klebstoffeinspritzung adhesive injection
Klebstoffentwicklungen adhesive developments
Klebstoffgrundstoff adhesive resin
Klebstoffhaftung adhesion
Klebstoffharz adhesive resin
Klebstoffhersteller adhesives manufacturer
Klebstoffilm adhesive film
Klebstoffindustrie adhesives industry
Klebstoffkennwerte adhesive properties
Klebstofflösung adhesive solution
Klebstoffmarkt adhesives market
Klebstoffmenge amount of adhesive
Klebstoffmerkmale adhesive properties/characteristics
Klebstoffmischung adhesive (mix)
Klebstoffolie film adhesive
Klebstofformulierung adhesive formulation
Klebstoffgeneration generation of adhesives
Klebstoffpalette range of adhesives
Klebstoffparameter adhesive properties
Klebstoffpaste adhesive paste
Klebstoffpatrone adhesive cartridge
Klebstoffproduktion adhesive production
Klebstoffprogramm range of adhesives
klebstoffrei free from adhesive
Klebstoffreste adhesive residues
Klebstoffrezeptur adhesive formulation
Klebstoffrohstoff adhesive resin/raw material
Klebstoffschicht adhesive coating/layer/film
Klebstoffschichtdicke adhesive (film) thickness
Klebstoffschwindung adhesive shrinkage
Klebstoffsortiment range of adhesives
klebstoffspezifisch adhesive-related
Klebstoffsystem adhesive system/formulation
Klebstofftechnik adhesive technology
Klebstofftechnologie adhesive technology
Klebstoffverarbeiter adhesive user
Klebstoffverarbeitung application of adhesive
Klebstoffverbrauch adhesive consumption
Klebstoffviskosität adhesive viscosity
Klebstoffzusammensetzung adhesive composition
Klebstreifen adhesive tape
Klebsystem adhesive
Klebtechnik bonding technique
klebtechnisch adhesive: **klebtechnische Eigenschaften** adhesive properties
Klebung 1. bonded joint. 2. bonding
Klebverbindung bonded joint
Klebverbindungsfestigkeit bond strength
Klebverbund bonded joint
Klebverhalten adhesive properties
Klebvermögen adhesive power
Klebvorgang bonding process
Klebwerkstoff adhesive
Klebwert bond strength
Klebwirkung adhesive effect
Kleinchargenfertigung small batch production
kleine Produktionsserien short runs
kleine Stückzahlen small numbers
kleinflächig small-area
Kleingewerbe small businesses
Kleinserien short runs
Kleinserienanwendungen small-scale applications
Kleinserienfertigung small scale production
Kleinstmaß minimum dimension
Kleinstmenge minimum amount/quantity
kleinstmöglich smallest possible
Kleinstserien extremely short runs
Kleinstwert minimum value
kleintechnisch small-scale, pilot plant scale
Klima climate, climatic conditions
Klimaauslagerung ageing under standard climatic conditions
klimabedingt due to climatic conditions
Klimabedingungen climatic conditions
Klimabeständigkeit resistance to climatic conditions
Klimaeinflüsse climatic influences
Klimakammer conditoning cabinet
Klimakatastrophe climatic disaster
Klimalagerung ageing under climatically controlled conditions
Klimaprüfschrank climatically controlled test cabinet/chamber
Klimaraum conditioning chamber
Klimaraumlagerung ageing under standard climatic conditions
Klimaschrank conditioning cabinet
Klimaschwankungen climatic variations
klimatisch climatic
klimatisiert climatically controlled
Klimatisierung climatic control
Klimawechselbeanspruchung exposure to changing climatic conditions
Klimawechselbeständigkeit resistance to changing climatic conditions
Klimawechsellagerung ageing under changing climatic conditions
Klimawechselprüfung testing under changing climatic conditions
Klinker clinker
klümpchenfrei free from lumps
Klumpen lump
Klumpenbildung lump formation
klumpenfrei free from lumps
KMK *abbr. of* **kritische Mizellenkonzentration**, critical micelle concentration, c.m.c.
Knackpunkt sticking point
Knäuelkonformation tangle configuration
knetbar kneadable
Knetelemente kneading discs
Kneter kneader, compounder
Kneterschaufeln mixing/kneader blades
Knetflügel mixing/kneader blades
Knetkammer mixing compartment/chamber
Knetmaschine kneader
Knetraum mixing chamber/compartment
Knetschaufel mixing/kneader blades
Knickbelastung buckling stress

knicken to buckle, to fold
Knickfestigkeit buckling resistance
Knicklast buckling stress
Knickpunkt kink *(in a curve)*
Knickversuch folding test
Koagulation coagulation
Koagulationserscheinungen signs of coagulation
koaguliert coagulated
Koaleszenz coalescence
Koaleszenzhilfsmittel coalescing agent
Koaleszenzmittel coalescing agent
Koalesziermittel coalescing agent
koaxial coaxial
Kobaltbeschleuniger cobalt accelerator
Kobaltblau cobalt blue
Kobaltnaphthenat cobalt naphthenate
Kobaltoctoat cobalt octoate
Kobaltpigment cobalt pigment
Kobaltsalz cobalt salt
Kobaltseife cobalt soap
Kobaltsikkativ cobalt drier
Kobalttrockenstoff cobalt drier
kochend boiling
kochfest resistant to boiling water
Kochsalz common salt
Kochversuch boiling test
Kochwasser boiling water
Kochwasserbeständigkeit resistance to boiling water
Koeffizient coefficient
koflockulieren to co-flocculate
kohärent coherent
Kohärenz coherence
Kohäsion cohesion
Kohäsionsarbeit cohesive energy
Kohäsionsbruch cohesive fracture/failure
Kohäsionseigenschaften cohesive properties
Kohäsionsenergie cohesive energy
Kohäsionsfähigkeit cohesion
Kohäsionsfestigkeit cohesive strength
Kohäsionskraft cohesive force
Kohäsionsriß cohesive failure
Kohäsionsversagen cohesive fracture/failure
kohäsiv cohesive
Kohäsivbruch cohesive fracture/failure
Kohäsivfestigkeit cohesive strength
Kohäsivversagen cohesive fracture/failure
Kohle 1. coal. 2. charcoal. 3. carbon
kohleartig carbon-like
Kohlebogenlampe carbon arc lamp
Kohlefaser carbon fibre
kohlefaserverstärkt carbon fibre reinforced
Kohlefilter activated charcoal filter
Kohlelichtbogen carbon arc
Kohlelichtbogenstrahlung carbon arc radiation
Kohlelichtlampe carbon arc lamp
Kohlendioxid carbon dioxide
Kohlendioxiddurchlässigkeit carbon dioxide permeability
Kohlendioxidgas carbon dioxide (gas)
Kohlenhydrat carbohydrate
Kohlenmonoxid carbon monoxide
Kohlensäure carbonic acid
Kohlenstoff carbon
Kohlenstoff-Kohlenstoffbindung carbon-carbon bond
kohlenstoffarm low-carbon
Kohlenstoffaser carbon fibre
Kohlenstoffaserlaminat carbon fibre laminate
Kohlenstoffatom carbon atom
Kohlenstoffelektrode carbon electrode
Kohlenstoffkette carbon chain
Kohlenstoffpigment carbon black
kohlenstoffrei carbon-free
Kohlenstoffring benzene ring
Kohlenteer coal tar
Kohlenwasserstoff hydrocarbon
Kohlenwasserstoffgruppe hydrocarbon group
Kohlenwasserstoffharz hydrocarbon resin
Kohlenwasserstoffkautschuk hydrocarbon rubber
Kohlenwasserstoffkette hydrocarbon chain
Kohlenwasserstoffpolymer hydrocarbon polymer
Kohlenwasserstoffrest hydrocarbon radical
Kohlenwasserstoffverbindung hydrocarbon compound
Kohlenwasserstoffwachs hydrocarbon wax
Kohlepartikel charcoal particle
Kohlepulver powdered charcoal
Kohleteer coal tar
Kolben 1. flask. 2. plunger, ram, piston
Kolbenpumpe piston/reciprocating pump
Kollodiumwolle collodion cotton, nitrocotton, nitrocellulose
Kolloid colloid
kolloidal colloidal
Kolloidalbereich colloidal range
kolloidchemisch colloidochemical
kolloiddispers colloidally dispersed
Kolloidmühle colloid mill
Kolloidstabilisator colloid stabiliser
Kolonne column
Kolophonium rosin, colophony
Kolophoniumester colophony/rosin ester
Kolophoniumharz rosin, colophony
koloristisch colouristic
Kombination combination
Kombinationsverhältnis mixing ratio
kombinierbar capable of being combined/blended
kompatibel compatible
kompatibilisieren to make compatible
Kompatibilität compatibility
Kompatibilitätsprobleme compatibility problems
kompensieren to compensate
komplementär complementary
Komplementärfarbe complementary colour
Komplexbildner chelator, chelating agent
Komplexbildung chelation
Komplexsalz complex salt

kompliziert complex
Kompliziertheit complexity
Komponente component
Kompression compression
Kondensat condensate
Kondensatharz condensation resin
Kondensation condensation
Kondensationsgrad degree of condensation
kondensationshärtend condensation-curing
Kondensationsharz condensation resin
Kondensationskunststoff condensation polymer
Kondensationsparameter condensation conditions
Kondensationspolymer condensation polymer
Kondensationspolymerisation condensation polymerisation
Kondensationsprodukt condensation product
Kondensationsreaktion condensation reaction
Kondensationsstufe condensation stage
kondensationsvernetzend condensation curing/crosslinking
Kondensationsvernetzung condensation curing/crosslinking
Kondensatiosfeuchtigkeit condensation
kondensierfähig capable of being condensed
kondensiert condensed
Kondenswasser condensation
kondenswasserbelastet exposed to condensation
Kondenswasserfestigkeit resistance to condensation
konfektioniert formulated, made-up, compounded; *this word is often used for the sake of convenience rather than clarity, as in this example:* **Es empfiehlt sich der Einsatz von konfektionierten Pigmentpasten** It is best to use ready-made pigment pastes
Konfektionierung *originally associated with the clothing industry, this word is now widely used to express whatever an author wants it to mean. Here are some interpretations:* manufacture, conversion, fabrication, formulation, compounding, solution. *When the word does not seem to make sense in a given context, translators should substitute whatever word they think fits the bill, even if it is not among those given above.*
Konformation configuration, arrangement
konisch conical
konjugiert conjugated
Konservendose can
Konservendosenlack can coating lacquer
Konservierer preservative
Konservierung preservation, conservation
Konservierungsmittel preservative
Konservierungsstoff preservative
Konsistenz consistency
Konsistenzgeber thickener, thickening agent
Konsolidierung consolidation
konstant constant
Konstante constant
Konstantspannungsquelle constant voltage source
Konstantstromquelle constant power source
Konstantwert constant figure
konstruieren 1. to construct. 2. to design
Konstrukteur designer
Konstruktion 1. design. 2. construction
Konstruktionsgrundsätze design principles
Konstruktionsklebstoff structural adhesive
Konstruktionsmerkmal 1. design feature. 2. structural feature
Konstruktionsprinzipien design principles
konstruktiv structural
konstruktive Gestaltung structural design
Konsumentenmarkt consumer market
Konsumentenzufriedenheit consumer satisfaction
Kontaktbluten bleeding *(of pigments or dyes)*
Kontaktdruck contact pressure
Kontaktfläche contact surface
Kontaktheizung contact heating
Kontaktkleben contact bonding
Kontaktkleber contact adhesive
Kontaktklebstoff contact adhesive
Kontaktkorrosion contact corrosion
Kontaktkühlung contact cooling
kontaktlos 1. non-contact, contactless. 2. solid state, electronic
Kontaktstelle contact point, point of contact
Kontaktthermometer contact thermometer
Kontamination contamination
kontaminiert contaminated
kontinuierlich continuous
Kontinuität continuity
Kontrastfarbe contrasting colour
kontrastierend contrasting
kontrastreich high-contrast
Kontrasttaster contrast scanner
Kontrastverhältnis contrast ratio
kontrollieren to check
Kontrollmuster control sample
Kontrollprüfung quality control test
Konvektion convection
konz. conc. *(acids and alkalis)*
Konzentrat concentrate
Konzentration concentration
konzentrationsabhängig depending on the concentration
Konzentrationsänderung change in concentration
Konzentrationsbereich concentration range
Konzentrationsgefälle concentration gradient/difference
Konzentrationsgradient concentration gradient/difference
Konzentrationssprung sudden increase in concentration
Konzentrationsstreuungen variations in concentration
Konzentrationsunterschied difference in concentration

Konzentrationsverhältnis concentration ratio
Konzentrationszunahme increase in concentration
konzentriert concentrated
Konzept concept, idea, notion, design, scheme
Koordinate coordinate
koordiniert coordinated
Kopal copal
Kopalharz copal
koppeln to couple
Kork cork
Korn particle
kornartig granular
körnig granular
Kornklasse particle size range
Kornoberfläche particle surface
Kornpartikel particle
Kornporosität particle porosity
Kornspektrum particle size range
Kornstruktur particle structure
Körnung 1. particle size. 2. grit *(of sand or emery paper)*
Körnungsverteilung particle size distribution
Kornveränderung changes in particle characteristics
Kornverteilung particle size distribution
Kornzusammensetzung particle size distribution
Korona- *see* **Corona-**
Körperfarbe body colour
Körperfarbenpigment body colour pigment
Körpergehalt solids content
körperreich high-solids
Korrektur correction
Korrekturfaktor correction factor
Korrekturflüssigkeit correcting fluid
Korrekturmaßnahmen corrective measures
Korrekturwert correction factor
Korrelation correlation
Korrelationskoeffizient correlation coefficient
korrelieren to correlate
korrigiert corrected, adjusted
korrodierend corroding, corrosive
Korrosion corrosion
Korrosion, elektrochemische electrolytic corrosion
Korrosion, elektrolytische electrolytic corrosion
Korrosion, galvanische electrolytic corrosion
Korrosion, lochfraßartige pitting
korrosionsanfällig susceptible to corrosion
Korrosionsanfälligkeit susceptibility to corrosion
Korrosionsangriff corrosive attack
Korrosionsbeginn start of corrosion
Korrosionsbelastbarkeit corrosion resistance
Korrosionsbelastung corrosive attack
korrosionsbeständig corrosion resistant
Korrosionsbeständigkeit corrosion resistance
korrosionsempfindlich corrosion-sensitive, susceptible to corrosion
Korrosionserscheinung sign of corrosion
korrosionsfest corrosion resistant
Korrosionsfestigkeit corrosion resistance
korrosionsfördernd corrosive
Korrosionsfreiheit freedom from corrosion
Korrosionsgefahr risk of corrosion
korrosionsgeschützt protected against corrosion
Korrosionsgeschwindigkeit corrosion rate
korrosionshemmend corrosion-inhibiting
korrosionshindernd corrosion-inhibiting
korrosionsinhibierend corrosion-inhibiting, anti-corrosive
Korrosionsinhibition corrosion inhibition
Korrosionsinhibitor corrosion inhibitor
Korrosionsneigung tendency to corrode
Korrosionsnester localised corrosion
Korrosionspotential corrosion potential
Korrosionsprodukte corrosion products
Korrosionsprozeß corrosion process
Korrosionsprüfung corrosion test
Korrosionsrate corrosion rate
Korrosionsreaktion corrosion reaction
Korrosionsrisiko risk of corrosion
Korrosionsschäden corrosive damage, damage due to corrosion *or*: caused by corrosion
Korrosionsschutz corrosion protection, anti-corrosive effect
Korrosionsschutzanstrich anti-corrosive coating
Korrosionsschutzanstrichfarbe anti-corrosive paint
Korrosionsschutzanstrichstoff anti-corrosive paint
Korrosionsschutzbeschichtung anti-corrosive coating
Korrosionsschutzeffekt anti-corrosive effect
Korrosionsschutzeigenschaften anti-corrosive properties, corrosion resistance properties
korrosionsschützend anti-corrosive
Korrosionsschutzfarbe anti-corrosive paint
Korrosionsschutzfilm anti-corrosive film/coating
Korrosionsschutzgrundfarbe anti-corrosive primer
Korrosionsschutzgrundierung 1. anti-corrosive primer. 2. anti-corrosive primer coat
Korrosionsschutzinhibitor corrosion inhibitor
Korrosionsschutzlack anti-corrosive paint
Korrosionsschutzlösung anti-corrosive solution
Korrosionsschutzmaßnahmen anti-corrosive measures
Korrosionsschutzmittel anti-corrosive agent
Korrosionsschutzpigment anti-corrosive pigment
Korrosionsschutzprimer anti-corrosive primer
Korrosionsschutzschicht anti-corrosive coating/film
Korrosionsschutzsektor anti-corrosive sector
Korrosionsschutzsystem anti-corrosive paint/formulation
Korrosionsschutzüberzug anti-corrosive film/coating
Korrosionsschutzverhalten anti-corrosive properties
Korrosionsschutzwirksamkeit (degree of) corrosion resistance

Korrosionsschutzwirkung anti-corrosive effect
Korrosionsspuren signs/traces of corrosion
korrosionsstabil corrosion resistant
Korrosionstechnologie corrosion technology
Korrosionsüberwachung corrosion monitoring, monitoring the corrosion process
Korrosionsverhalten corrosion behaviour
korrosionsverhindernd corrosion-preventing
Korrosionsverlust material lost through corrosion
Korrosionsversuch corrosion test
Korrosionsverursacher corrosive agent
korrosionsverursacht due to corrosion
Korrosionsvorgang corrosion process
korrosionswehrend anti-corrosive
Korrosionswirkung, elektrolytische electrolytic corrosion
korrosiv corrosive
korrosive Belastung corrosive attack
Korrosivität corrosiveness
Korrosivwirkung corrosive effect
Korund corundum
Kosten-Leistungsverhältnis cost-performance ratio
Kosten-Nutzenrechnung cost-benefit calculation
Kosten-Nutzenüberlegungen cost-benefit considerations
Kosten-Nutzenverhältnis cost-benefit ratio
Kostenabschätzung cost estimate
Kostenanalyse cost analysis
kostenaufwendig costly, expensive
Kostenbetrachtung cost appraisal
Kostenbewertung cost appraisal
kosteneffektiv cost-effective
Kosteneffizienz cost-effectiveness
Kosteneinsparung cost saving
Kostenersparnis cost saving
Kostenfaktor cost factor
Kostengegenüberstellung cost comparison
kostengerecht inexpensive, reasonably priced, economical, low-cost
kostengünstig inexpensive, reasonably priced, economical, low-cost
kostenintensiv cost-intensive
Kostenrechnung cost calculation
Kostenreduzierung cost reduction
Kostensenkung cost reduction
kostensparend cost-saving
kostenspezifisch cost-related
Kostensteigerung cost increase
Kostenstruktur cost structure
kostenträchtig costly, expensive
Kostenvergleich cost comparison
Kostenvorteil cost advantage
kostspielig costly, expensive
Kotflügel bumper
kovalent covalent
KPVK *abbr. of* **kritische Pigmentvolumenkonzentration**, critical pigment volume concentration, c.p.v.c.
Kraft 1. force. 2. power. 3. energy. 4. stress
Kraft, aufgelegte applied force
Kraft-Dehnungsdiagramm stress-strain diagram/curve
Kraft-Durchbiegungskurve force-deflection curve
Kraft-Längerungsänderungsdiagramm stress-strain diagram/curve
Kraft-Verformungsdiagramm force-deformation curve/diagram, stress-strain diagram
Kraft-Verformungskurve stress-strain/force-deformation curve
Kraft-Verformungsverhalten stress-strain behaviour
Kraft-Wegdiagramm force-deformation diagram, stress-strain diagram
Kraft-Wegkurve 1. force-deformation curve *(general term)*. 2. stress-strain curve *(if referring to tensile stress)*
Kraftabbau energy/stress dissipation
Kraftamplitude force amplitude
Kraftangriff application of load(s)/stress(es)
Kraftaufnehmer force transducer
Kraftaufwand amount of force needed
Kräftegleichgewicht force equilibrium
Krafteinleitung application of load(s)/stress(es)
Krafteinwirkung force acting on something
Kraftfahrzeug (motor) vehicle
Kraftfahrzeugdecklack automotive paint/enamel/finish
Kraftfahrzeugindustrie motor industry
Kraftfahrzeugsektor motor industry
Kraftfluß stress transmission/distribution, transmission/distribution of forces/stresses
Kraftfortpflanzung stress propagation
Kraftmaximum maximum force
Kraftmeßdose force transducer
Kraftmeßeinrichtung force transducer
Kraftmeßplatte force transducer
Kraftpapier Kraft paper
Kraftschluß *connection where power is transmitted by frictional contact*: **Die Kraft wird durch Kraftschluß übertragen** The power is transmitted by friction
kraftschlüssig non-positive *(connection or joint)*
Kraftstoff fuel, petrol
Kraftübertragung power/force transmission
Krakelierung crackling
Krater crater
kraterähnlich crater-like
Kraterbildung cratering
kraterförmig crater-like/-shaped
Kraterfreiheit freedom from cratering
kratzbeständig mar resistant, scratchproof
Kratzbeständigkeit mar/scratch resistance
kratzempfindlich easily scratched
Kratzempfindlichkeit susceptibility to scratching
Kratzer scratch
kratzfest mar resistant, scratchproof
Kratzfestbeschichtung mar/scratch resistant coating
Kratzfestigkeit mar/scratch resistance
kratzfestmachendes Additiv scratchproofing additive

Kratzfestmittel scratchproofing additive
Kratzspur scratch
kratzunempfindlich mar resistant, scratchproof
Kräusellack wrinkle finish
Krebs cancer
krebserregend carcinogenic, cancer-causing
krebserzeugend carcinogenic, cancer-causing
krebsverdächtig suspected of causing cancer
Kreide chalk, whiting
kreidend chalking
Kreidung chalking
Kreidungsbeständigkeit resistance to chalking
Kreidungseffekt chalking (effect)
Kreidungsergebnisse results of chalking tests
Kreidungsneigung tendency to chalk
Kreidungsresistenz resistance to chalking
Kreiselpumpe centrifugal pump
kreisförmig circular
Kresol cresol
Kresol-Formaldehydkondensat cresol-formaldehyde condensate
Kresolharz cresol resin
kresolisch cresolic
Kresolnovolak cresol novolak
Kresolresol cresol resol
Kresylglycidylether cresyl glycidyl ether
Kriechbeanspruchung creep stress
Kriechbelastung creep stress
Kriechbeständigkeit creep strength
Kriechdehnung creep strain
Kriecheffekt creep effect
Kriechen creep
Kriechfaktor creep factor
Kriechfestigkeit creep strength
Kriechgeschwindigkeit creep rate
Kriechkurve creep curve
Kriechmodul creep modulus
Kriechmodulkurve creep modulus curve
Kriechmodullinie creep modulus curve
Kriechneigung tendency to creep
Kriechrate creep rate
Kriechspannung creep stress
Kriechspur creep stress
Kriechstrecke leakage current, tracking
Kriechstrom leakage current, tracking
Kriechstrombeständigkeit tracking resistance
kriechstromfest tracking resistant
Kriechstromfestigkeit tracking resistance
Kriechstromprüfung tracking resistance test
Kriechstromsicherheit tracking resistance
Kriechstromzeitbeständigkeit long-term tracking resistance
Kriechverformung creep deformation
Kriechverhalten creep behaviour
Kriechversuch creep test
Kriechwegbildung tracking
Kriechwiderstand creep resistance
Kristall crystal
kristallartig crystal-like
Kristallaufbau crystal structure
Kristallbildung crystallisation
Kristallform crystal shape
Kristallgitter crystal lattice
kristallin crystalline
Kristallinität crystallinity
Kristallinitätsänderung change in crystallinity
Kristallinitätsgrad degree of crystallinity
Kristallisation crystallisation
Kristallisationsbedingungen crystallising conditions
kristallisationsbeständig non-crystallising
kristallisationsfähig crystallisable
Kristallisationsfähigkeit crystallisability
kristallisationsfrei non-crystallising
Kristallisationsgeschwindigkeit rate of crystallisation
Kristallisationsneigung tendency to crystallise
Kristallisationspunkt crystallisation temperature
kristallisationsstabil non-crystallising
Kristallisationstemperatur crystallisation temperature
Kristallisationstendenz tendency to crystallise
Kristallisationszustand crystalline state
kristallisierbar crystallisable
kristallisieren to crystallise
kristallisierend crystallising
Kristallit crystallite
Kristallitorientierung crystallite orientation
Kristallitphase crystallite phase
Kristallitschmelzbereich crystallite melting range
Kristallitschmelzpunkt crystallite melting point
Kristallitschmelztemperatur crystallite melting point
kristallographisch crystallographic
Kristallöl white spirit
Kristallstruktur crystal structure
Kristallviolett crystal violet
Kristallwachstum crystal growth
Kristallwachstumsgeschwindigkeit rate of crystal growth
Kristallwasser water of crystallisation
Kriterium criterion *(plural:* criteria*)*
kritisch critical
kritische Mizellkonzentration critical micelle concentration, c.m.c.
kritische Pigmentvolumenkonzentration critical pigment volume concentration, c.p.v.c.
kryogen cryogenic
Kryostat cryostat, low-temperature thermostat
Kryptonfluorid-Excimerlaser krypton fluoride excimer laser
KTL 1. *abbr. of* **kathodische/kataphoretische Tauchlackierung**, cathodic/electrophoretic/electrodeposition painting 2. *abbr. of* **kathodische/kataphoretischer Tauchlack**, cataphoretic/electrophoretic paint
KTL-beschichtet cataphoretically/electrophoretically painted
KTL-Beschichtung cataphoretic/electrophoretic coating
KTL-Formulierung cataphoretic/electrophoretic/

electrodeposition paint (formulation)
KTL-Grundierung cataphoretic/electrophoretic/ electrodeposition primer
KTL-Grundschicht cataphoretic/electrophoretic primer coat
KTL-System cathodic electrophoretic/electrodeposition/electrodeposition paint
kubisch cubical
kubischer Ausdehnungskoeffizient coefficient of cubical/volume expansion
kugelähnlich spherical
Kugelblase spherical bubble
Kugelfallviskosimeter falling sphere viscometer
Kugelform spherical shape
kugelförmig spherical
kugelig spherical
Kugelkorn spherical particle
Kugelmahlung ball milling
Kugelmühle ball mill
Kugelschaum foam consisting of spherical bubbles
Kugelschreiber ballpoint (pen)
Kugelschreibertinte ballpoint ink
kühl cool
Kühleffekt cooling effect
Kühlflüssigkeit coolant
Kühlkreis cooling circuit
Kühlkreislauf cooling circuit
Kühlleistung cooling efficiency/capacity
Kühlmantel cooling jacket
Kühlmedium cooling medium, coolant
Kühlmittel coolant
Kühlmöbel freezers and refrigerators
Kühlschlange cooling coil
Kühlschmierstoff low-temperature lubricant
Kühlschrank refrigerator
Kühlsystem cooling system
Kühltruhe freezer, deep-freeze
Kühlungsrate cooling rate
Kühlungsverhältnisse cooling conditions
Kühlwasser cooling water
Kühlwasserkreislauf cooling water circuit
Kühlwassernetz cooling water supply
Kühlwasserzufluß cooling water inlet/supply
Kulturmedium culture medium
Kumaron coumarone
Kumaron-Indenharz coumarone-indene resins
Kumaronharz coumarone resin
Kundendienst after-sales/customer service
Kunstgewerbe arts and crafts
Kunstharz synthetic resin
kunstharzbeschichtet synthetic resin coated
Kunstharzbeton polymer concrete
Kunstharzbindemittel synthetic resin binder
Kunstharzestrich synthetic resin screed
kunstharzgebunden synthetic resin bound
Kunstharzgrundierung synthetic resin primer
Kunstharzkleber synthetic resin adhesive
Kunstharzklebstoff synthetic resin adhesive
Kunstharzlack synthetic resin-based paint
Kunstharzleim synthetic resin adhesive
Kunstharzpulver powdered synthetic resin
Kunstharzputz synthetic resin modified plaster
kunstharzvergütet synthetic resin modified
Kunstkautschuk synthetic rubber
kunstkautschukbeschichtet synthetic rubber coated/covered
Kunstleder (PVC) leathercloth, (PVC) coated fabric
Künstlerfarben artists' colours
künstlich artificial, synthetic
künstliche Alterung artificial ageing
künstliche Bewitterung artificial weathering
Kunststein artifical stone
Kunststoff plastic, polymer
Kunststoffbranche plastics industry
Kunststoffdispersionskleber dispersion/emulsion/latex adhesive
Kunststoffdispersion synthetic resin dispersion, polymer dispersion
Kunststoffemballage plastics pack
Kunststoffgebinde plastics container
Kunststoffgrundierung 1. primer for plastics. 2. polymer-based primer
Kunststoffindustrie plastics industry
Kunststofflack 1. paint for plastics. 2. polymer-based paint
Kunststofflaminat synthetic resin laminate
Kunststofflatex synthetic resin latex, polymer latex
Kunststofflösung polymer solution
kunststoffmodifiziert synthetic resin modified
Kunststoffmolekül polymer molecule
Kunststoffoberfläche plastics surface
Kunststoffolie plastics film/sheeting
Kunststoffphase polymer phase
Kunststoffpulver finely ground plastics waste
Kunststoffscheibe plastics glazing sheet
Kunststofftechnologie plastics/polymer technology
Kunststoffteilchen polymer particle
Kunststoffußboden plastics floor
Kunststoffüberzug plastics coating
kunststoffvergütet synthetic resin modified
Kunststoffverklebung 1. bonded plastics joint. 2. bonding of plastics
Kunststoffwerkstoff plastics material
Kunststoffwirtschaft plastics industry
Kunststoffzementfaktor resin-cement factor, resin content
Kupfer copper
Kupferblech copper sheet
Kupferblechelektrode copper sheet electrode
Kupferfolie copper foil
Kupferion copper ion
kupferkaschiert copper-clad
Kupferkaschierung copper cladding
kupferlaminiert copper-clad
Kupferlegierung copper alloy
Kupfermetall metallic copper
Kupfernaphthenat copper naphthenate

Kupferoberfläche copper surface
Kupferoxid copper oxide
Kupferphthalocyanin copper phthalocyanine
Kupferpulver copper powder
Kupfersalz copper salt
Kupfersulfat copper sulphate
Kupfertiefdruck rotogravure (printing)
Kupferverbindung copper compound
Kuppplungswirkung coupling effect
Kurve curve
Kurvenast part of the curve
Kurvenschar group of curves
Kurventeil part of the curve
Kurvenzug curve
Kurzbewitterung accelerated weathering
Kurzbewitterungsgerät accelerated weathering apparatus
Kurzbewitterungsversuch accelerated weathering test
Kurzbezeichnung abbreviation
kurzfristig short-term, brief(ly), quickly, immediately, at short notice, for a short time
Kurzhalskolben short-neck flask
Kurzkettenverzweigung short-chain branching
kurzkettig short-chain
Kurzölalkydharz short-oil (modified) alkyd (resin)
kurzölig short-oil
Kurzprüfmethode accelerated test method
Kurzprüfung accelerated test
Kurzschluß short circuit
Kurzschreibweise abbreviated form
Kurztest accelerated test
kurzwellig short-wave
Kurzzeitbeanspruchung short-term stress/loading
Kurzzeitbelastung short-term stress
Kurzzeitbewitterungsverhalten short-term weathering resistance
Kurzzeitbewitterungsversuch accelerated weathering test
Kurzzeitbruchlast short-term breaking stress
Kurzzeitdurchschlagfestigkeit short-term dielectric strength
Kurzzeiteigenschaften short-term properties
kurzzeithärtend fast curing
kurzzeitig short-term, briefly, for a short time
Kurzzeitkriechversuch accelerated creep test
Kurzzeitlagerung 1. short-term ageing. 2. short-term immersion
Kurzzeitmessung accelerated determination/test
Kurzzeitprüfung accelerated test
Kurzzeitverhalten short-term performance/behaviour
Kurzzeitversuch accelerated test
Kurzzeitwerte 1. short-term test results. 2. accelerated test results
Kurzzeitzugbeanspruchung short-term tensile stress
Küstenbereich coastal region
küstennah 1. near the coast. 2. near the sea
KV *abbr. of* **Kontrastverhältnis**, contrast ratio
KV-Wert contrast ratio
KW *abbr. of* **Kohlenwasserstoff**, hydrocarbon
KW-Harz hydrocarbon resin

L

labil unstable
Labor laboratory
Laboranlage laboratory equipment
Laboransatz laboratory batch
Laborbeanspruchung exposure under laboratory conditions
Laborbedingungen laboratory conditions
Laborbeschichtungsanlage laboratory coating machine/line
Laborbewitterungsversuch laboratory weathering test
Labordissolver high-speed laboratory stirrer/mixer *(see explanatory note under* **Dissolver***)*
Laborgerät laboratory instrument
Laborkneter laboratory kneader/compounder
Labormaßstab laboratory scale
Labormengen laboratory quantities
Laborofen laboratory oven
Laborperlmühle laboratory pearl mill
Laborprobe laboratory sample/specimen
Laborprodukt laboratory product
Laborprüfung laboratory test
Laborraum laboratory
Laboruntersuchung laboratory investigation/test
Laborversuch laboratory test/experiment
Laborwaage laboratory balance
Lack paint, lacquer, varnish, enamel
Lack- und Farbenindustrie paint/coatings industry
Lack, lösemittelarmer high-solids paint
Lackabfälle paint residues
lackabweisend paint-repellent
Lackadditiv paint additive
Lackadhäsion paint (film) adhesion
Lackansatz paint formulation
Lackanstrich paint film
Lackapplikation paint application
Lackaufbau coating composition
Lackaufzug paint film
Lackbad paint dipping bath
Lackbehälter paint tin/can/container
Lackbenzin white spirit
Lackbereich paint sector
Lackbindemittel paint resin, surface coating resin
Lackbranche coatings/paint industry
Lackchemiker paint chemist
lackchemisch *adjective relating to paint chemistry:* **lackchemischer Aufsatz** article

devoted to *or:* dealing with paint chemistry
Lackdämpfe paint (solvent) vapours
Lackdispersion emulsion paint
Lackdose paint can
Lackdraht coated/varnished wire
Lackeigenschaften paint properties/characteristics
Lackeinbrenntemperatur (paint) stoving/baking temperature
Lackemulsion emulsion paint
Lackentfernung paint removal/stripping
Lackfabrik paint factory
Lackfachleute paint experts
Lackfachmann paint expert
Lackfarbe paint
Lackfarbenbestandteil paint constituent/component
Lackfehler paint defect
Lackfestkörper paint solids
Lackfeststoffgehalt paint solids content
Lackfilm paint film
Lackfilmaufbau paint film structure
Lackfilmelastizität paint film flexibility/elasticity
Lackformulierung paint formulation
Lackgewebe varnished fabric
Lackglasgewebe varnished glass cloth
Lackhaftung paint adhesion
Lackharz paint resin
lackherstellend paint producing *(e.g. industry)*
Lackhersteller paint manufacturer
Lackherstellung paint manufacture/production
Lackhilfsmittel paint additive
Lackierbarkeit paintability
Lackiereinbrenntemperatur stoving/baking temperature
lackierfähig paintable
Lackierhalle painting shop
Lackierkabine paint spraying booth
Lackierobjekt 1. article being painted. 2. article to be painted
Lackierofen paint drying/stoving oven
Lackierstraße painting line
lackiert varnished, painted
Lackierung 1. paint film, finish, coat *(of paint)*. 2. painting
Lackierungsdefekt paint film defect
Lackierungseigenschaften paint film properties
Lackierungsfehler paint film defect
Lackierungsqualität paint film quality
Lackierzyklus painting cycle
Lackindustrie paint/coatings industry
Lackkonsistenz paint consistency
Lackkunstharz synthetic paint resin
Lacklabor paint (research) laboratory
Lacklaborant paint chemist
Lackleinöl refined linseed oil
Lacklieferant paint supplier
Lacklösemittel paint solvent
Acklöser paint solvent
Lacklösung lacquer/varnish solution
Lackmuspapier litmus paper
Lackoberfläche paint film surface
Lackpapier varnished paper
Lackpolyester polyester paint resin
Lackpolymer paint resin, surface coating resin
Lackprimer primer
Lackprobe paint sample
Lackqualität paint quality BUT: **...können mit allen gängigen Lackqualitäten überlackiert werden** ...can be overpainted with all standard paints
Lackreste paint residues
Lackrezept paint formulation
Lackrezeptierung 1. paint formulation. 2. developing a paint formulation
Lackrezeptur paint formulation
Lackrohstoff paint resin, surface coating resin
Lackrückstände paint residues
Lackschicht paint film
Lackschichtablösung detachment/lifting of the paint film
Lackschichtdicke paint film thickness
Lackschlamm paint sludge
Lackschlammanfall amount of paint sludge produced
Lackschlammaufkommen production of paint sludge
Lackschlammfestkörper paint sludge solids
Lackschutzschicht protective (paint) film
Lacksektor paint/coatings industry
Lackspezifikation paint specification
Lackspritznebel spray mist
Lackstörung paint film defect
Lacksystem paint (formulation), surface coating system
Lacktechnik paint technology
Lacktechniker paint technologist
lacktechnisch *adjective relating to paints*: **lacktechnische Eigenschaften** paint properties
Lacktechnologie paint technology
lacktechnologisch *adjective relating to paints:* **lacktechnologische Eigenschaften** paint properties
Lacküberzug paint film
Lackverarbeitung paint application
Lackverbrauch paint consumption
Lackvernetzer paint hardener/catalyst
lackverwendend paint using *(e.g. industry)*
Lackverwender paint user
Lackviskosität paint viscosity
Lackzusammensetzung paint composition
Lackzusatz paint additive
Lactam lactam
Lactamschmelze lactam melt
Lacton lactone
Lactonharz lactone resin
Ladung 1. charge. 2. load, cargo
Ladungsdichte charge density
Ladungsverteilung charge distribution
Lager warehouse, storeroom
Lagerbedingungen storage conditions

Lagerbehälter storage tank/container
Lagerbestand stock in hand
lagerbeständig having a good shelf life
Lagerbeständigkeit shelf/storage life, storage stability
Lagercontainer storage tank/container
Lagerdauer 1. ageing period. 2. time of immersion, immersion period *(see explanatory note under* **Lagerung***)*. 3. storage period
Lagerdauer 1. storage period. 2. ageing period
Lagereigenschaften storage properties/characteristics/stability, shelf/storage life
Lagerfähigkeit shelf/storage life, storage stability
Lagerfähigkeitsgarantie guaranteed shelf life
Lagerfläche storage area
Lagergebäude store, warehouse
Lagerhalle warehouse, storeroom
Lagerhaltung storage, warehousing
Lagerhaltungskosten warehousing costs
Lagerkapazität storage capacity
Lagerkonservierungsmittel preservative
Lagerkosten storage/warehousing costs
Lagerplatz storeroom
lagerstabil having a long shelf life
Lagerstabilität shelf/storage life, storage stability
Lagertank storage tank
Lagertemperatur storage temperature
Lagerung 1. immersion *(of test specimen in a liquid)*. 2. ageing. 3. storage. **Lagerungsbedingungen** 1. storage conditions. 2. ageing conditions
Lagerungsbehälter storage tank
Lagerungsdauer 1. ageing period. 2. time of immersion, immersion period. 3. storage period *(see explanatory note under* **Lagerung***)*
Lagerungseigenschaften shelf/storage life, storage stability
Lagerungstemperatur 1. ageing temperature. 2. immersion temperature. 3. storage temperature *(see explanatory note under* **Lagerung***)*
Lagerungsversuch 1. ageing test. 2. immersion test *(see explanatory note under* **Lagerung***)*
Lagerungszeit 1. ageing period. 2. time of immersion, immersion period. 3. shelf/storage life
Lagerzeit 1. storage period/time. 1. ageing period. 2. time of immersion, immersion period. 3. storage period/time
LAKW *abbr. of* **leichtflüchtiger aromatischer Kohlenwasserstoff**, highly volatile aromatic hydrocarbon
lamellar lamellar, platelet-like
Lamelle lamella
lamellenartig lamellar
lamellenförmig lamellar
Lamellenklebung lamination
Lamellenstruktur lamellar structure
lamelliert laminated
laminar laminar
laminare Schichtenströmung laminar flow
laminare Strömung laminar flow
laminares Fließen laminar flow
Laminarfluß laminar flow
Laminat laminate
Laminataufbau laminate structure
Laminatdicke laminate thickness
Laminateigenschaften laminate properties
Laminatoberfläche laminate surface
Laminator laminating unit
Laminatpapier laminating paper
Laminieranlage laminating plant
laminierbar capable of being laminated
Laminierdruck laminating pressure
laminieren to laminate
Laminierharz laminating resin
Laminierkleber laminating adhesive
Laminiermaschine laminator, laminating machine
Laminierpresse laminating press
Laminiersystem laminating adhesive
Laminierung lamination
Lammfellwalze lambswool roller
Längenausdehnung linear expansion
Längenausdehnungskoeffizient, thermischer coefficient of linear expansion
Längenkontraktion longitudinal shrinkage/contraction
Längenschrumpf longitudinal shrinkage/contraction
Längenschwindung longitudinal shrinkage/contraction
Längenschwund longitudinal shrinkage/contraction
längerkettig long-chain
längerwellig long-wave
langfristig long-term
Langfristprognose long-term forecast
langkettig long-chain
langlebig long-lasting/-life,
Langlebigkeit longevity, long (working/service) life
Langölalkydharz long-oil (modified) alkyd (resin)
Langölalkydharzlack long-oil alkyd paint
langölig long-oil
längs longitudinal(ly)
Längsachse longitudinal axis
langsamflüchtig low-volatility
Langsamrührer slow-speed stirrer/mixer
langsamverdunstend slow-evaporating
Längsausdehnung longitudinal expansion
Längsdehnung longitudinal expansion
längsgerichtet longitudinally oriented
Längskontraktion longitudinal shrinkage
Längsleimung longitudinal glueing
Längsrichtung longitudinal direction
Längsscherung longitudinal shear
Längsschrumpf longitudinal shrinkage
Längsschrumpfung longitudinal shrinkage
Längsschwindung longitudinal shrinkage
Längsspannung longitudinal stress
langwellig long-wave

Langzeitbeanspruchung long-term stress/loading
Langzeitbelastbarkeit long-term resistance *(e.g. to loads, chemicals etc.)*
Langzeitbelastung long-term stress
Langzeitbelastungseigenschaften long-term load bearing properties
Langzeitbeständigkeit long-term resistance/stability
Langzeitbestrahlung long-term irradiation
Langzeitbewitterungsbeständigkeit long-term weathering resistance
Langzeitdruckbeanspruchung long-term compressive stress
Langzeiteffekt long-term effect
Langzeiteigenschaften long-term properties
Langzeiteinsatz long-term use
Langzeiterfahrung long-term experience
Langzeitfestigkeit long-term strength
Langzeitgebrauch long-term use
langzeitig long-term
Langzeitkontakt long-term contact
Langzeitkriechverhalten creep behaviour
Langzeitlagerung 1. long-term ageing. 2. prolonged immersion *(of test piece in a liquid)*
Langzeitprüfung 1. long-term test. 2. creep test
Langzeitschutz long-term protection
Langzeitstabilität long-term stability
Langzeittemperaturbeständigkeit long-term heat resistance
Langzeituntersuchung creep test, long-term test
Langzeitverhalten long-term behaviour/performance
Langzeitversuch 1. long-term test. 2. creep test
Langzeitwasserlagerung prolonged immersion in water
Lanthan lanthanum
Lanthanid lanthanide
Lärm noise
Lärmdämmung sound insulation
Lärmemission noise/sound emission
Lärmemissionswert 1. sound emission value *(in a scientific context)*. 2. noise level *(general term)*
lärmgedämpft soundproofed
Lärmminderung noise abatement
Lärmminderungsmaßnahmen noise abatement measures
Lärmschutzwand acoustic barrier
Lasche, doppelte double strap
Lasche, einfache single strap
Laschung, abgeschrägte zweischnittige tapered double strap lap joint
Laschung, abgesetzte einschnittige recessed single strap joint
Laschung, abgesetzte zweischnittige recessed double strap joint
Laschung, einschnittige single strap joint, butt strap joint
Laschung, zweischnittige double strap joint, double butt-strap joint
Laserabtastgerät laser scanning analyser
Laseranlage laser equipment
Laserbeschrifter laser marker, laser marking/etching device
Laserbeschriftung laser marking/etching
Laserbeschriftungsprogramm laser marking/etching program
Laserbestrahlung laser irradiation
Laserdruck laser printing
Laserdrucker laser printer
Lasereinstellung laser setting
Laserenergie laser energy
Laserfrequenz laser frequency
lasergesteuert laser controlled
Laserhärtung laser (beam) curing
Laserinstrument laser instrument
Laserkleben laser bonding
Laserleistung laser output
Laserlicht laser light
Laserlichtabsorption laser light absorption
Laserlichtpolarisation laser beam polarisation
Laserlöten laser soldering
Lasermikrolöten laser micro-soldering
Laserpyrolyse laser pyrolysis
Laserschneideverfahren laser beam cutting
Laserschreiber laser marker
lasersensibel laser sensitive
Laserstrahl laser beam
Laserstrahlführung laser (beam) control
Laserstrahlhärtung laser (beam) curing
Laserstrahlleistung laser (beam) output
Laserstrahlparameter laser beam parameter
Laserstrahlschneiden laser (beam) cutting
Laserstrahlung laser (beam) radiation
Lasersystem laser system
Lasertechnik laser technology
Lasertyp laser type, type/kind of laser
lasierend transparent
lasiert glazed
Last load, stress
Last, aufgelegte applied stress/load
Last-Verformungsdiagramm 1. force-deformation diagram *(general term)*. 2. stress-strain diagram *(if referring to tensile stresses)*
lastabhängig depending on the load
Lastamplitude stress amplitude
Lastanlegung application of stress
Lastaufbringung application of stress
Lastaufnahmevermögen load-bearing/-carrying capacity
lastbeaufschlagt under stress
Lastdruck load pressure
Lastenheft list of requirements, (technical) specification
Lastfrequenz load frequency
lastkompensiert load-compensated
Lastspiel stress/load cycle
Lastspielfrequenz stress/load cycle frequency
Lastspielzahl number of stress/load cycles

Laststandzeit time under stress, stress duration
Laststeigerung 1. load increment *(in tensile testing)*. 2. load increase
Lasttragevermögen load-bearing capacity
Lastübertragung stress/load transfer
lastunabhängig independent of the load
Lastverteilung stress/load distribution
Lastwechsel stress/load cycle
Lastwechselzahl number of stress/load cycles
Lastzyklus stress/load cycle
Lasur glaze, varnish, lacquer
Lasuranstrich glaze coat
Lasurfilm glaze coat
Lasurharz glazing resin
Latenthärter latent hardener
Latex latex, dispersion, emulsion
Latexfarbe emulsion paint
Latexgrundierung emulsion primer
Latexharz dispersion resin
Latexpolymer polymer latex
Latexteilchen latex particle
Laubholz hardwood
Läufer curtaining, sagging
Läuferbildung curtaining, sagging
Lauge alkali solution
Laurinlactam laurinlactam
Laurinsäure lauric acid
Lauroylperoxid lauroyl peroxide
LC-Pigment liquid crystal pigment
LCD liquid crystal display
LCD-Anzeige liquid crystal display
LCD-Flüssigkristallanzeige liquid crystal display
Lebensdauer (working/service) life
Lebensmittel foodstuffs
Lebensmittelbedarfsgegenstände consumer goods for food contact applications
Lebensmittelbehälter food container/pack
Lebensmittelbereich food sector
Lebensmittelbranche food industry/sector
lebensmittelecht suitable for food contact applications
Lebensmittelgesetz food regulations
Lebensmittelindustrie food industry
lebensmittelnah in contact with foodstuffs
lebensmittelrechtlich *adjective relating to food regulations*: **lebensmittelrechtliche Bestimmungen** food regulations
lebensmittelrechtliche Unbedenklichkeit suitability for food contact applications
Lebensmittelverpackung 1. food packaging. 2. food pack/container
Lecithin lecithin
Leckage leakage
leckagefrei leak-proof, non-leaking
Leckprüfgerät leak tester
Leder leather
Lederartikel leathergoods
Lederfaserwerkstoff reconstituted leather
Lederindustrie leather industry
Ledernarbung leather grain effect
Leerstelle void
Legierung alloy
Leichtbau lightweight construction
Leichtbaukonstruktion lightweight construction
Leichtbaustruktur lightweight structure
Leichtbauteil lightweight part/component
Leichtbeton lightweight concrete
leichtbrennbar readily flammable
leichtdispergierbar easily dispersed
leichtentflammbar readily flammable
leichtentzündbar readily flammable
leichtentzündlich readily flammable
leichtfließend free-flowing, easy-flow
leichtflüchtig readily volatile
Leichtmetall light metal
Leichtpigment lightweight pigment
leichtpigmentierbar easily pigmented
Leichtspat gypsum
leichtzugänglich easily accessible
Leim glue
Leimauftrag application of glue/adhesive
Leimauftragsmenge amount of glue/adhesive applied
Leimauftragungsgerät glue applicator
Leimdispersion adhesive dispersion
Leimfarbe distemper
Leimflotte adhesive solution
Leimfuge glued/bonded joint
Leimharz adhesive resin
Leimmischung adhesive mix
Leimstrich adhesive film/coating
Leimzylinder glueing cylinder/roll
Leinöl linseed oil
Leinölalkyd linseed oil alkyd
Leinölalkydharz linseed oil alkyd (resin)
Leinölfettsäure linseed oil fatty acid
Leinölkitt linseed oil based mastic
Leistung 1. performance, efficiency, capacity *(e.g. of a machine)*. 2. output *(of power)*. 3. input *(of power)*. 4. rating *(electrical)*. 5. energy, power. 6. service *(provided by a company)*
Leistung, abgegebene power output
Leistung, aufgenommene power input, power/energy used
Leistung, elektrische power
Leistung, installierte connected/installed load
Leistung, mechanische 1. mechanical power/output *(of a machine)*. 2. power, horsepower *(of an engine)*
Leistung, thermische thermal efficiency
Leistung, zugeführte power input
Leistungsabgabe power/energy output
Leistungsangaben performance/output data
Leistungsangebot services *(offered by a company)*
Leistungsaufnahme 1. energy used/consumed. 2. power input
Leistungsbedarf power/energy requirement, power/energy consumption
Leistungsbewertung assessment of performance
Leistungsbilanz energy balance
Leistungsdaten performance data

Leistungsdichte power density
Leistungserhöhung increased output/efficiency
leistungsfähig powerful, efficient, capable, high-output, high-performance
Leistungsfähigkeit efficiency, effectiveness, performance, output
Leistungsgrenze performance limit
Leistungskraft efficiency
Leistungsmerkmale performance features
Leistungsminderung reduced output/efficiency
Leistungsniveau (level of) performance
Leistungsprofil overall performance
Leistungsreserven power/energy reserves
leistungsschwach inefficient
leistungssparend power/energy saving
leistungsstark powerful, efficient, high-output, high-performance
Leistungssteigerung increased output/efficiency
Leistungssteuerung output control
Leistungsvergleich performance comparison
Leistungsvermögen performance, efficiency, effectiveness
Leistungswert efficiency, effectiveness
leitend conductive
Leiter conductor *(of heat or electricity)*
Leiterplatte printed circuit board
Leiterplattenindustrie printed circuit board industry
Leiterplattenmaterial printed circuit board material
Leiterpolymer ladder polymer
leitfähig conductive
Leitfähigkeit conductivity
Leitfähigkeitslack conductive varnish
Leitfähigkeitspigment conductive pigment
Leitfähigkeitsruß conductive carbon black
Leitfähigkeitsschicht conductive coating
Leitfähigkeitssensor conductivity sensor/probe
Leitfähigkeitsverbesserung improvement in conductivity
Leitfähigkeitszelle conductivity cell
Leitgrundierfilm conductive primer coat
Leitklebschicht conductive adhesive film
Leitklebstoff conductive adhesive
Leitlinie guideline
Leitrechner central/master computer
Leitruß conductive carbon black
Leitstand control console/desk/panel
Leitungswasser tap water
leuchtend luminous
Leuchtfarbe fluorescent paint
Leuchtkraft luminosity
Leuchttinte fluorescent ink
Lewisbase Lewis base
Lewissäure Lewis acid
Licht light
Lichtabbau photodegradation
Lichtabsorber light absorbing agent
lichtabsorbierend light absorbent
Lichtabsorption light absorption
Lichtabsorptionskoeffizient light absorption coefficient
lichtaktivierbar capable of being photo-activated
Lichtalterung light ageing
Lichtausbeute luminous efficiency
lichtbeständig light resistant
Lichtbeständigkeit lightfastness
Lichtbeugung light refraction
Lichtbogen arc
lichtbogenbeständig arc resistant
Lichtbogenbeständigkeit arc resistance
lichtbogenfest arc resistant
Lichtbogenfestigkeit arc resistance
Lichtbrechungsindex refractive index
Lichtdurchgang light transmission
lichtdurchlässig translucent
Lichtdurchlässigkeit light transmission
lichtecht lightfast
Lichtechtheit lightfastness
Lichteinwirkung exposure to light
Lichtemission light emission
lichtempfindlich light sensitive
Lichtempfindlichkeit light sensitivity
Lichtenergie light energy
Lichtfleck light spot
Lichtgeschwindigkeit velocity of light
lichthärtbar light curing
lichthärtend light curing
lichtinduziert light induced
Lichtintensität light intensity
Lichtinterferenz light interference
Lichtleiter optic(al) fibre
Lichtleiterfaser optic(al) cable
Lichtleitfaser optic(al) fibre
Lichtmikroskop optical microscope
Lichtmikroskopie optical microscopy
lichtmikroskopische Aufnahme photomicrograph
Lichtquelle light source
lichtreflektierend light reflecting
Lichtreflexion light reflection
Lichtremission light remission
lichtresistent light resistant
Lichtschranke light beam guard
Lichtschutz light stabilisation
Lichtschutzadditiv light stabiliser
Lichtschutzausrüstung light stabilisation
Lichtschutzeffekt light stabilising effect
Lichtschutzmittel light stabiliser
Lichtschutzsystem light stabliser
Lichtschutzwirkung light stabilisation
Lichtschutzzusatz light stabiliser
Lichtspektrum (light) spectrum
lichtstabil lightfast
Lichtstabilisator light stabiliser
Lichtstabilisatorwirkung light stabilising effect
lichtstabilisierend light stabilising
lichtstabilisiert light stabilised
Lichtstabilisierung light stabilisation
Lichtstabilisierwirkung light stabilising effect
Lichtstabilität light stability
Lichtstrahl light beam

Lichtstrahloszillograph light beam oscillograph
Lichtstreukoeffizient light scattering coefficient
Lichtstreuung light scattering
Lichtstreuvermögen light scattering power
Lichtstreuwirkung light scattering effect
Lichttaster light scanner
Lichttransmission light transmission
Lichtumwandlungsenergie light transformation energy
lichtundurchlässig opaque, impermeable to light
Lichtundurchlässigkeit opacity, impermeability to light
Lichtwellenlänge wavelength (of light)
Lieferant supplier
Lieferantenauswahl choice of supplier
Lieferantennachweis list of suppliers
Lieferantenverzeichnis list of suppliers
Lieferantenwechsel change of supplier
lieferbar available
Lieferbedingungen delivery conditions, terms of delivery
Lieferform form in which supplied
Lieferkonzentration concentration as delivered
Lieferprogramm range *(e.g. of products)*
Lieferschein delivery note
Lieferschwierigkeiten delivery problems
Lieferspezifikation delivery specification
Lieferung 1. delivery. 2. consignment
Lieferzeit delivery date
Ligand ligand
Lignin lignin
Ligninleim lignin glue
Ligninsulfonat lignin sulphonate
Ligninsulfonsäure lignin sulphonic acid
linear linear
linear-elastisch linear-elastic
Lineardichroismus linear dichroism
linearer Ausdehnungskoeffizient coefficient of linear expansion
linearer Wärmeausdehnungskoeffizient coefficient of linear expansion
Linearität linearity
Linienspektrum line spectrum
Linolensäure linolenic acid
Linoleum linoleum
Linolsäure linoleic acid
lipohil lipophilic
Lipophilie lipophilic character
Lippenstift lipstick
Literaturdaten figures quoted in the literature
Literaturwert figure quoted in the literature
Lithiumcarbonat lithium carbonate
Lithiumhydroxid lithium hydroxide
lithiumorganisch organo-lithium
Lithiumoxid lithium oxide
Lithiumwasserglas lithium silicate waterglass
Lithographie lithography
Lithographielack litho varnish
Lithoponpigment lithophone pigment
lizensiert licensed
Lizenzgeber licensor
Lizenznehmer licensee
LKZ *abbr. of* **Luftkontaktzeit**, air contact time
Loch hole
Lochfraß pitting
lochfraßartige Korrosion pitting
Lochfraßkorrosion pitting
Lochkorrosion pitting
Lochplatte perforated sheet
Lockerung loosening
logarithmisch logarithmic
lohnintensiv wage intensive
Lohnkosten labour costs
Lohnlackierung contract painting
lokal local
Los batch
lösbar 1. soluble. 2. temporary *(joint)*
lösbare Verbindung temporary joint
Löseeigenschaften dissolving properties
Lösegeschwindigkeit dissolving speed/rate
Lösekessel dissolving vat
Lösekraft dissolving power
Lösemittel solvent
Lösemittelabdunstung solvent evaporation
Lösemittelabgabe solvent emission/evaporation/release
Lösemittelabreicherung drop in solvent content
Lösemittelacrylat acrylic resin solution
Lösemittelanteil solvent content *When used in the plural,* **-anteile** *can usually be ignored, e.g.* **Da keine Lösemittelanteile verdunsten müssen...** since there is no solvent which has to evaporate...
lösemittelarm low-solvent, high-solids
Lösemittelausstoß solvent emission
lösemittelbasierend solvent-based
Lösemittelbasis, auf solvent-based
lösemittelbeladen containing solvent
lösemittelbeständig solvent resistant
Lösemittelbeständigkeit solvent resistance
Lösemitteldampf solvent vapour
Lösemitteldiffusion solvent diffusion
Lösemittelechtheit solvent resistance
Lösemitteleinfluß effect of solvent
Lösemitteleinsparung solvent saving(s)
Lösemitteleinstellung solvent composition
Lösemittelemission solvent emission
lösemittelempfindlich affected by solvents
Lösemittelempfindlichkeit susceptibility to solvent attack
lösemittelerfordernd requiring the addition of solvent
lösemittelfest solvent resistant
Lösemittelfestigkeit solvent resistance
lösemittelfrei solvent-free/-less
Lösemittelfreiheit freedom from solvents
lösemittelfreundlich solvent-compatible
Lösemittelgehalt solvent content
Lösemittelgemisch solvent blend
lösemittelgetränkt solvent impregnated
lösemittelhaltig 1. containing solvent. 2. solvent-based

Lösemittelkleber solvent-based adhesive
Lösemittelklebstoff solvent-based adhesive
Lösemittelkombination solvent blend
Lösemittelkonzentration solvent concentration
Lösemittellack solvent-based paint
lösemittellöslich soluble in solvents, solvent-soluble
Lösemittellöslichkeit solubility in solvents
Lösemittelmischung solvent blend
Lösemittelphase solvent phase
Lösemittelproblematik solvent problem
Lösemittelreaktivierung solvent activation *(of a dry adhesive film)*
lösemittelreduziert with a reduced solvent content, containing less solvent
Lösemittelregenerat recovered solvent
lösemittelreich high-solvent, with a high solvent content
Lösemittelresistenz solvent resistance, resistance to solvent attack
Lösemittelreste solvent residues
Lösemittelrestgehalt residual solvent content
Lösemittelretention solvent retention
Lösemittelrückgewinnung solvent recovery
lösemittelverdünnbar 1. solvent-based. 2. thinnable with solvent
lösemittelverdünnt thinned with solvent
Lösemittelverdunstung solvent evaporation
Lösemittelverflüchtigung solvent evaporation
Lösemittelverlust solvent loss
lösemittelverträglich solvent-compatible
Lösemittelverträglichkeit solvent compatibility
Lösemittelzugabe addition of solvent
Lösemittelzusammensetzung solvent composition
Lösemittelzusatz addition of solvent
lösen to dissolve
Löser solvent
Lösergemisch solvent blend
Lösetemperatur dissolving temperature
Löseverhalten dissolving/solubility characteristics
lösevermittelnd solubilising
Lösevermögen dissolving power
löslich 1. soluble. 2. temporary *(bond)*
löslich, begrenzt sparingly soluble
löslich, beschränkt sparingly soluble
löslich, gut readily soluble
löslich, schlecht difficult to dissolve
löslich, weitgehend largely soluble
Löslichkeit solubility
Löslichkeitsbereich solubility range
Löslichkeitseigenschaften solubility characteristics
Löslichkeitsinhibitor solubility inhibitor
Löslichkeitsvermittler solubiliser
Losnummer batch number
Lösung solution
Lösung, ethanolische ethanol solution
Lösung, methanolische methanol solution
Lösungseffekt dissolving effect
Lösungsfreiheit freedom from solvent(s)
Lösungsklebung solution bonding
Lösungsmittel solvent
Lösungsmittel-, lösungsmittel- *see* **Lösemittel-, lösemittel-**
Lösungsmöglichkeit way of solving: **Eine Lösungsmöglichkeit wäre...** One way of solving the problem would be...
Lösungsphase solution phase
Lösungspolymer solution polymer
Lösungspolymerisat solution polymer
Lösungspolymerisation solution polymerisation
lösungspolymerisiert solution polymerised
Lösungsverhalten solubility/dissolving characteristics
Lösungsvermittler solution aid
Lösungsvermögen dissolving power
Lösungsviskosität solution viscosity
Lösungsvorgang dissolving process
Lösungsvorschlag suggested solution
Lötbad solder bath
Lötbadbeständigkeit solder bath rsistance
Lötbadfestigkeit solder bath resistance
Lötbadlagerung solder bath immersion
löten to solder
lötfest solder bath resistant
Lötverbindung soldered joint
Lötwelle soldering iron
Luft air
Luftabschluß, unter in the absence of air
Luftaktivierungsphase air activation phase
Luftausschluß exclusion of air
Luftbelastung air pollution
Luftbläschen air bubbles
Luftblase air bubble
luftblasenfrei free from air bubbles
Luftbürste airbrush
luftdicht airtight
Luftdruck 1. atmospheric pressure. 2. air pressure
Luftdurchlässigkeit air permeability
Luftdurchwirbelung forced air circulation
Lufteinschlüsse entrapped air
Lufteintrag introduction of air *(e.g. into a paint by stirring)*
Luftfahrt aircraft/aviation industry
Luftfahrtindustrie aircraft/aviation industry
Luftfeuchte 1. relative humidity. 2. atmospheric humidity/moisture, moisture in the atmosphere
Luftfeuchte, relative relative humidity
Luftfeuchteempfindlichkeit sensitivity to atmospheric humidity
Luftfeuchtegehalt atmospheric moisture content
Luftfeuchtigkeit relative/atmospheric humidity
Luftfeuchtigkeit, relative relative humidity
Luftfilter air filter
luftforciert trocknend force drying
luftfrei free from air
lufthärtend air drying/curing
Luftkontakt, in in contact with air
Luftkontaktzeit air contact time
luftlos airless

luftloses Sprühverfahren airless spraying
Luftmenge amount of air
Luftnetz air supply system
Luftreinhaltemaßnahmen measures to keep the air clean
Luftreinhaltung keeping the air clean
Luftreinhaltungsbestimmung clean air act
Luftreinhaltungsgesetz clean air act
Luftsauerstoff atmospheric oxygen
Luftschall airborne sound
Luftspalt air gap
Luftstrom current of air
lufttrockenbar air drying
lufttrocknend air drying
Lufttrocknung air drying
Luftumwälzofen air circulating oven
Luftumwälzung 1. air circulation. 2. air circulating system
luftundurchlässig airtight, impermeable to air
luftverfilmend forming a film on exposure to air
Luftverschmutzung air/atmospheric pollution
Luftversorgung air supply
Luftverunreinigung air/atmospheric pollution
Luftwege airways
Luftzirkulation air circulation
Luftzutritt access of air
Lumineszenz luminescence
Lumineszenzspektroskopie luminescence spectroscopy
Lumineszenztaster luminescence scanner
Lunker void
lyophil lyophilic
lyophob lyophobic
lyotrop lyotropic
Lysin lysine

M

Machbarkeit feasibility
mager short-oil
Magnesium magnesium
Magnesiumlegierung magnesium alloy
Magnesiumoxid magnesium oxide
Magnesiumsilikat magnesium silicate
Magnesiumstearat magnesium stearate
Magnesiumsulfat magnesium sulphate
magnetisch magnetic
Mahlaggregat grinder, grinding unit
Mahlansatz millbase
Mahlbehälterform grinding compartment shape
Mahldauer grinding time
Mahlergebnis grinding result
Mahlgut millbase, material being ground, material to be ground
Mahlgutansatz millbase batch
Mahlgutformulierung millbase formulation
Mahlgutzusammensetzung millbase composition
Mahlkammer grinding compartment
Mahlkörper grinding medium
Mahlkörperdichte grinding medium density
Mahlkörperdurchmesser grinding medium diameter
Mahlkörperform grinding medium shape
Mahlkörperfüllgrad amount of grinding medium
Mahlkörpergröße grinding medium size
Mahlkörperhärte grinding medium hardness
Mahlkörperrauhigkeit grinding medium roughness
Mahlkörperschüttung grinding medium
Mahlmedium grinding medium
Mahlorgan grinding element
Mahlpaste millbase
Mahlprozeß grinding process/operation
Mahlraum grinding compartment
Mahlscheibe grinding disc
Mahlstamm millbase
Mahltopf grinding compartment
Mahlung grinding
Mahlwalzwerk grinding rolls
Mahlwirkung grinding effect/efficiency
MAK-Wert maximum allowable concentration, MAC
Makroaufbau macrostructure
Makrobereich macro-range/-region
Makroblase macro-bubble
makrobrownsche Bewegung macro-Brownian movement
makrocyklisch macrocyclic
Makromolekül macromolecule
makromolekular macromolecular
Makromolekülkette macromolecule chain
Makromonomer macromonomer
Makroradikal macroradical
Makroriß macrocrack
Makroschaum macrofoam
makroskopisch macroscopic
Makrostruktur macrostructure
Maleinanhydrid maleic anhydride
Maleinat maleate
Maleinatharz maleic resin
maleiniert maleated
Maleinimid maleinimide
maleinisiert maleated
Maleinsäure maleic acid
Maleinsäureanhydrid maleic anhydride
Maleinsäurederivat maleic acid derivative
Maleinsäureester maleic ester
Maler painter, decorator
Malerbürste paint brush
Malerfarbe decorator's/decorative paint
Malerlack decorator's/decorative paint
Malerpinsel paint brush
Malonsäure malonic acid
Malzextrakt malt extract
Mangan manganese

Manganchlorid manganese chloride
manganhaltig containing manganese
Manganitpigment manganite pigment
Manganviolet manganese violet
Manometer manometer
Markenname trade/brand name
Markenzeichen trade mark
markieren to mark
Markierfarbe 1. marking paint. 2. road marking paint
Markierungsfarbe 1. marking paint. 2. road marking paint
Markierstift marking pen
Markiertinte marking ink
Marktakzeptanz market acceptance
Marktanteil market share
Marktaufteilung market breakdown
Marktausweitung market expansion
Marktchancen market opportunities
Marktdurchdringung market penetration
Marktführer market leader
marktgängig currently available, (available) on the market
Marktgliederung market breakdown
Marktlücke market gap
marktorientiert market-orientated
Marktpotential market potential
Marktprognose market forecast
Marktsättigung market saturation
Marktstudie market study
marktüblich standard, ordinary
Marktverhältnisse market conditions
Marktwirtschaft market economy
Marmor marble
Martens-Wärmeformbeständigkeit Martens heat distortion temperature
Martensgrad Martens temperature
Martenswert Martens temperature
Martenszahl Martens temperature
maschinell mechanical
Maschinenausfallzeit machine downtime
Maschinenausrüstung equipment
Maschinenbau mechanical engineering
Maschinenbediener machine operator
Maschinenbedienpersonal machine operators
Maschinengeschwindigkeit machine speed
Maserung grain *(of wood)*
Maske mask
maskiert masked
Massekonstanz constant weight
Massen% percent/% by weight
Massenabnahme weight decrease, decrease in weight
Massenanstieg weight increase, increase in weight
Massendurchsatz material throughput
Masseneinheit unit mass/weight
Massenflußrate mass flow rate
Massenprozente percent/% by weight
massenselektiver Detektor mass selective detector
Massenspektrometer mass spectrometer
Massenspektrometrie mass spectrometry
massenspektrometrisch mass-spectrometric
Massenspektroskopie mass spectroscopy
Massenspektrum mass spectrum
Massenteile parts by weight, p.b.w.
Massenverhältnis proportion by weight
Massenverlust loss of weight, weight decrease, decrease in weight
Massenzunahme weight increase, increase in weight
Massenzuwachs weight increase, increase in weight
Massepolymerisation bulk polymerisation
Masseverlustrate rate of weight decrease
maßgeschneidert tailor-made
Materialaufwand amount of material (needed)
Materialbruch material failure
Materialeigenschaften material properties
Materialfehler material defect
Materialfluß material flow
materialgegeben material-related
Materialkonstante material constant
Materialprüfung material testing
materialspezifisch material-related, specific to the material
materialtypisch typical for the material
Matrix matrix
Matrixharz matrix resin
matt matt
mattierend flatting
mattiert matt
Mattierung flatting
Mattierungseffekt flatting effect
Mattierungsinseln dull spots/areas
Mattierungsmittel flatting agent
Mattierungsmittelanteil flatting agent content
Mattierungswirkung flatting effect
Mattlack matt finish paint
Mattlackierung matt finish
Mattlacküberzug matt finish
Mauer wall
Mauerfläche masonry surface
Mauerwerk masonry, brickwork
maximale Arbeitsplatzkonzentration maximum allowable concentration, MAC
Maximalkonzentration maximum concentration
Maximalkraft maximum force
Maximalschrumpfung maximum shrinkage
Maximalspannung maximum stress
Maximaltemperatur maximum temperature
Maximalviskosität maximum viscosity
Maximalwert maximum value
MDI diphenylmethane diisocyanate
Mechanik 1. mechanism. 2. mechanical properties
mechanisch mechanical
mechanische Beanspruchung mechanical loading/stress
mechanische Bearbeitung machining
mechanische Eigenschaften mechanical

properties
mechanische Leistung mechanical power/output *(of a machine)*
mechanische Werte mechanical properties
mechanischer Verlustfaktor loss factor
mechanisches Verhalten mechanical performance
mechanisiert mechanised
Mechanismus mechanism
Medienbeständigkeit chemical resistance
Medium medium
Medizin medicine
medizinisch medical
Meeresnähe, in in a marine environment
Meeresorganismus marine organism
Meerwasser seawater
mehrachsig multi-axial
mehrbasisch polybasic
mehrdimensional multi-dimensional
mehrdirektional multi-directional
mehrfarbig multi-coloured
mehrfunktionell multifunctional
mehrkernig multi-nuclear
Mehrkomponentenklebstoff multi-component adhesive
mehrkomponentig multi-component/-pack
mehrlagig multi-layer
Mehrphasensystem multi-phase system
mehrphasig multi-phase
Mehrscheibenisoliereinheit sealed unit
Mehrscheibenisolierglas sealed unit
Mehrschichtauftrag multi-film application
Mehrschichtdispersionspartikel multi-shell dispersion particle
Mehrschichtenlackierung multi-coat finish
mehrschichtig multi-layer/-coat
Mehrschichtlackierung 1. multi-coat paint application. 2. multi-coat paint film
Mehrschichtsystem multi-coat paint
Mehrschichtüberzug multi-coat finish
mehrstufig multi-stage
Mehrwegbehälter multi-trip/returnable container
Mehrweggebinde multi-trip/returnable container
Mehrwegverpackung multi-trip/returnable packaging
mehrwertig 1. polyvalent. 2. polyhydric *(if referring to an alcohol)*
Melamin melamine
Melamin-Formaldehydharz melamine/melamine-formaldehyde/MF resin
Melamin-Formaldehydleim melamine-formaldehyde adhesive
Melamin-Methylol-Butylether melamine methylolbutyl ether
Melamin-Methylol-Methylether melamine methylolmethyl ether
Melamincyanurat melamine cyanurate
Melaminhärter melamine hardener
Melaminharz melamine/melamine-formaldehyde/MF resin
melaminharzgetränkt melamine resin impregnated
Melaminharzmenge amount of melamine resin
Melasse molasses
Membran 1. membrane. 2. diaphragm
Membrandicke membrane thickness
Membranfläche membrane surface
Membranmodul membrane unit
Membrantrenntechnik membrane separating technology
Membrantrennverfahren membrane separating process
menügesteuert menu-controlled
Mercaptan mercaptan
Mercaptanendgruppe terminal mercaptan group
Mercaptangruppe mercaptan group
Mercaptanhärter mercaptan hardener
Mercaptid mercaptide
Mercapto-Butylzinnstabilisator mercaptobutyltin stabiliser
Mercapto-Octylzinnstabilisator mercaptooctyltin stabiliser
Mercaptobenzimidazol mercaptobenzimidazole
Mercaptobenzothiazol mercaptobenzothiazole
Mercaptocarbonsäure mercaptocarboxylic acid
Mercaptoether mercaptoether
Mercaptogruppe mercapto group
Mercaptoharz mercapto resin
mercaptomodifiziert mercapto-modified
Mercaptopropylgruppe mercaptopropyl group
Mercaptopropylsilan mercaptopropyl silane
Mercaptosilan mercaptosilane
Mercaptoverbindung mercapto compound
Merkapto-, merkapto- *see* **Mercapto-, mercapto-**
Merkblatt data sheet
Merkblattwerte figures given in technical data sheets
Merkmal feature, characteristic
Mesometrie mesometry
mesomorph mesomorphous
Meß- und Regeltechnik measuring and control technology
Meß- und Steuereinheit measuring and control unit
Meßanlage measuring equipment
Meßapparatur measuring instrument/equipment
Meßaufnehmer sensor, transducer, probe
meßbar measurable
Meßbedingungen test conditions
Meßbereich measuring range
Meßdose sensor, transducer, probe
Meßeinrichtung measuring equipment
Meßelektrode measuring electrode
Meßergebnis test result
Meßfehler measuring error
Meßflüssigkeit test liquid
Meßfrequenz measuring frequency
Meßfühler sensor, transducer, probe
Meßgeometrie experimental/test setup
Meßgerät measuring instrument
Messing brass

Meßkolben measuring flask
Meßkurve curve
Meßlösung test solution
Meßmethode method of determination
Meßmethodik method of determination
Meßparameter test conditions
Meßprinzip measuring principle
Meßpunkt measuring point
Meßreihe test series
Meßresultat test result
Meßsystem measuring system
Meßtemperatur test temperature
Messung determination, measurement
Meßwert measured value
Meßwinkel measuring angle
Meßzelle measuring cell
Metall metal
Metallabfall scrap metal, metal scrap
Metallacetylacetonat metal acetyl acetonate
Metallack metal finish
Metallalkoxid metal alkoxide
Metallalkoxyverbindung metal alkoxy compound
Metallanstrich metal coating
Metallatom metal atom
Metallbau metal construction
metallbedampft metallised
Metallbedampfungsverfahren metallisation
metallbeschichtet metallised
Metallcarboxylat metal carboxylate
Metallchelat metal chelate
Metallchlorid metal chloride
Metalldampf metal vapour, vaporised metal
Metalldithiocarbamat metal dithiocarbamate
Metalldithiophosphat metal dithiophosphate
Metalleffekt metal effect
Metalleffektlack metallic finish
Metalleffektpigment metallic pigment
Metallegierung metal alloy
Metallentfettung metal degreasing
Metallfläche metal surface
Metallfolie metal foil
metallfrei metal-free, free from metals
Metallgrundierung metal primer
Metallhaftung adhesion to metals
Metallhalogenid metal halide
metallhaltig containing metal
Metallhydroxid metal hydroxide
Metallicdecklack metallic finish
Metalliclack metallic paint
Metallicpigment metallic pigment
Metallion metal ion
metallionenfrei free from metal ions
metallisch metallic
Metallisierbad metal plating bath
metallisiert metallised
Metallisierung metallisation
Metallkatalysator metal catalyst
Metallkation metal cation
Metallklebstoff metal adhesive
Metallklebung 1. bonding of metals. 2. bonded metal joint
Metallklebverbindung metal bond, bonded metal joint
Metallkomplex metal complex
Metallkonstruktion metal structure/construction
Metallkontakt contact with metal
Metallmercaptid metal mercaptide
Metallmöbel metal furniture
Metallnaphthenat metal naphthenate
Metalloberfläche metal surface
Metalloctoat metal octoate
metallorganisch organo-metallic
Metalloxid metal oxide
metalloxidbeschichtet metal oxide coated
Metalloxidpulver metal oxide powder
Metalloxidschicht metal oxide film
Metallpentadionat metal pentadionate
Metallpigment metallic pigment
Metallprimer metal primer
Metallpulver metal powder
Metallresinat metal resinate
Metallsalz metal salt
Metallschicht metal film/layer
Metallschrott scrap metal, metal scrap
Metallschuppen metal flakes
Metallseife metal soap
Metallseifenstabilisator metal soap stabiliser
Metallstearat metal stearate
Metallstruktur metal structure
Metallsubstrat metal substrate/surface
Metallsulfat metal sulphate
Metallteil metal component
Metalluntergrund metal substrate
Metallverbindung metal compound
Metallvergiftung metal poisoning *(of catalyst)*
Metallverklebung 1. bonding of metal. 2. metal bond, bonded metal joint
Metallverpackung metal pack
Metaphosphat metaphosphate
metastabil metastable
Methacroylgruppe methacroyl group
Methacrylalkylester alkyl methacrylate
Methacrylat methacrylate
Methacrylatgruppe methacrylate group
Methacrylatharz methacrylic resin
Methacrylatklebstoff methacrylate adhesive
Methacrylatoligomer methacrylate oligomer
Methacrylatpolymer polymethacrylate
Methacrylbindemittel acrylic binder
Methacrylester methacrylate
Methacrylfestharz solid acrylic resin
Methacrylgruppe methacrylic group
Methacrylharz acrylic resin
Methacrylimid methacrylimide
Methacrylnitril methacrylonitrile
Methacrylsäure methacrylic acid
Methacrylsäurealkylester alkyl methacrylate
Methacrylsäureallylester allyl methacrylate
Methacrylsäureester methacrylate
Methacrylsäuremethylester methyl methacrylate
Methanol methanol, methyl alcohol

methanolfrei free from methanol
methanolhaltig containing methanol
methanolisch methanolic
methanolische Lösung methanol solution
methanollöslich methanol soluble
methanolverethert methanol etherified
Methanolyse methanolysis
Methinharz methine resin
Methoxygruppe methoxy group/radical
methoxyliert methoxylated
Methoxymethylmelamin methoxymethyl melamine
Methoxypropanol methoxypropanol
Methoxypropylacetat methoxypropyl acetate
Methylacetat methyl acetate
Methylacetylacetonat methyl acetyl acetonate
Methylacrylat methyl acrylate
Methylalkohol methyl alcohol
Methylalkylsiliconharz methyl alkyl silicone resin
Methylalkylsiloxan methyl alkyl siloxane
Methylbenzylalkohol methyl benzyl alcohol
Methylbutylglykol methyl butyl glycol
Methylcellulose methyl cellulose
Methylchlorid methyl chloride
Methylchloroform methyl chloroform
Methylchlorsilan methyl chlorosilane
Methylcyanacrylat methyl cyanoacrylate
Methyldiethanolamin methyl diethanolamine
Methylenbrücke methylene bridge
Methylenchlorid methylene chloride
Methylengruppe methylene group
Methylester methyl ester
Methylether methyl ether
Methylethergruppe methyl ether group
Methylethylketon methyl ethyl ketone, MEK
Methylethylketonperoxid methyl ethyl ketone peroxide
Methylethylketoxim methyl ethyl ketoxime
Methylglykol methyl glycol
Methylglykolacetat methyl glycol acetate
Methylglykoxygruppe methyl glycoxy group/radical
Methylgruppe methyl group/radical
methyliert methylated
Methylierung methylation
Methylinden methyl indene
Methylisobutylketon methyl isobutyl ketone, MIBK
Methylmethacrylat methyl methacrylate
Methylmethacrylatbasis, auf methyl methacrylate-based
Methylolacrylamid methylolacrylamide
Methylolamid methylolamide
Methylolether nethylol ether
Methylolgruppe methylol group
Methylolharnstoff methylol urea
methyloliert methylolised
Methylolmelamin methylolmelamine
Methylolphenol methylol phenol
Methylolphenolharz methylol phenol resin
Methyloxazolin methyl oxazoline
Methylpentandiol methyl pentane diol
Methylphenylpolysiloxan methylphenyl polysiloxane
Methylphenylsiliconharz methylphenyl silicone resin
Methylpolysiloxan methyl polysiloxane
Methylpyrrolidon methyl pyrrolidone
Methylrest methyl radical/group
Methylrot methyl red
Methylseitengruppe methyl side group
Methylsilanrest methyl silane group
Methylsiliconfestharz solid methyl silicone resin
Methylsiliconharz methyl silicone resin
Methylsiliconöl methyl slicone fluid
Methylstyrol methyl styrene
Methyltetrahydrophthalsäureanhydrid methyl tetrahydrophthalic anhydride
Methyltrichlorsilan methyl trichlorosilane
methylverethert methyl etherified
Methylzinnmercaptid methyl-tin mercaptide
Methylzinnstabilisator methyl-tin stabiliser
Methylzinnverbindung methyl-tin compound
MF-Harz melamine/melamine-formaldehyde/MF resin
MF-Leimharz melamine/melamine-formaldehyde adhesive resin
MF-Tränkharz melamine/melamine-formaldehyde impregnating resin
MFT *abbr. of* **Mindestfilmbildetemperatur**, minimum film forming temperature
MG *abbr. of* **Molekulargewicht**, molecular weight
Mg-Legierung magnesium alloy
MIBK methyl isobutyl ketone
Micell- *see* **Mizell-**
Migration migration
migrationsarm low-migration
migrationsbeständig migration resistant
Migrationsbeständigkeit migration resistance
migrationsecht migration resistant
Migrationsechtheit migration resistance
migrationsfrei non-migrating
Migrationsneigung migration tendency
Migrationsrate migration rate
Migrationsverhalten migration behaviour
Migrieren migration
Mikroagglomerat microagglomerate
mikroanalytisch microanalytical
Mikrobe microbe
Mikrobenbefall microbial attack
Mikrobereich microrange, microregion
mikrobiell microbial
mikrobiologisch microbiological
mikrobiologisch abbaubar microbiodegradable
Mikrobiozid microbiocide
Mikroblase microbubble, microblister
mikrobrownsche Bewegung micro-Brownian movement
Mikrobruch microfracture
Mikrobürette microburette
Mikrodispersion microdispersion
Mikroeindringhärte micropenetration hardness

Mikroelektronik microelectronics
Mikroemulsion microemulsion
mikrofein microfine
Mikrogefüge microstructure
Mikrogel microgel
Mikrogravimetrie microgravimetry
mikrogravimetrisch microgravimetric
Mikrohärte microhardness
Mikrohohlglaskugeln hollow glass microbeads/microspheres
Mikrohohlraum microvoid
Mikrokalorimeter microcalorimeter
Mikrokontakt microcontact
mikrokristallin microcrystalline
Mikrolöten micro-soldering
Mikroluftbläschen microscopic air bubbles
Mikromaßstab microscale
Mikromorphologie micromorphology
mikronisiert micronised
Mikroorganismus microorganism
Mikropartikel microparticle
Mikroperlen microbeads
Mikrophase microphase
Mikropipette micropipette
Mikroporen pinholes, micropores
mikroporös microporous
Mikroporosität microporosity
Mikroquerriß transverse microcrack
Mikrorauheit microroughness
Mikrorauhigkeit microroughness
Mikroriß microcrack
Mikroschaum microfoam
Mikroschnitt microtome section
Mikroskop microscope
Mikroskopie microscopy
mikroskopisch microscopic
Mikrospalte microcrack
Mikrospektroskopie microspectroscopy
Mikrostruktur microstructure
mikrostrukturell microstructural
mikrosuspendiert microsuspended
Mikroteilchen microparticle
Mikrotitaniumdioxid microtitanium dioxide
Mikrotom microtome
Mikrotomschnitt microtome section
Mikrotomschnittaufnahme microtome section photomicrograph
Mikrotropfen microdroplet
Mikroverkapselung microencapsulation
Mikroverklebung microbond
Mikrowaage microbalance
Mikrowelle microwave
Mikrowellenbereich microwave range
mikrowellendurchlässig permeable to microwaves
Mikrowellendurchlässigkeit microwave permeability
Mikrowellenfeld microwave field
Mikrowellenheizung microwave heating
Mikrowellenplasmaanlage microwave plasma plant
milchig milky
milchigtrüb milky
milchigweiß milky-white
Milchsäure lactic acid
mindergiftig slightly toxic, harmful. *This word has now been officially replaced by* **gesundheitsschädlich**, *q.v.*
Mindestanforderungen minimum requirements
Mindestfilmbildetemperatur minimum film forming temperature
Mindesthärtetemperatur minimum curing temperature
Mindestkonzentration minimum concentration
Mindestmenge minimum amount/quantity
Mindestmolmasse minimum molecular weight
Mindestreaktionszeit minimum reaction time
Mindesttemperatur minimum temperature
Mindesttrockenzeit minimum drying time
Mindestverfilmungstemperatur minimum film forming temperature
Mindestwert minimum figure/value
Mineralfarbe silicate paint
Mineralfüllstoff mineral filler
mineralgefüllt mineral filled
mineralisch mineral
Mineralöl petroleum, mineral oil
mineralölbasiert petroleum-based
Mineralsäure mineral acid
mineralverstärkt mineral filled
Mineralwolle rock/mineral wool
minimale Filmbildetemperatur minimum film forming temperature
Minimalwert minimum value
Minimierung minimisation
Minimumgrenzwert minimum value
Minimumlosgröße minimum batch size
Minustemperaturen sub-zero temperatures
Mischaggregat mixer, mixing unit
Mischanlage mixing equipment/unit
mischbar miscible
Mischbarkeit miscibility, compatibility
Mischbedingungen mixing conditions
Mischbehälter mixing tank/vessel
Mischbindemittel binder blend
Mischcopolymer copolymer
Mischcopolymerisat copolymer
Mischeffekt mixing effect/performance
Mischeinrichtung mixing unit/equipment
Mischen mixing
Mischer mixer
Mischerinhalt mixer contents
Mischether mixed ethers
Mischfehler mixing error
Mischflügel rotor blade
Mischgefäß mixing vessel
Mischgerät mixer
Mischgewebe mixed fabric
Mischgrad mixing efficiency, degree of mixing
Mischgut 1. mixer contents. 2. material being mixed. 3. material to be mixed
Mischgüte mixing quality

Mischhärter hardener blend
Mischharz resin blend
Mischkatalysator catalyst blend, mixed catalyst
Mischkessel mixing vessel
Mischkondensat co-condensate
Mischleistung mixing efficiency
Mischmaschine mixer
Mischoxide mixed oxides
Mischperoxid peroxide blend
Mischphase mixed phase
Mischpolyester mixed polyesters
Mischpolyether mixed polyethers
Mischpolymer copolymer
Mischpolymerisat copolymer
Mischpolymerisation copolymerisation
Mischprozeß mixing process/operation
Mischsäure mixed acid
Mischteilchen mixed particle
Mischtemperatur mixing temperature
Mischungsansatz batch *(e.g. of resin-catalyst mix)*
Mischungsgüte mixing efficiency
Mischungsstabilität mix stability
Mischungsverhältnis mixing ratio
Mischungsvorschriften mixing instructions
Mischvorgang mixing process/operation
Mischwalzwerk mixing rolls
Mischwerk mixer
Mischwirkung mixing effect/performance
Mischzeit mixing time
Mitarbeiter, gewerbliche blue collar workers
Mittelbettkleber medium-bed adhesive
mitteldispers medium-particle
mitteleleastisch semi-flexible, moderately flexible
mittelfett medium-oil
mittelflexibel semi-flexible, moderately flexible
mittelflüchtig medium-volatility
mittelkettig medium-chain
mittellang medium-length
mittelmolekular medium-molecular weight
Mittelölalkydharz medium-oil modified alkyd (resin)
Mittelölalkydharzlack medium-oil alkyd paint
mittelölig medium-oil
mittelreaktiv moderately reactive
mittelsiedend medium-boiling
Mittelsieder medium-boiling solvent
mittelständisch medium-size *(business, company etc.)*
mittelviskos medium-viscosity
Mittelwert mean/average value
mittlere Teilchengröße mean particle size
mittlerer Teilchendurchmesser mean particle size
mittleres Molekulargewicht average molecular weight
Mizellbildung micelle formation
Mizelle micelle
Mizellkonzentration, kritische critical micelle concentration, c.m.c.
MMA methyl methacrylate
mμ-TiO_2 micro-titanium dioxide
Möbelausrüstungsteile furniture fittings
Möbelbau furniture making
Möbelbranche furniture industry
Möbelfolie furniture covering film
Möbelindustrie furniture industry
Möbellack furniture varnish/lacquer/paint
Möbellackierung furniture finish
Möbelstück piece of furniture
Modellflüssigkeit model liquid
Modelliermasse modelling compound
Modellrechnung model calculation
Modellsystem model system/formulation
modifizierend modifying
Modifizierharz modifying resin
Modifiziermittel modifying agent
modifiziert modified
Modifizierung modification
Modifizierungsgrad degree of modification
Modifizierungsgruppe modifying group
Modifizierungsmittel modifying agent, modifier
Modifizierungsmöglichkeiten modification possibilities/options
Modifizierungsrest modifying radical/group
Modul modulus
modular modular
Modularität modular character
modulartig modular
Modulationsfrequenz modulation frequency
Modulierung modulation
Modulwert modulus
Mohnsamenöl poppyseed oil
Molarität molarity
Molekül molecule
molekular molecular
Molekularanordnung molecular arrangement/configuration
Molekularaufbau molecular structure
Molekularbewegung molecular movement
molekulardynamisch molecular-dynamic
Molekulargewicht molecular weight
Molekulargewicht, gewichtsmittleres weight-average molecular weight
Molekulargewicht, mittleres average molecular weight
Molekulargewichtsabnahme decrease in molecular weight
Molekulargewichtsabbau drop in molecular weight
Molekulargewichtsänderung change in molecular weight
Molekulargewichtsbestimmung molecular weight determination
Molekulargewichtserhöhung increase in molecular weight
Molekulargewichtsmittelwert average molecular weight
Molekulargewichtsverlust drop/decrease in molecular weight
Molekulargewichtsverteilung molecular weight distribution

Molekulargewichtszahlenmittel number-average molecular weight
Molekulargewichtszunahme increase in molecular weight
Molekularmasse molecular weight
Molekularorientierung molecular orientation
Molekularschicht molecular layer
Molekularsieb molecular sieve
Molekularstruktur molecular structure
Molekülart type/kind of molecule
Molekularzustand molecular state
Molekülaufbau molecular structure
Molekülbaustein molecule unit
Molekülbeweglichkeit molecule mobility
Molekülbewegung molecule movement
Molekülbruchstück molecule segment/fragment
Moleküleigenschaften molecule properties/characteristics
Molekülfragment molecule segment/fragment
Molekülgebilde molecular structure
Molekülgestalt molecular shape
Molekülgröße molecule size
Molekülgruppe molecule group
Molekülgruppierung molecule group
Molekülkette molecule chain
Molekülkettenbeweglichkeit molecule chain mobility
Molekülkettenlänge molecule chain length
Molekülkettenorientierung molecule chain orientation/alignment
Molekülkettensegment molecule chain segment
Molekülkonformation molecular arrangement/configuration
Moleküllage molecule layer
Moleküllänge molecule length
Molekülmasse molecular weight
Molekülorientierung molecular orientation
Molekülschicht molecular layer
Molekülsegment molecule segment
Molekülspaltung molecule fission
Molekülstruktur molecular structure
Molekülvergrößerung molecule enlargement
Molekülvernetzung molecule crosslinkage
Molekülverschiebung molecular displacement
Molekülzusammensetzung molecule constitution
Molenbruch molar fraction
Molenfluß molecular flow
Molgewicht molecular weight
Molgewichtsverteilung molecular weight distribution
Molmasse molecular weight
Molmassenverteilung molecular weight distribution
Molmassenzunahme increase in molecular weight
Molmassenzuwachs increase in molecular weight
Molverhältnis molar ratio
Molwärme molar heat
Molybdän molybdenum
Molybdändisulfid molybdenum disulphide
Molybdänsulfid molybdenum sulphide
Molybdäntrioxid molybdenum trioxide
Molybdat molybdate
Molybdatorange molybdenum orange
Molybdatrot molybdenum red
Monatsproduktion monthly production/output
Monatstonnen tonnes per month
Monoalkohol monohydric alcohol
Monoalkylzinnchlorid monoalkyl tin chloride
Monoalkylzinnstabilisator monoalkyl tin stabiliser
Monoalkylzinnverbindung monoalkyl tin compound
Monoamin monoamine
monoaxial uniaxial
Monocarbonsäure monocarboxylic acid
Monocarbonsäureester monocarboxylate
Monocarbonsäurevinylester vinyl monocarboxylate
Monochlorphenol monochlorophenol
monochromatisch monochromatic
Monofluortrichlormethan monofluorotrichloromethane
monofunktionell monofunctional
Monohydroxyverbindung monohydroxy compound
Monoisocyanat monoisocyanate
monoklin monoclinic
Monolage monolayer
monolithisch monolithic
monomer monomeric
Monomer monomer
Monomeransatz monomer blend
Monomeranteil monomer content
monomerarm low-monomer, with a low monomer content
Monomerbaugruppe monomer unit
Monomerbaustein monomer unit
Monomereinheit monomer unit
Monomerengemisch monomer blend
Monomerenverhältnis monomer ratio
Monomerenzusammensetzung monomer composition
monomeres Vinylchlorid vinyl chloride monomer
monomerfrei monomer-free, free from monomer
Monomergemisch monomer blend
monomerhaltig containing monomer
Monomerkonzentration monomer concentration
monomerlöslich monomer soluble
Monomermischung monomer blend
Monomermolekül monomer molecule
Monomerradikal monomer radical
Monomerreste monomer residues
Monomerweichmacher monomeric plasticiser
Monomethylolharnstoff monomethylol urea
monomolekular monomolecular
monoolefinisch monoolefinic
Monoorganosilan monoorganosilane
Monooxazolin monooxazoline
Monosaccharid monosaccharide

Monosäure monoacid
monosubstituiert monosubstituted
Monosulfid monosulphide
Monstyrol styrene monomer
Montage 1. assembly, mounting, fitting, installation. 2. assembly shop
Montageband assembly line
montagefreundlich easy to fit/instal
Montageklebstoff assembly adhesive
montageleicht easy to fit/instal
Montageleim assembly glue
Montansäure montanic acid
Montansäureester montanate
Montanwachs montan wax
montieren to assemble, to mount, to fit, to instal
Montmorillonit montmorillonite
Moosgummi foam rubber
Morpholin morpholine
Morphologie morphology
morphologisch morphological
Mörtel mortar
Mörtelbelag mortar screed
Mörtelbett layer of mortar
Mörtelfuge mortar joint
Mörtelmischung mortar mix
Motorblock engine block
Motorenöl engine oil
Motorhaube bonnet
Motorraum engine compartment
MSD mass selective detector
Mühle mill
Müll refuse, waste, rubbish, garbage
multiaxial multiaxial
Multifunktionalität multifunctionality
multifunktionell multifunctional
multimolekular multimolecular
multivalent polyvalent
Muscheln barnacles
Muster sample
mutagen mutagenic
Mutagen mutagen
Mutagenität mutagenity
mutieren to mutate

N

Nabe hub
Nachbarmolekül neighbouring/adjacent molecule
Nachbarpolymersegment adjacent polymer segment
nachbehandelt post-treated
Nachbehandlung post-treatment
nachchloriert post-chlorinated
nachdosieren to add more **...so daß ein Nachdosieren des Amins nicht notwendig war.** ...so that no more amine had to be added.
nachgeschaltet downstream
nachhärtbar post-curing
Nachhärtung post-curing
Nachhärtungsbedingungen post-curing conditions
Nachhärtungstemperatur post-curing temperature
Nachkondensation post-condensation
Nachkristallisation post-crystallisation
Nachpolymerisation post-polymerisation
Nachschwindung post-/after-shrinkage
Nachsintern post-sintering
Nachtempern 1. annealing. 2. post-curing
Nachtrocknung post-drying
Nachverbrennung afterburning
Nachverbrennungsanlage afterburning plant
nachvernetzend post-curing
Nachvernetzung post-curing
nachwachsend renewable *(i.e. raw materials)*
Nadel needle, pin
nadelförmig needle-shaped
Nadelholz softwood
Nadelstichbildung pinholing
Nadelstiche pinholes
Nagellack nail varnish
Näherung approximation
Nährlösung nutrient solution
Nährmedium nutrient medium
Nährstoff nutrient medium
Nahrungsmittel foodstuffs
nahtlos seamless
Naphthalin naphthalene
Naphthalinderivat naphthalene derivative
Naphthochinondiazid naphthoquinone diazide
Nasenbildung curtaining, sagging
naß wet
naß-auf-naß wet-on-wet
Naß-auf-Naß-Auftrag wet-on-wet application
naß-in-naß wet-in-wet
Naß-in-Naß-Lackierung wet-in-wet painting
Naß-in-Naß-Verfahren wet-in-wet painting
Naßabriebfestigkeit wet scrub resistance
Naßbruchwiderstand wet tear strength
naßchemisch wet-chemical
naßchemische-Bestimmung wet analysis
Naßdeckkraft wet hiding power
Naßdeckvermögen wet hiding power
Nässe wet
Naßenthaftung detachment of wet paint film
Naßetikettierung wet labelling
Naßfestharz wet strength resin
Naßfestigkeit wet strength
Naßfestmittel wet strength agent
Naßfilm wet film
Naßfilmauflage wet film weight
Naßfilmdicke wet film thickness
Naßfilmstärke wet film thickness
Naßfilmviskosität wet film viscosity

Naßgemisch wet mix
Naßhaftfestigkeit wet adhesion, wet bond strength
Naßhaftung wet (film) adhesion
Naßhaftungsproblematik wet adhesion problems
Naßhaftungsprobleme wet adhesion problems
Naßhelligkeit wet brightness
Naßklebekraft wet adhesive strength
Naßkleben wet bonding
Naßklebstoff wet adhesive
Naßklebzeit wet bonding time
Naßlack liquid paint
Naßmahlung wet grinding
Naßmörtel wet mortar
Naßquerzugfestigkeit wet transverse tensile strength
Naßraum wet-process room
Naßschichtdicke wet film thickness
Naßtack wet tack
Naßverklebung wet bonding
Naßwischfestigkeit wet rub resistance
Natriumbenzoat sodium benzoate
Natriumbisulfit sodium bisulphite
Natriumcarbonat sodium carbonate
Natriumcarboxymethylcellulose sodium carboxymethyl cellulose
Natriumchlorid sodium chloride
Natriumdichromat sodium dichromate
Natriumdodecylbenzolsulfonat sodium dodecylbenzene sulphonate
Natriumdodecylsulfat sodium dodecyl sulphate
Natriumfluorid sodium fluoride
Natriumhydroxid sodium hydroxide
Natriumhypochlorit sodium hypochlorite
Natriumlaurat sodium laurate
Natriumlaurylsulfonat sodium lauryl sulphonate
Natriumnitrit sodum nitrite
Natriumphosphat sodium phosphate
Natriumpolyacrylat sodium polyacrylate
Natriumsalz sodium salt
Natriumsilikat sodium silicate
Natriumsulfit sodium sulphite
Natriumverbindung sodium compound
Natriumwasserglas sodium silicate waterglass
Natronlauge caustic soda solution
natur natural
Naturbewitterung natural/outdoor weathering
Naturgummi natural rubber
Naturharz natural resin
naturharzmodifiziert natural resin modified
Naturkautschuk natural rubber
Naturkautschukvulkanisat natural rubber vulcanisate, vulcanised natural rubber
Naturkleber natural adhesive
Naturlatex natural rubber latex
natürlich natural
natürliche Bewitterung natural weathering
Naturrohstoff natural raw material
Naturstein natural stone
Natursteinplatte natural stone slab
Naturstoff natural substance
Naturwachs natural wax
NC-Lack nitrocellulose lacquer
NCO-Gruppe isocyanate group
Nebel mist
Nebeneffekt side effect
Nebenprodukt by-product
nebenproduktfrei free from by-products
Nebenreaktion side/secondary reaction
nebenvalent secondary valency
Nebenvalenzbindung secondary valency bond
Nebenvalenzkräfte secondary valency forces
Nebenwirkungen side effects
negativ negative
negativ geladen negatively charged
Neigung 1. slope *(of a curve)*. 2. tendency
Neigungswinkel angle of inclination
nematisch nematic
NE-Metall non-ferrous metal
Nennviskosität nominal viscosity
Neopren neoprene
Netzfehler wetting defect
Netzmittel wetting agent
Netzmittelkonzentration wetting agent concentration
Netzmittellösung wetting agent solution
Netzwerk network
Neuanstrich 1. new coat of paint. 2. re-painting
Neubauten new buildings
neuentwickelt new, newly/recently developed
Neuentwicklung recent/new development
neutral neutral, inert
Neutralisation neutralisation
Neutralisationsgrad degree of neutralisation
Neutralisationsmittel neutralising agent
neutralisiert neutralised
Neutralisierung neutralisation
Neutralisierungsmittel neutralising agent
Neutron neutron
Neutronaktivierung neutron activation
Newtonsch Newtonian
Newtonsche Flüssigkeit Newtonian liquid
Newtonsches Fließen Newtonian flow
Newtonsches Fließverhalten Newtonian flow
Newtonsches Verhalten Newtonian behaviour
NH_3-Plasma ammonia plasma
nichtablaufend non-sag
nichtabreagiert unreacted
nichtabsorbierend non-absorbent
Nichtakzeptanz non-acceptance
nichtbelastet unstressed
nichtbeschichtet uncoated
nichtbewittert unweathered
nichtblockend non-blocking
nichtblockiert unblocked
nichtbrennbar non-flammable, incombustible
Nichtbrennbarkeit non-flammability, flame/fire resistance
nichtdeckend transparent
nichtdispergiert undispersed
Nichteignung unsuitability
Nichteisenmetall non-ferrous metal

nichtelastisch non-elastic, plastic
nichtentflammbar non-flammable
Nichtentflammbarkeit non-flammability
nichtenzymatisch non-enzymatic
nichtextrahierbar non-extractable
nichtflüchtig non-volatile
Nichtflüchtigkeit non-volatility
nichtfunktionell non-functional
nichtgilbend non-yellowing
nichthaftend 1. non-stick. 2. non-adhering
nichthydrolisierbar non-hydrolysable
nichtimprägniert unimpregnated
nichtionisch non-ionic
nichtionisierfähig non-ionisable
nichtionisiert non-ionised
nichtionogen non-ionic
nichtisotherm non-isothermal
nichtklebrig non-tacky
nichtkorrosiv non-corrosive
nichtkristallin non-crystalline
nichtkristallisierend non-crystallising
nichtleitend non-conductive
Nichtleiter non-conductor
nichtleitfähig non-conductive
nichtlinear non-linear
Nichtlinearität non-linearity
Nichtlösemittel non-solvent
Nichtlöser non-solvent
nichtmagnetisch non-magnetic
nichtmetallisch non-metallic
nichtmischbar immiscible
Nichtmischbarkeit immiscibility
nichtmodifiziert unmodified
nichtnewtonsch non-Newtonian, pseudoplastic
nichtnewtonsche Flüssigkeit non-Newtonian liquid
nichtnewtonsches Verhalten non-Newtonian/pseudoplastic behaviour/flow
nichtnukleiert non-nucleated
nichtoxidiert unoxidised
nichtpigmentiert unpigmented
nichtplastifiziert unplasticised
nichtpolar non-polar
nichtpolymerisiert unpolymerised
nichtquellbar non-swellable
nichtradikalisch non-radical
nichtreagierend inert, neutral
nichtreagiert unreacted
nichtreaktiv inert
nichtreflektierend non-reflecting
nichtrostender Stahl stainless steel
nichtrutschend non-slip
nichtsaugend non-absorbent
nichtsaugfähig non-absorbent
nichtschäumend non-foaming
nichtschmelzend infusible
nichtschrumpfend non-shrink(ing)
nichtsphärisch non-spherical
nichtstabilisiert unstabilised
nichtthixotrop non-thixotropic
nichttoxisch non-toxic
nichttransparent opaque, non-transparent
nichttrocknend non-drying
nichttropfend non-drip
nichtumgesetzt unreacted
nichtverdampfbar non-volatile
nichtvergilbend non-yellowing
nichtvernetzt uncured, uncrosslinked
nichtversprödend non-embrittling
nichtverträglich incompatible
nichtwassermischbar water immiscible
nichtwäßrig 1. non-aqueous (solution). 2. non-waterborne/water-based *(paint or adhesive)*
nichtweichgemacht unplasticised
nichtzerstörend non-destructive
Nickelborlegierung nickel-boron alloy
Nickelkatalysator nickel catalyst
Nickelpigment nickel pigment
Nickelpulver nickel powder
Nickeltitangelb nickel titanium yellow
Niederdruckdestillation low-pressure distillation
Niederdruckplasma low-pressure plasma
Niederdruckpolymerisation low-pressure polymerisation
Niederdruckreaktor low-pressure reactor
Niederdruckspritzen low-pressure spraying
Niederdruckverfahren low-pressure process
niederenergetisch low-energy
niederfrequent low-frequency
niedergeschlagen precipitated
niedermolekular low-molecular weight
Niederschlag 1. rain(fall). 2. precipitate. 3. condensation
Niederschlagsfeuchtigkeit condensed moisture
niedershorig with a low Shore hardness
niedertourig slow-speed
niederviskos low-viscosity
niedrigaktiv low-reactivity
niedrigdicht low-density
niedrigenergetisch low-energy
niedrigflüchtig low-volatility
Niedrigfrequenzbereich low-frequency range
niedriglegiert low-alloyed
niedrigmolekular low-molecular weight
niedrigpigmentiert with a low pigment content
niedrigpreisig low-price, inexpensive
niedrigreaktiv low-reactivity
niedrigschmelzend low-melting
niedrigsiedend low-boiling
Niedrigsieder low-boiling solvent
Niedrigtemperaturflexibilität low-temperature flexibility
Niedrigtemperaturplasmabehandlung low-temperature plasma treatment
niedrigtourig slow-speed
niedrigviskos low-viscosity
NIR near-infra-red
nitriergehärtet nitrided
Nitrierschicht nitrided layer
Nitrierstahl nitrided steel
Nitrilgehalt nitrile content
Nitrilgruppe nitrile group

Nitrilkautschuk nitrile rubber
Nitrillatex nitrile rubber latex
Nitrilotriessigsäure nitrilotriacetic acid
Nitrilseitengruppe nitrile side group
Nitrit nitrite
Nitrobenzoyloxysilan nitrobenzoyl oxysilane
Nitrocellulose nitrocellulose
Nitrocelluloselack nitrocellulose/NC lacquer
Nitrolack nitrocellulose/NC lacquer
Niveau level
NMR nuclear magnetic resonance
NMR-Spektroskopie NMR spectroscopy, nuclear magnetic resonance spectroscopy
NMR-Spektrum NMR spectrum, nuclear magnetic resonance spectrum
Nockenwelle camshaft
Nominaldurchmesser nominal diameter
Nominallänge nominal length
Nominalradius nominal radius
Nominalwert nominal value/figure
nominell nominal
Nonyloxazolin nonyl oxazoline
Norm standard specification
normalaktiv normally active
normalbrennbar flammable
normalentflammbar flammable
Normalentflammbarkeit normal flammability
normalisiert standard, standardised
Normalklima standard conditioning atmosphere
Normalklimahärtung room temperature curing
Normalspannung normal stress
Normalstahl ordinary steel
Normaltemperatur room temperature
Normaltemperaturaushärtung room temperature curing
normalviskos normal-viscosity
Normbedingungen standard conditions
Normenvorschrift standard specification
normgerecht standard
normiert standardised
Normierung standardisation
Normklima standard conditioning atmosphere
Normprobe standard test piece/specimen
Normprobekörper standard test piece/specimen
Normprüfkörper standard test piece/specimen
Normstab standard test piece
Normung standardisation
Normvorschrift standard specification
Normzugstab standard tensile test piece
Novelle amendment *(to a law or regulation)*
novelliert amended
Novolak novolak resin
Novolakharz novolak resin
NTA nitrilotriacetic acid
Nuancieren tinting
Nuancierpaste tinting paste
nukleierend nucleating
nukleiert nucleated
Nukleierung nucleation
Nukleierungshilfsmittel nucleating agent
Nukleierungsmittel nucleating agent
Nukleierungsvorgang nucleating process
Nukleierungswirkung nucleating effect
nukleophil nucleophilic
Nulleinstellung zero setting/adjustment
Nullprobe reference sample
Nullpunkt zero (point)
Nullpunkteinstellung zero adjustment/setting
Nullpunktkorrektur zero correction
Nullstellung zero setting/position/adjustment
Nullwert zero value
Nut-Federverbindung tongue and groove joint
Nut-Federverbindung, abgeschrägte scarf tongue and groove joint
Nutverbindung, winkelförmige slip recessed joint
Nutzen use, benefit
Nutzen-Aufwandbetrachtung cost-benefit considerations
Nutzenergie usable energy
Nutzfahrzeug commercial vehicle
Nutzinhalt effective capacity
Nutzung 1. utilisation, exploitation. 2. recycling, recovery
Nutzung, thermische energy recovery/recycling
Nutzungsdauer useful/service/working life
Nutzungsgrad productivity
Nutzvolumen effective volume

O_2-Plasma oxygen plasma
oberer Grenzwert upper limit
Oberfläche surface
Oberfläche, orangenschalenartige orange peel effect
Oberfläche, spezifische specific surface area
Oberfläche, wirksame effective surface
Oberflächenabdunstung surface evaporation
Oberflächenabrieb surface abrasion
Oberflächenabtrag surface wear
Oberflächenadditiv wetting agent, surface active agent
Oberflächenadhäsionskräfte surface adhesive forces
oberflächenaktiv surface active
oberflächenaktivierend surface activating
Oberflächenaktivierung surface activation
Oberflächenaktivität surface activity
Oberflächenalterung surface ageing
Oberflächenanalytik surface analysis
oberflächenanalytisch surface-analytical
Oberflächenangriff surface atack
Oberflächenausrüstung 1. surface finish. 2. surface finishing, painting
Oberflächenbeeinträchtigungen surface defects/blemishes

oberflächenbehandelt 1. surface treated. 2. painted. 3. coated
Oberflächenbehandlung 1. coating. 2. surface treatment
Oberflächenbehandlungsanlage surface treatment plant
Oberflächenbehandlungssystem surface coating system/formulation
Oberflächenbelegung surface deposit
Oberflächenbenetzung surface wetting
Oberflächenbenetzungsmittel surface wetting agent
Oberflächenbeschaffenheit surface finish
oberflächenbeschichtet coated
Oberflächenbeschichtung surface coating
Oberflächenbeständigkeit surface durability
Oberflächenbindung surface bonding
Oberflächenbrillanz (surface) sparkle
Oberflächencharakter surface character
Oberflächenchemie surface chemistry
Oberflächendefekt surface mark/blemish/defect
Oberflächendelamination surface delamination
Oberflächendiffusion surface diffusion
Oberflächeneffekt surface effect
Oberflächeneigenschaften surface characteristics/properties
oberflächenenergetisch surface energy
Oberflächenenergie surface energy
Oberflächenenergieanteil surface energy content
Oberflächenentropie surface entropy
Oberflächenerosion surface erosion
Oberflächenfehler surface mark/blemish/defect
Oberflächenfeuchte surface moisture
Oberflächenfilm surface film
Oberflächenglanz 1. surface polish. 2. surface gloss
Oberflächenglätte surface smoothness
Oberflächengriff handle: **mit trockenem Oberflächengriff** with a dry handle, dry to the touch
Oberflächengüte surface finish
Oberflächenhaftung adhesion
oberflächenhart with a hard surface
Oberflächenhärte surface hardness
Oberflächenhaut surface skin
Oberflächenklebekraft surface adhesive force
Oberflächenklebrigkeit surface tack
Oberflächenkontur surface contour
Oberflächenkorrosion surface corrosion
Oberflächenkraft surface energy
Oberflächenkratzer surface scratch
Oberflächenkratzfestigkeit scratch/mar resistance
Oberflächenladung surface charge
Oberflächenleitfähigkeit surface conductivity
Oberflächenmarkierungen surface marks/blemishes/defects
Oberflächenmittel wetting agent, surface active agent
oberflächenmodifiziert surface coated *(e.g. filler particles)*
Oberflächenmodifizierung surface modification
oberflächenmontierbar surface mountable
Oberflächenmorphologie surface morphology
oberflächennah near the surface
Oberflächennähe, in near the surface
Oberflächennarbe surface texture/grain
Oberflächenneubildung surface renewal
Oberflächenoptik surface appearance
oberflächenoptimiert with a high-quality surface finish
oberflächenorientiert surface-orientated
Oberflächenoxidation surface oxidation
Oberflächenphysik surface physics
Oberflächenpolarität surface polarity
Oberflächenqualität surface finish
Oberflächenrauheit surface roughness
Oberflächenrauhigkeit surface roughness
Oberflächenreflexion surface reflection
Oberflächenreibung surface friction
Oberflächenreibwiderstand surface abrasion resistance
Oberflächenreinigungsanlage surface cleaning plant
Oberflächenriß surface crack
Oberflächenschäden, thermische surface charring
Oberflächenschädigung surface damage
Oberflächenschicht surface layer
Oberflächenschlieren surface streaks
Oberflächenschrumpfung surface shrinkage
Oberflächenschutz surface coating/protection
Oberflächenspannung surface tension
Oberflächenspannungsgradient surface tension gradient
Oberflächenstörungen surface defects/blemishes
Oberflächenstruktur surface texture
Oberflächentechnik surface finishing, painting
oberflächentechnisch surface: **oberflächentechnische Untersuchung** surface test
Oberflächentemperatur surface temperature
Oberflächentemperaturfühler surface thermocouple
Oberflächentrocknung surface drying
oberflächenunbehandelt 1. not surface treated. 2. uncoated
Oberflächenunebenheit surface irregularity
Oberflächenunruhe surface irregularities
Oberflächenveränderungen surface changes
Oberflächenveredelung 1. surface treatment. 2. painting
Oberflächenvergrößerung surface enlargement
Oberflächenverletzung surface damage
Oberflächenversprödung surface embrittlement
oberflächenverstärkt surface enhanced
oberflächenverstärkte Ramanstreuung surface enhanced Raman scattering, SERS
Oberflächenvertiefung surface depression
Oberflächenverunreinigungen surface

contaminants/dirt
Oberflächenviskosität surface viscosity
Oberflächenvorbehandlung surface preparation/pre-treatment
Oberflächenvorbehandlungsmethode method of surface preparation
Oberflächenvorbereitung surface preparation/pre-treatment
Oberflächenwiderstand surface resistance
Oberflächenwiderstand, spezifischer surface resistivity
oberflächenzäh with a hard surface
Oberflächenzustand condition of the surface
oberflächlich surface: **oberflächlich behandelt** surface coated **oberflächliche Klebrigkeit** surface tack
Octadecylalkohol octadecyl alcohol
octadecylsubstituiert octadecyl-substituted
Octansäure octanoic acid
octasubstituiert octa-substituted
Octylacrylat octyl acrylate
Octylfettsäureester octyl fatty acid ester
Octylmethacrylat octyl methacrylate
Octylsilylgruppe octyl silyl group
Octylzinncarboxylat octyl tin carboxylate
Octylzinnmercaptid octyl tin mercaptide
Octylzinnstabilisator octyl tin stabiliser
Octylzinnverbindung octyl tin compound
Ofen oven
Ofenhärtung oven curing
Ofentemperatur oven temperature
ofentrocknend oven drying
Ofentrocknung oven drying
offene Wartezeit open assembly time
offene Zeit open assembly time
offener Becher open cup *(in flash point determination)*
offenzellig open-cell *(foam)*
Offsetdruckfarbe offset printing ink
OH-Gehalt hydroxyl content
OH-Gruppe hydroxyl group
OH-gruppenhaltig containing hydroxyl groups
OH-haltig containing hydroxyl groups
OH-Radikal hydroxyl group
OH-substituiert hydroxyl-substituted
OH-Zahl hydroxyl value
Ohmscher Widerstand Ohm's resistance
Ohmsches Gesetz Ohm's law
OHZ *abbr. of* **OH-Zahl**, hydroxyl value
Ökoaudit ecological/environmental audit
Ökobericht eco-report, environmental report
Ökobilanz 1. environmental audit. 2. eco-balance, ecological equilibrium
Ökobilanzierung 1. environmental auditing, preparation of an environemtal audit. 2. life cycle assessment, LCA
Ökokatastrophe ecological/environmental disaster
Ökologie ecology
ökologisch ecological
ökologisch bedenklich ecologically critical/doubtful/risky
ökologisch relevant ecologically/environmentally important
ökologisch sinnvoll ecologically sound/acceptable
ökologisch unbedenklich ecologically sound/safe/acceptable
ökologisches Handeln to act with due consideration for the environment
ökonomisch economic(al)
ökonomisch sinnvoll making economic sense
ökopolitisch ecopolitical
Ökoprofil ecoprofile
Ökosystem ecosystem
ökosystemar ecosystematic
Ökotoxikologie ecotoxicology
ökotoxikologisch ecotoxicological
ökotoxisch ecotoxic
Ökotoxizität ecotoxicity
Ökozeichen environmental symbol
Oktyl- *see* **Octyl-**
Öl oil
Öl-Harzlack oleoresinous paint
Öl-in-Wasser-Emulsion oil-in-water emulsion, O/W emulsion
Ölabsorption oil absorption value
Ölabsorptionszahl oil absorption value
ölabweisend oil repellent
Ölanstrich oil paint film
Ölanteil oil content
Ölaufnahme oil absorption
Ölaufnahmezahl oil absorption value
Ölbasis, auf oil based
ölbeheizt oil heated
ölbeständig oil resistant
Ölbeständigkeit oil resistance
Olefin olefin
Olefinbasis, auf olefin-based
Olefincopolymer olefin copolymer
Olefinharz olefin resin
olefinisch olefinic
Olefinoberfläche polyolefin surface
Olefinpolymer polyolefin
Olefinwachs olefin wax
Ölfarbe oil paint
ölfest oil resistant
ölfrei oil free
Ölgehalt oil content
ölgelöst 1. oil based *(e.g. printing ink)*. 2. dissolved in oil
ölgestreckt oil extended
Ölharz oleoresin
ölig oily
Oligoalkylenglykol oligoalkylene glycol
Oligoalkylenterephthalat oligoalkylene terephthalate
Oligobutylenglykolether oligobutylene glycol ether
Oligobutylenterephthalat oligobutylene terephthalate
Oligoester oligoester

Oligoesteracrylat oligoester acrylate
Oligoether oligoether
Oligoglykol oligoglycol
oligomer oligomeric
Oligomer oligomer
Oligomerisationsgrad degree of oligomerisation
Oligomerkette oligomer chain
Oligomerpolyester oligomeric polyester
Oligopolmarkt oligopolistic market
Oligourethan oligourethane
Ölkrise oil crisis
Ölländer oil producing countries
Öllänge oil length
öllöslich oil soluble
ölmodifiziert oil modified
Ölphase oil phase
Ölpreiskrise oil crisis
Ölpreisschock oil crisis
ölreich long-oil
Ölreste oil residues
Ölrückstände oil residues
Ölsäure oleic acid
Ölsäureester oleate
ölverstreckt oil extended
Ölverteuerung oil price increase
ölverträglich oil compatible
Ölvorräte oil reserves
Ölzahl oil absorption value
Oniumsalz onium salt
Onlinelackierbarkeit on-line paintability
Onlinelackierung on-line painting
Onlinemessung on-line determination/measurement
opak opaque
Opaleszenz opalescence
opaleszierend opalescent
Opazität opacity
Optik 1. optics, optical industry. 2. appearance
optimal optimum, best (possible), ideal, most suitable
Optimalbedingungen ideal/optimum conditions
optimieren to perfect, to upgrade, to improve
Optimierung upgrading, improvement
Optimum, optimum optimum
optisch optical
optische Faser optic(al) fibre
optischer Fehler surface defect
optoelektronisch optoelectronic
orange orange
Orangenhauteffekt orange peel effect
orangenschalenartige Oberfläche orange peel effect
Orangenschaleneffekt orange peel effect
Ordnungszahl atomic number
Ordnungszustand state of order
organisch organic
organisch modifiziert organically modified
Organismus organism
Organoalkoxysilan organoalkoxysilane
Organobleiverbindung organolead compound
Organochloralkoxysilan organochloroalkoxysilane
organofunktionell organofunctional
Organogruppe organic group/radical
organometallisch organometallic
Organometallverbindung organometallic compound
organomodifiziert organomodified
Organooligosiloxan organooligosiloxane
organophil organophilic
organophob organophobic
Organophosphat organophosphate
Organophosphit organophosphite
Organopolysiloxan organopolysiloxane
Organorest organic radical/group
Organosilan organosilane
Organosiliciumverbindung organosilicon compound
Organosiloxan organosiloxane
Organosol organosol
Organotitanat organotitanate
Organozinnchlorid organotin chloride
Organozinnmercaptid organotin mercaptide
Organozinnstabilisator organotin stabiliser
Organozinnverbindung organotin compound
orientieren to align, to orient
Orientierung alignment, orientation
orientierungsfrei non-oriented, free from orientation
Orientierungsgrad degree of orientation
Orientierungshilfe guide
Orientierungspunkt point of reference
Orientierungsspannungen frozen-in stresses
Originalabfüllung original contents
Originalgebinde original container
Originalverpackung original packaging/pack/container
Originalviskosität original/initial viscosity
Originalzustand original condition
Orthodiphenol ortho-diphenol
Orthophosphorsäure ortho-phosphoric acid
Orthophthalsäure ortho-phthalic acid
orthoständig in the ortho position
Orthostellung ortho position
orthosubstituiert ortho-substituted
orthotrop orthotropic
Osmometrie osmometry
Osmose osmosis
osmotisch osmotic
OSW *abbr. of* **Ozonschutzwachs**. anti-ozonant wax
Oszillation oscillation, vibration
Oszillationsversuch oscillation/vibration test
oszillierend oscillating
Oszillograph oscillograph
oval oval
oxalkyliert oxalkylated
Oxalsäure oxalic acid
Oxazolidin oxazolidine
Oxazolidonharz oxazolidone resin
Oxazolin oxazoline
Oxazolinring oxazoline ring

oxethyliert ethoxylated
Oxialkylengruppe oxyalkylene group
Oxialkylgruppe oxyalkyl group
Oxicarbonsäure oxycarboxylic acid
Oxid oxide
Oxidation oxidation
Oxidationsanfälligkeit oxidation sensitivity
Oxidationsbeständigkeit oxidation resistance, resistance to oxidation
oxidationsempfindlich oxidation-sensitive, susceptible to oxidation
oxidationsfähig oxidation-sensitive, susceptible to oxidation
Oxidationsgeschwindigkeit oxidation rate, rate of oxidation
Oxidationsgrad degree of oxidation
Oxidationskatalysator oxidation catalyst
Oxidationsmittel oxidising agent
Oxidationsprodukt oxidation product
Oxidationsschutzmittel antioxidant
Oxidationsstabilisator antioxidant
Oxidationsstabilität oxidation resistance
oxidationsunempfindlich not susceptible to oxidation
oxidativ oxidative
oxidativ trocknend oxidation drying
oxidativ vernetzend oxidation curing
oxidbeschichtet oxide coated
Oxidfilm oxide film/layer
Oxidhaut oxide layer
Oxidhydrat hydroxide
oxidierbar oxidisable
oxidierend oxidising
oxidisch oxidic
Oxidpigment oxide pigment
Oxidschicht oxide film/layer
Oxiethylen oxyethylene
Oxim oxime
oximblockiert oxime blocked
Oximethylen oxymethylene
Oxin oxine
Oxipropylen oxypropylene
Oxoverbindung oxo compound
Oxy-, oxy- *see* **Oxi-, oxi-**
Ozon ozone
Ozonabbau ozone depletion
Ozonalterung ozone ageing
Ozonalterungsgerät ozone ageing apparatus
Ozonalterungsschrank ozone ageing cabinet
Ozonalterungsversuch ozone ageing test
Ozonangriff ozone attack
Ozonbeständigkeit ozone resistance
Ozonbewitterung ozone weathering
Ozoneinfluß effect of ozone
Ozonfestigkeit ozone resistance
Ozongefährdung risk to the ozone layer
ozongeschützt ozone stabilised
ozoninduziert ozone induced
ozonisiert ozonised
Ozonisierung ozonisation
Ozonkonzentration ozone concentration
Ozonloch ozone hole, hole in the ozone layer
Ozonprüfkammer ozone test chamber/cabinet
Ozonriß ozone crack
Ozonrißbildung ozone cracking
ozonschädigend ozone depleting, ozone damaging
Ozonschädigung depletion of the ozone layer, damage to the ozone layer, ozone depletion
Ozonschicht ozone layer
ozonschichtschädigend ozone depleting
ozonschichtunschädlich harmless for the ozone layer
Ozonschutz protection against ozone
Ozonschutzmittel antiozonant
Ozonschutzsystem antiozonant
Ozonschutzwachs antiozonant wax
Ozonschutzwirkung antiozonant effect
Ozonwiderstandsfähigkeit ozone resistance

P

p-Toluolsulfonsäure p-toluene sulphonic acid
PA-Intensität photoacoustic intensity
Packmittel (packaging) container, pack
Packmittelabfall packaging waste
Packstoff packaging material
Packungsdichte packing density
Packungsfaktor packing factor
PAI-Verfahren pre-application irradiation
Palette 1. range. 2. pallet
Palmitat palmitate
Palmitinsäure palmitic acid
Papier paper
Papierbahn paper web
Papierbeschichtung paper coating
Papierchromatographie paper chromatography
papierchromatographisch paper chromatographic
papiererzeugende Industrie paper producing industry
Papierfabrik paper mill
Papierfasern paper fibres
Papierimprägniermittel paper impregnating agent
Papierkleber paper adhesive
Papierklebstoff paper adhesive
Papierlack paper varnish
Papierlaminat paper based laminate
Papiermasse paper pulp
Papiersack paper sack
Papierschichtstoff paper based laminate
Papierstreichmasse paper coating compound
Papiertuch paper towel
Papierveredlung paper coating/finishing
Pappdose cardboard container

Pappe cardboard
Pappkarton cardboard box
Papptrommel cardboard drum
parabelähnlich parabolic
parabelförmig parabolic
parabolisch parabolic
Paradichlorbenzol p-dichlorobenzene
Paradiphenol para-diphenol
Paraffin paraffin
paraffinisch paraffinic
Paraffinkette paraffin chain
Paraffinkohlenwasserstoff aliphatic hydrocarbon
Paraffinwachs paraffin wax
Paraformaldehyd paraformaldehyde
Parallelerscheinungen side effects
parallelgeschaltet connected in parallel
parallelorientiert aligned/arranged in parallel
Paramethylstyrol p-methyl styrene
Paraphenylendiamin p-phenylene diamine
paraständig in the para position
Parastellung para position
parasubstituiert para-substituted
Paratoluolsulfonsäure p-toluene sulphonic acid
Paraxylol p-xylene
Parkett parquet (floor)
Parkettklebstoff parquet flooring adhesive
Parkettlack parquet floor varnish
Parkettversiegelung parquet floor sealant
Partialdruck partial pressure
Partialdruckdifferenz partial pressure difference, difference in partial pressure
Partialgasdruck partial gas pressure
partialgeladen partially charged
Partiekontrolle batch control
partiell partly, partially
partiellmethyliert partly methylated
partiellverethert partly etherified
Partieprüfung batch control
Partikel particle
Partikelagglomeration particle agglomeration
Partikelform particle shape
Partikelgrenzfläche particle interface
Partikelgrenzwert maximum particle size
Partikelgröße particle size
Partikelgrößenverteilung particle size distribution
Partikelladungsdetektor particle charge detector
Partikeloberfläche particle surface
Partikelverschweißung fusion of the particles
PAS 1. photoacoustic spectroscopy. 2. *abbr. of* **Polyacrylsäure**, polyacrylic acid
PAS-Haftschicht polyacrylic acid primer coat
PAS-Lösung polyacrylic acid solution
PASI-Verfahren preassembly irradiation
passiv passive
Passivation passivation
passivierend passivating
Passivierung passivation
Passivität passivity
Paste paste
Pastellnuance pastel colour/shade
Pastellton pastel shade
pastös paste-like
Patentanmeldung patent application
Patentanspruch patent claim
Patentantrag patent application
Patentanwalt patent agent
Patentblatt patent journal
Patenterstellung granting of a patent
Patenterteilung granting of a patent
patentfähig patentable
patentiert patented
Patentinhaber patentee
Patentliteratur patent literature
Patentlösung panacea
Patentrechte patent rights
patentrechtlich geschützt patented
Patentschrift patent specification
Patrone cartridge
Patronenheizkörper cartridge heater
Pb-frei lead free
Pb-Gehalt lead content
PCD particle charge detector
PCS photon correlation spectroscopy
PCS-Untersuchung photon correlation spectroscopy
PDMS *abbr. of* **Polydimethylsiloxan**, polydimethyl siloxane
PEA *abbr. of* **Polyesteracrylat**, polyester acrylate
PEM photo-elastic modulator
Pendel pendulum
Pendeldämpfung pendulum damping
Pendelhammer pendulum
Pendelhammergerät pendulum impact tester
Pendelhärte pendulum hardness
Pendelhärtegerät pendulum hardness instrument
Pendelschlagversuch pendulum impact test
Pendelschlagwerk pendulum impact tester
penetrant penetrating
Penetration penetration
Penetrationsvermögen penetrating power
penetrierfähig capable of good penetration *(e.g. primers, impregnating agents etc.)*
Penetrierfähigkeit ability to penetrate *(e.g. impregnating agent into substrate)*
Penetrierung penetration
penetrometrisch penetrometric
Pentachlorphenol pentachlorophenol
Pentaerythrit pentaerythritol
Pentaerythritester pentaerythritol ester
Pentan pentane
Pentanol pentanol
Pentenharz pentene resin
Per perchloroethylene, tetrachloroethylene
Peradipinsäure peradipic acid
Perameisensäure performic acid
Perbenzoesäure perbenzoic acid
Perbernsteinsäure persuccinic acid
Perbuttersäure perbutyric acid
Perchlorethylen perchloroethylene, tetrachloroethylene
perchloriert perchlorinated
Perchlorsäure perchloric acid

Perdampf tetrachloroethylene vapour/fumes
Peressigsäure peracetic acid
Perester per-ester
Perfluoralkohol perfluoroalcohol
perfluoralkylmodifiziert perfluoroalkyl modified
Perfluorcarbonverbindung perfluorocarbon compound
Perfluorharz perfluororesin
perfluoriert perfluorinated
Perfluorogruppe perfluoro group
Perfluorpolyether perfluoropolyether
Perfluorvinylether perfluorovinyl ether
Perhydrosiloxanharz perhydrosiloxane resin
Perhydroxyradikal perhydroxy radical/group
Periodensystem periodic system
periodisch periodic
Peripheriegerät peripheral instrument/equipment
Perkolations-PVK percolation p.v.c.
Perkolationsschwelle percolation threshold
Perkolationstheorie percolation theory
Perleffekt pearlescent effect, pearlescence
Perlen beads
Perlglanzeffekt pearlescent effect
Perlglanzlackierung pearlescent finish
Perlglanzpigment pearlescent pigment
perlglanzpigmentiert pearlescent
Perlglanzplättchen pearlescent pigment flakes
Perlmühle pearl mill
Perlmutteffekt pearlescent effect
Perlpolymerisation bead/suspension polymerisation
Permanenz permanence
Permanganat permanganate
Permeabilität permeability
Permeabilitätskoeffizient permeability coefficient
Permeabilitätsverhalten permeability behaviour
Permeabilitätswert permeability coefficient
Permeation permeation, diffusion
Permeationsgeschwindigkeit permeation rate
Permeationskoeffizient permeation/diffusion coefficient
Permeationsrate permeation rate
Permeationsverhalten permeation behaviour
Permeationswert permeability coefficient
permeiert permeated
Permittivitätszahl permittivity
Peroxid peroxide
Peroxidbedarf peroxide requirement
Peroxiddosierung amount of peroxide (added)
peroxidfrei peroxide-free
Peroxidgruppe peroxide group
peroxidhärtend peroxide curing
peroxidisch peroxide
peroxidisch aushärtbar peroxide curing
peroxidisch vernetzend peroxide curing
peroxidische Vernetzung peroxide cure/crosslinkage
Peroxidkonzentration peroxide concentration
Peroxidmenge amount of peroxide
Peroxidmolekül peroxide molecule
Peroxidradikal peroxide radical
Peroxidsuspension peroxide suspension
peroxidvernetzbar peroxide curing/crosslinking
Peroxidvernetzer peroxide curing/crosslinking agent
peroxidvernetzt peroxide cured/crosslinked
Peroxidvernetzung peroxide cure/crosslinkage
Peroxidzerfall peroxide decomposition
Peroxidzersetzung peroxide decomposition
Peroxydicarbonat peroxydicarbonate
Peroxyradikal peroxide radical
Peroxyverbindung peroxy compound
Persäure per-acid
Personal staff, employees
Personalaufwand 1. personnel expenditure, wages and salaries. 2. number of personnel
personalaufwendig labour intensive
personalintensiv labour intensive
Personalkosten labour costs, wages and salaries
Personalmangel staff shortage, shortage of staff
Personalschwierigkeiten staff problems
Personenautomobil (motor) car
Personenkraftwagen (motor) car
Personenwagen (motor) car
Perspektiven possibilities, prospects
Persulfat persulphate
Perverbindung per compound, peroxide
PES-Pulverlack polyester powder coating
Pestizid pesticide
PET polyethylene terephthalate
Petrischale Petri dish
Petrochemie petrochemistry
petrochemisch petrochemical
Petrolether petroleum ether
Petroleumfraktion petroleum fraction
Petroleumharz hydrocarbon resin
PF-Harz phenolic/phenol-formaldehyde resin
Pflanzenöl vegetable oil
pflanzenölbasiert vegetable oil based
Pflanzenölfarbe vegetable oil based paint
pflanzlich vegetable
pflegeleicht easy-care
Pflegeleichtausrüstung easy-care finish
Pflichtenheft specification
pfropfen to graft
Pfropfcopolymer graft (co)polymer
Pfropfcopolymerisat graft (co)polymer
Pfropfharz grafting resin
Pfropfmischpolymer graft (co)polymer
Pfropfmischpolymerisat graft (co)polymer
Pfropfpolymer graft (co)polymer
Pfropfpolymerisat graft (co)polymer
Pfropfpolymerisation graft (co)polymerisation
pfropfpolymerisiert graft (co)polymerised
Pfropfreaktion grafting reaction
Pfropfung grafting
PGC *abbr. of* **Pyrolyse-Gas-Chromatographie** pyrolysis gas chromatography
pH-Abfall drop in pH
pH-Abhängigkeit pH dependence
pH-Absenkung reduction in pH
pH-Änderung change in pH

pH-Anhebung pH increase, increase in pH
pH-Bereich pH range
pH-Bestimmung pH determination
pH-Einfluß effect of pH
pH-Einstellung pH setting/adjustment
pH-Gerät pH meter
pH-geregelt pH controlled
pH-Gradient pH difference
pH-Meßgerät pH meter
pH-Puffersystem pH buffer system
pH-Regler pH regulator
pH-Regelung pH control
pH-Schwankungen variations in pH
pH-Sensor pH sensor
pH-Verschiebung shift in pH
pH-Wert pH (value)
pH-Wert- *see* pH-
phänomenologisch phenomenological
Phase phase
Phase, dispergierte disperse phase
Phase, disperse disperse phase
Phasenauftrennung phase separation
Phasendiagramm phase diagram
Phasengleichgewicht phase equilibrium
Phasengrenze phase boundary
Phasengrenzfläche phase interface
Phaseninversion phase inversion
Phasenmorphologie phase morphology
Phasenstabilität phase stability
Phasenstruktur phase structure
Phasentransfer phase transfer
Phasentrennung phase separation
Phasenübergang phase transition
Phasenumkehr phase inversion
Phasenumschlag phase reversal
Phasenumwandlung phase reversal
Phasenversatz phase displacement
Phasenverschiebung phase displacement
Phasenverteilung phase distribution
phasenverträglich phase compatible
Phasenwinkel phase angle
Phenanthren phenanthrene
Phenol phenol
Phenol-Formaldehydharz phenolic/phenol-formaldehyde resin
Phenol-Formaldehydharzklebstoff phenolic/phenol-formaldehyde adhesive
Phenol-Formaldehydkondensat phenol-formaldehyde condensate
Phenol-Formaldehydleim phenolic/phenol-formaldehyde adhesive
Phenol-Furfuralharz phenol-furfural resin
Phenoladditiv phenolic additive
Phenoladdukt phenol adduct
phenolarm low-phenol
Phenolcarbonsäure phenol carboxylic acid
Phenolderivat phenol derivative
Phenolester phenol ester
Phenolethergruppierung phenol-ether group
Phenoletherrest phenol-ether group
Phenolflüssigharz liquid phenol resin
Phenolgeruch phenolic smell, smell of phenol
Phenolgruppe phenol group
Phenolharz phenolic/phenol-formaldehyde resin
Phenolharzbasis, auf phenolic based
phenolharzgetränkt phenolic resin impregnated
Phenolharzhartpapier phenolic paper laminate
Phenolharzkleber phenolic adhesive
Phenolharzklebstoff phenolic adhesive
Phenolharzlack phenolic varnish/lacquer
Phenolharzlaminat phenolic laminate
Phenolharzpapierlaminat phenolic paper laminate
Phenolharzschichtstoff phenolic laminate
phenolisch phenolic
phenolmodifiziert phenol modified
Phenolnovolak phenol novolak
Phenoloxidase phenol oxidase
Phenolphthalein phenolphthalein
Phenolresol phenol resol
Phenolrest phenolic group
Phenoltriazin phenol triazine
Phenoplaste phenolics
Phenoplastharz phenolic resin
Phenoxyether phenoxy ether
Phenoxyethylacrylat phenoxyethyl acrylate
Phenoxyradikal phenoxy radical/group
Phenoxyrest phenoxy radical/group
Phenylendiamin phenylene diamine
Phenylethylrest phenyl ethyl radical/group
Phenylglycidylether phenylglycidyl ether
Phenylgruppe phenyl group/radical
Phenylharnstoff phenyl urea
Phenylisocyanat phenyl isocyanate
Phenylisopropylrest phenyl isopropyl radical/group
Phenylmethylsilicon phenylmethyl silicone
Phenylmethylsiliconharz phenylmethyl silicone resin
Phenylmethylsiliconöl phenylmethyl silicone fluid
Phenylnaphthylamin phenyl naphthylamine
Phenyloxazolin phenyl oxazoline
Phenylpropylsiloxan phenylpropyl siloxane
Phenylrest phenyl radial/group
Phenylring phenyl ring
Phenylseitengruppe phenyl side group
Phenylsiliconharz phenyl silicone resin
Phosgen phosgene
Phosgenierung phosgenisation
Phosphat phosphate
Phosphatierbad phosphating bath
phosphatiert phosphated
Phosphatierung phosphating
Phosphation phosphate ion
Phosphatpigment phosphate pigment
Phosphatschicht phosphate film
Phosphatweichmacher phosphate plasticiser
Phosphinsäure phosphinic acid
Phosphit phosphite
Phosphonsäure phosphonic acid
Phosphor phosphorus

Phosphor-Halogenverbindung phosphorus-halogen compound
Phosphoreszenz phosphorescence
phosphoreszierend phosphorescent, fluorescent
phosphorhaltig containing phosphorus
Phosphorigsäureester phosphite
phosphororganisch organophosphorus
Phosphorpentoxid phosphorus pentoxide
Phosphorsäure phosphoric acid
Phosphorsäureester phosphate
Phosphorverbindung phosphorus compound
Photoabbau photodegradation
photoabbaubar photodegradable
photoaktiv photoreactive
Photoaktivität photoreactivity
photoakustisch photoacoustic
Photoätzen photoetching, photoengraving, process engraving
photochemisch photochemical
photochrom photochromic, photosensitive
Photodegradation photodegradation
Photodiode photodiode
Photoelektron photoelectron
photoempfindlich light sensitive, photosensitive
Photoempfindlichkeit photosensitivity
photogehärtet light cured
photogenarbt photoetched, photoengraved
photographisch photographic
photohärtbar light curing/curable
photoinduziert light induced, photoinduced
Photoinduzierung photoinduction
Photoinitiator photoinitiator
photoinitiiert light initiated, photoinitiated
Photoinitiierung photoinitiation
Photoionisation photoionisation
Photoionisationsdetektor photoionisation detector
photokatalytisch photocatalytic
Photolack resist, photoresist (material)
photolytisch photolytic
photometrisch photometric
Photon photon
Photonenenergie photon energy
Photonenkorrelationsspektroskopie photon correlation spectroscopy
Photooxidation photooxidation
Photopolymer photoreactive polymer
Photopolymerisation photopolymerisation
photopolymerisierbar photopolymerisable
photopolymerisiert photopolymerised
Photoreaktion photoreaction
photoreaktiv photoreactive
Photoresistenzentwickler photoresist developer
Photoresistmaterial photoresist material
Photosensibilisator photosensitiser
Photospaltung photodegradation
Photostabilität light stability
photothermisch photothermal
Photovervielfältiger photocopier
Photozelle photoelectric cell
Phthalat phthalate
Phthalatester phthalate
Phthalatharz phthalate resin
Phthalatweichmacher phthalate plasticiser
Phthalocyanin phthalocyanine
Phthalocyaninblau phthalocyanine blue
Phthalocyaninfarbstoff phthalocyanine pigment
Phthalocyaningermaniumdihydroxid phthalocyanine germanium dihydroxide
Phthalocyaningerüst phthalocyanine skeleton
Phthalocyaningrün phthalocyanine green
Phthalocyaninoctacarbonsäure phthalocyanine octacarboxylic acid
Phthalocyaninpigment phthalocyanine pigment
Phthalocyaninring phthalocyanine ring
Phthalopigment phthalo pigment
Phthalsäure phthalic acid
Phthalsäureanhydrid phthalic anhydride
Phthalsäurediallylester diallyl phthalate
Phthalsäuredibutylester dibutyl phthalate, DBP
Phthalsäurediethylhexylester diethylhexyl/dioctyl phthalate, DOP
Phthalsäurediglycidylester diglycidyl phthalate
Phthalsäureester phthalate
physikalisch physical
physikalisch abbindend physically setting
physikalisch trocknend physically drying
physikalische Eigenschaften physical properties
physikalische Trocknung physical drying
physikalisches Trocknen physical drying
physikochemisch physicochemical
physiologisch physiological
physiologisch einwandfrei non-toxic
physiologisch indifferent non-toxic
physiologisch inert non-toxic, physiologically inert
physiologisch nicht einwandfrei toxic
physiologisch unbedenklich non-toxic
physiologische Unbedenklichkeit non-toxicity
physiologisches Verhalten toxicological properties
Physisorption physisorption
PID *abbr. of* **Photoionisationsdetektor**, photoionisation detector
PIDS polarisation intensity differential scattering
Piezoeffekt piezoelectric effect
piezoelektrisch piezoelectrical
Piezoelektrizität piezoelectricity
piezoresistiv piezoresistive
Pigment pigment
Pigment-Kunststoffkonzentrat pigment masterbatch
Pigment-Volumenkonzentration pigment volume concentration, p.v.c.
Pigment-Volumenkonzentration, kritische critical pigment volume concentration, c.p.v.c.
Pigmentabbau pigment degradation
Pigmentagglomerat pigment agglomerate
Pigmentanreibung 1. pigment paste. 2. pigment grinding
Pigmentanreicherung pigment enrichment
Pigmentansatz pigment mix

Pigmentanteil pigment content
pigmentär pigmentary
Pigmentaufnahme pigment wetting properties
Pigmentaufnahmevermögen pigment wetting properties
Pigmentaufschlämmung pigment slurry
Pigmentausrichtung pigment (particle) alignment
Pigmentausschwimmen, horizontales floating
Pigmentausschwimmen, vertikales flooding
pigmentbenetzend pigment wetting
Pigmentbenetzung pigment wetting
Pigmentbenetzungseigenschaften pigment wetting properties
Pigmentbindevermögen pigment binding power
Pigmentcharge pigment batch
Pigmentdichte pigment density
Pigmentdispergiermittel pigment dispersing agent
Pigmentdispergierung pigment dispersion
Pigmentdispersion pigment dispersion
Pigmentdispersionsadditiv pigment dispersing agent
Pigmentdosierung 1. addition of pigment. 2. amount of pigment (used or added)
Pigmentfarbstoff pigment
Pigmentfeinheit pigment fineness
pigmentfrei unpigmented
Pigmentgebiet pigment sector
Pigmentgehalt pigment content
Pigmentgranulat pigment granules
Pigmentgröße pigment particle size
Pigmenthärte pigment hardness
Pigmentierbarkeit pigmentability
pigmentiert pigmented
Pigmentierung 1. pigmentation. 2. pigment content
Pigmentierungsart type of pigment
Pigmentierungsgrad pigment content
Pigmentierungshöhe pigment content, amount of pigment
Pigmentklasse pigment group
Pigmentkonzentrat pigment concentrate
Pigmentkonzentration pigment concentration
Pigmentkonzentratpaste pigment masterbatch
Pigmentkorn pigment particle
Pigmentkörnchen pigment particle
Pigmentmassenkonzentration pigment mass concentration, p.m.c.
Pigmentmenge pigment content, amount of pigment
Pigmentmischung pigment blend
Pigmentnester (local) pigment accumulations
Pigmentnetzeigenschaften pigment wetting properties
Pigmentnetzer pigment wetting agent
Pigmentoberfläche pigment (particle) surface
Pigmentoberflächenkonzentration pigment surface concentration
Pigmentopazität pigment opacity
Pigmentpackungsfaktor pigment packing factor
Pigmentpartikel pigment particle
Pigmentpartikeldurchmesserverteilung pigment particle size distribution
Pigmentpartikelgrößenverteilung pigment particle size distribution
Pigmentpartikelverteilung pigment particle size distribution
Pigmentpaste pigment paste
Pigmentpastenharz pigment paste resin
Pigmentplättchen pigment platelets
Pigmentpräparation pigment paste
Pigmentpulver powdered pigment
Pigmentruß carbon black
Pigmentschuppen pigment flakes
Pigmentsedimentation pigment sedimentation, settling out of pigment
Pigmentstabilisator pigment stabiliser
Pigmentstabilisierung pigment stabilisation
Pigmentstrich pigmented coating
Pigmentstruktur pigment structure
Pigmentsuspension pigment suspension
pigmenttechnisch *adjective relating to pigments:* **pigmenttechnische Eigenschaften** pigment properties
Pigmentteilchen pigment particle
Pigmenttragevermögen pigment wetting power/properties
Pigmentverbrauch pigment consumption
Pigmentverhalten pigment behaviour
Pigmentvermahlen pigment grinding
Pigmentverteiler pigment dispersing agent
Pigmentverteilung pigment dispersion
Pigmentverträglichkeit pigment compatibility
Pigmentvolumenkonzentration pigment volume concentration, p.v.c.
Pigmentvolumenkonzentration, kritische critical pigment volume concentration, c.p.v.c.
Pigmentzerstörung destruction of pigment particles
Pilotanlage pilot plant
Pilotproduktionsanlage pilot production plant
Pilotprojekt pilot project
Pilotserie pilot run
Pilotstadium pilot stage
Pilzbefall attack by mould
pilzbeständig resistant to mould
Pilze mould
pilzhemmend fungicidal
Pilzkultur mould culture
Pilzschutzmittel fungicide
Pinholebildung pinholing
Pinsel brush
Pinselapplikation brush application
Pinseln brushing, brush application
Pinselreiniger brush cleaner
Pinselstruktur brush marks
Piperazin piperazine
Piperidin piperidine
Pistole spraygun
Plädoyer plea
plan plane
Planetenmischer planetary mixer

planparallel plane parallel
Planungsstadium planning stage
Plasma plasma
plasmabehandelt plasma treated
Plasmabehandlung plasma treatment
plasmabeschichtet plasma coated
Plasmaentladung plasma discharge
plasmageschnitten cut with a plasma arc
Plasmakammer plasma chamber
plasmamodifiziert plasma treated
Plasmamodifizierung plasma treatment
Plasmamodul plasma treatment unit
Plasmapolymerisation plasma polymerisation
plasmapolymerisiert plasma polymerised
Plasmaprozessor plasma processor
Plasmaschicht plasma film/coating
Plasmaspritzen plasma spraying
Plasmateilchen plasma particle
Plasmaveredelungsverfahren plasma treatment
Plasmaverfahren plasma treatment process
Plasmavorbehandlung plasma pretreatment
Plastbeton polymer concrete
Plastifizierer plasticiser
Plastifizierharz plasticising resin
plastifiziert plasticised
Plastifizierung plasticisation
Plastifizierung, interne internal plasticisation
Plastifizierungsgrad degree of plasticisation
Plastifizierungskomponente 1. plasticiser. 2. plasticising resin
Plastifizierungsmittel plasticiser
Plastifizierungsvermögen plasticising capacity
plastisch plastic, soft
plastische Viskosität plastic viscosity
Plastisol plastisol, PVC paste
Plastisolmasse plastisol, PVC paste
Plastizität plasticity
Platinkatalysator platinum catalyst
platinkatalysiert platinum catalysed
Plättchen platelet
plättchenförmig platelet-like
Platte-Kegelviskosimeter plate-cone viscometer
Platte-Platte-Rheometer plate-plate rheometer
platzaufwendig requiring a lot of space
Platzbedarf (amount of) space required
platzintensiv requiring a lot of space
platzsparend requiring little space, space saving
PMMA-Platte acrylic sheet
PMS *abbr. of* **Polymethacrylsäure**, polymethacrylic acid
pneumatisch pneumatic
POK *abbr. of* **Pigmentoberflächenkonzentration**, pigment surface concentration
polar polar
Polarisation polarisation
Polarisationskräfte polarisation forces
Polarisationskurve polarisation curve
Polarisationsmikroskop polarising microscope
Polarisationswiderstand polarisation resistance
Polarisator polariser
polarisierbar polarisable
Polarisierbarkeit polarisability
polarisiert polarised
Polarisierung polarisation
Polarität polarity
Polarographie polarography
polarographisch polarographic
polieren to polish
polierfähig capable of being polished
Polierscheibe polishing wheel
poliert polished
Politur polish
Polstermaterial upholstery material
Polstermöbel upholstered furniture
Polyacetal polyacetal
Polyacetylen polyacetylene
Polyacrylamid polyacrylamide
Polyacrylat polyacrylate
Polyacrylatbasis, auf polyacrylate-based
Polyacrylatdichtstoff acrylic sealant
Polyacrylatdispersion polyacrylate/acrylic dispersion
Polyacrylatfarbe acrylic paint
Polyacrylatharz acrylic resin
Polyacrylester polyacrylate
Polyacrylnitril polyacrylonitrile
Polyacrylpolyol polyacrylic polyol
Polyacrylsäure polyacrylic acid
Polyacrylsäureester polyacrylate
Polyacrylverdicker polyacrylate thickener
Polyaddition polyaddition
Polyadditionsharz polyaddition resin
Polyadditionsklebstoff polyaddition-curing adhesive
Polyadditionsprodukt polyaddition product
Polyadditionsreaktion polyaddition reaction
Polyaddukt polyadduct
Polyadipat polyadipate
Polyadipinsäureester polyadipate
Polyalkohol polyalcohol
Polyalkoxysilan polyalkoxysilane
Polyalkylenglykol polyalkylene glycol
Polyalkylenoxidgruppe polyalkylene oxide group
Polyalkylenterephthalat polyalkylene terephthalate
Polyalkylmethacrylat polyalkyl methacrylate
Polyalkyloxygruppe polyalkyloxy group
Polyallylmethacrylat polyallyl methacrylate
Polyamid polyamide, nylon
Polyamidcarbonsäure polyamide carboxylic acid
Polyamidharz polyamide resin
Polyamidimid polyamide imide
Polyamidkette polyamide chain
Polyamidoamin polyamidoamine
Polyamidsäure polyamide acid
Polyamidwachs polyamide wax
Polyamin polyamine
Polyaminhärter polyamine hardener/catalyst
Polyaminoamid polyaminoamide
Polyaminoamidhärter polyaminoamide hardener/catalyst
Polyanhydrid polyanhydride

Polyanhydridharz polyanhydride resin
Polyanilin polyaniline
Polyanionenharz polyanionic resin
Polyaramid polyaramide
Polyarylamid polyarylamide
Polyarylat polyarylate
Polyarylether polyaryl ether
Polyaryletherketon polyaryl ether ketone
Polyarylsulfon polyaryl sulphone
Polyäth- *see* **Polyeth-**
Polyaziridin polyaziridine
Polyazomethin polyazomethine
polybasisch polybasic
Polybenzimidazol polybenzimide azol
Polybeton polymer concrete
Polyborsilikonharz polyborosilicone resin
polybromiert polybrominated
Polybutadien polybutadiene
Polybutadienkautschuk polybutadiene rubber
Polybuten polybutene
Polybutylenterephthalat polybutylne terephthalate, PBTP
Polybutylmethacrylat polybutyl methacrylate
Polybutylthiophen polybutyl thiophen
Polybutyltitanat polybutyl titanate
Polycaprolactam polycaprolactam
Polycaprolacton polycaprolactone
Polycarbonat polycarbonate
Polycarbonsäure polycarboxylic acid
Polycarbonsäureanhydrid polycarboxylic anhydride
Polycarbonsäureanhydridchlorid polycarboxylic anhydride chloride
polycarboxyliert polycarboxylated
Polychlorbutadien polychlorobutadiene, polychloroprene
Polychlorbutadienkautschuk polychlorobutadiene/polychloroprene rubber
polychloriert polychlorinated
Polychloropren polychloroprene
Polychloroprenkautschuk polychloroprene rubber
Polychlortrifluorethylen polychlorotrifluoroethylene
polycyklisch polycyclic
Polydialkylsiloxankette polydialkyl siloxane chain
Polydiallylphthalat polydiallyl phthalate
Polydicyclopentadien polydicyclopentadiene
Polydien polydiene
Polydimethylsiloxan polydimethyl siloxane
Polydiorganosiloxan polydiorganosiloxane
polydispers polydisperse
Polyederschaum macrofoam
Polyelektrolyt polyelectrolyte
Polyen polyene
Polyepoxid polyepoxide
Polyepoxidverbindung polyepoxy compound
Polyester polyester
Polyesteracrylat polyester acrylate
Polyesteracrylatemulsion polyester acrylate emulsion
Polyesteracrylatlack polyester acrylate paint
Polyesteracrylatlackierung polyester acrylate finish
Polyesteramid polyester amide
Polyesterbaustein polyester unit
Polyesterbeschichtung polyester coating
Polyesterbeschichtungspulver polyester powder coating
Polyesterbeton polyester concrete
Polyesterendgruppe terminal polyester group
Polyesterfaser polyester fibre
Polyesterglykol polyester glycol
Polyesterharz polyester resin
Polyesterharzlack polyester varnish/lacquer
Polyesterimid polyester imide
Polyesterimidharz polyester imide resin
Polyesterketon polyester ketone
Polyesterlack polyester paint/varnish
Polyesteroligomer polyester oligomer
Polyesterpolyol polyester polyol
Polyesterurethan polyester urethane
Polyether polyether
Polyetheracrylat polyether acrylate
Polyetheramid polyether amide
Polyetheramin polyether amine
Polyetherbasis, auf polyester-based
Polyetherdiamin polyether diamine
Polyetherdiol polyether diol
Polyetherester polyether ester
Polyetheretherketon polyetherether ketone, PEEK
Polyetheretherketonketon polyether ether ketone ketone, PEEKK
Polyetherglykol polyether glycol
Polyethergruppe polyether group
Polyetherharnstoffharz polyether urea resin
Polyetherimid polyether imide, PEI
Polyetherketon polyether ketone, PEK
Polyetherketonetherketonketon polyether ketone ether ketone ketone, PEKEKK
polyethermodifiziert polyether modified
Polyetherpolycarbonat polyether polycarbonate
Polyetherpolyol polyether polyol
Polyetherpolysiloxan polyether polysiloxane
Polyetherseitengruppe polyether side group
Polyethersiloxan polyether siloxane
Polyethersulfon polyether sulphone, PES
Polyethertriacrylat polyether triacrylate
Polyetherurethan polyether urethane
Polyethoxylat polyethoxylate
polyethoxyliert polyethoxylated
Polyethylen polyethylene, polythene
Polyethylenglykol polyethylene glycol
Polyethylenglykolacrylat polyethylene glycol acrylate
Polyethylenglykolterephthalat polyethylene glycol terephthalate
Polyethylenoxid polyethylene oxide
Polyethylenschaum polyethylene foam
Polyethylenterephthalat polyethylene terephthalate

Polyethylenwachs polyethylene wax
Polyfettalkyloxazolin polyfatty acid alkyl oxazoline
Polyfluoracrylat polyfluoroacrylate
Polyfluoralkylen polyfluoroalkylene
Polyfluorethylenpropylene polyfluoroethylene propylene
Polyfluorsiloxan polyfluorosiloxane
polyfunktionell multi-functional
Polyglycidylmethacrylat polyglycidyl methacrylate
Polyglycidylurethan polyglycidyl urethane
Polyglykol polyglycol
Polyglykolether polyglycol ether
Polyglykolfettsäureester polyglycol fatty acid
Polyglykolgruppe polyglycol group
Polyglykolterephthalsäureester polyglycol terephthalate
Polyharnstoff polyurea
Polyharnstoffamid polyurea amide
Polyharnstoffurethan polyurea urethane
Polyhexafluorpropylen polyhexafluoropropylene
Polyhydantoin polyhydantoin
Polyhydrazid polyhydrazide
Polyhydrochinon polyhydroquinone
Polyhydroxid polyhydroxide
Polyhydroxybenzoat polyhydroxybenzoate
Polyhydroxybutyrat polyhydroxybutyrate
Polyhydroxyethylacrylat polyhydroxyethyl acrylate
Polyhydroxypolyether polyhydroxypolyether
Polyhydroxystearinsäure polyhydroxystearic acid
Polyhydroxystyrol polyhydroxystyrene
Polyhydroxyvalerat polyhydroxyvalerate
Polyhydroxyverbindung polyhydroxy compound
Polyimid polyimide
Polyisobutylen polyisobutylene
Polyisocyanat polyisocyanate
Polyisocyanatharz polyisocyanate resin
Polyisocyanatreste polyisocyanate residues
polyisocyanatvernetzbar polyisocyanate curing
polyisocyanatvernetzt polyisocyanate cured
Polyisocyanurat polyisocyanurate
Polyisopren polyisoprene
Polykarbamid polyurea
Polykation polycation
Polykationenharz polycationic resin
Polyketon polyketone
Polykieselsäure polysilicic acid
Polykondensat polycondensate
Polykondensatharz polycondensation resin
Polykondensation polycondensation
Polykondensationsharz polycondensation resin
Polykondensationsklebstoff polycondensation-curing adhesive
Polykondensationsreaktion polycondensation reaction
polykonjugiert polyconjugated
polykristallin polycrystalline
Polylacton polylactone
polymer polymeric
Polymer polymer
Polymerabbau polymer degradation
Polymeranalytik polymer analysis
Polymeranalytiker polymer analyst
polymeranalytisch polymer-analytical
Polymeraufbau polymer structure
polymerbeschichtet polymer coated
Polymerbeton polymer concrete
Polymerbildung polymerisation
Polymerblend polymer blend/alloy, polyblend
Polymercharakterisierung polymer characterisation
Polymerchemie polymer chemistry
Polymerdispersion polymer dispersion
Polymeremulsion polymer dispersion
Polymeren- *see* **Polymer-**
Polymerfeststoff solid polymer
Polymerfilm polymer film
Polymerforschung polymer research
Polymerfüllstoff polymer filler
Polymergehalt polymer content
Polymergemisch polymer blend/alloy, polyblend
Polymergerüst polymer framework
Polymerharz synthetic resin
Polymerhauptkette main polymer chain
Polymerhohlkugeln hollow polymer beads
Polymerisat polymer
Polymerisat- *see* **Polymer-**
Polymerisation polymerisation
Polymerisationsbedingungen polymerising conditions
polymerisationsfähig polymerisable
Polymerisationsfähigkeit polymerisability, ability to polymerise
Polymerisationsgeschwindigkeit polymerisation rate
Polymerisationsgrad degree of polymerisation
polymerisationshemmend polymerisation inhibting
Polymerisationshilfsmittel polymerisation aid
Polymerisationsinitiator polymerisation initiator
Polymerisationskatalysator polymerisation catalyst
Polymerisationskessel polymerising vat
Polymerisationsklebstoff polymerisation curing adhesive
Polymerisationsmechanismus polymerisation mechanism
Polymerisationsneigung tendency to polymerise
Polymerisationsproduct polymer product
Polymerisationsreaktion polymerisation reaction
Polymerisationsstarter polymerisation initiator
Polymerisationstemperatur polymerisation temperature
Polymerisationsumsatz percentage polymerisation
Polymerisationsverfahren polymerisation (process)
polymerisationsverzögernd polymerisation retarding

Polymerisationsverzögerung polymerisation delay
Polymerisationswärme heat of polymerisation
Polymerisatlösung polymer solution
Polymerisatschmelze polymer melt
polymerisierbar polymerisable
polymerisieren to polymerise
Polymerkette polymer chain
Polymerkettenabbau polymer degradation
Polymerkettenlänge polymer chain length
Polymerknäuel polymer tangle
Polymerkonformation polymer configuration
Polymerkonzentration polymer concentration
Polymerkügelchen polymer bead
Polymerlegierung polymer blend/alloy, polyblend
Polymerlösung polymer solution
Polymermaterial polymer material
Polymermatrix polymer matrix
Polymermischung polymer blend/alloy, polyblend
polymermodifiziert polymer modified
Polymermolekül polymer molecule
Polymermörtel polymer mortar
Polymernetz polymer network
Polymernetzwerk polymer network
Polymeroberfläche plastics/polymer surface
polymeroptische Faser optic(al) fibre
Polymerpartikel polymer particle
Polymerperlen polymer beads
Polymerphase polymer phase
Polymerpulver polymer powder
Polymerrückgrat polymer backbone
Polymersalz polymer salt
Polymerschädigung polymer damage
Polymerschicht polymer film
Polymerschmelze polymer melt
Polymersegment polymer segment
Polymerstruktur polymer structure
Polymertechnik polymer engineering/technology
Polymertechnologie polymer technology
Polymerteilchen polymer particle
Polymerüberzug polymer coating
Polymerverbund polymer composite
polymerveredelt polymer/resin coated
Polymerverschnitt polymer blend/alloy, polyblend
Polymerweichmacher polymeric plasticiser
Polymerwerkstoff polymer material/substance
Polymerzusammensetzung polymer composition
Polymethacrylamid polymethacrylamide
Polymethacrylat polymethacrylate
Polymethacrylatharz acrylic/polymethacrylate resin
Polymethacrylimid polymethacrylimide, PMI
Polymethacrylnitril polymethacrylonitrile
Polymethacrylsäure polymethacrylic acid
Polymethacrylsäureester polymethacrylate
Polymethylenbrücke polymethylene bridge
Polymethylenharnstoff polymethylene urea
Polymethylenkette polymethylene chain
Polymethylmethacrylat polymethyl methacrylate
Polymethylpenten polymethyl pentene
Polymethylstyrol polymethyl styrene
Polynomansatz polynomial theorem
Polynomfunktion polynomial function
Polyol polyol
Polyolefin polyolefin
Polyolefinharz polyolefin resin
polyolefinisch polyolefinic
Polyolefinkautschuk polyolefin rubber
polyolgehärtet polyol cured
Polyolhärter polyol hardener/catalyst
Polyolharz polyol resin
Polyolkombination polyol blend
Polyolkomponente polyol component
Polyorganosiloxan polyorganosiloxane
Polyoxazolin polyoxazoline
Polyoxyalkyleneinheit polyoxyalkylene unit
Polyoxyalkylengruppe polyoxyalkylene group
Polyoxymethylen polyoxymethylene, POM
Polyoxymethylmelamin polyoxymethyl melamine
Polyphenol polyphenol
polyphenolisch polyphenolic
Polyphenylenether polyphenylene ether, PPE
Polyphenylenetherharz polyphenylene ether resin
Polyphenylenoxid polyphenylene oxide, PPO
Polyphenylensulfid polyphenylene sulphide, PPS
Polyphenylensulfidharz polyphenylene sulphide resin
Polyphenylensulfidsulfon polyphenylene sulphide sulphone, PPSS
Polyphenylsiloxan polyphenyl siloxane
Polyphosphat polyphosphate
Polyphosphorsäure polyphosphoric acid
Polyphthalamid polyphthalamide
Polyphthalocyanin polyphthalocyanine
Polypropen polypropene
Polypropylen polypropylene, PP
Polypropylenglykol polypropylene glycol
Polypropylenoxid polypropylene oxide
Polypropylenwachs polypropylene wax
Polypyrrol polypyrolle
Polyreaktion polymerisation
Polysaccharid polysaccharide
Polysäure polyacid
Polysilanverbindung polysilane compound
Polysiloxan polysiloxane
Polysiloxanbasis, auf polysiloxane based
Polysiloxanharz polysiloxane resin
Polysiloxanimid polysiloxane imide
polysiloxanmodifiziert polysiloxane modified
Polysiloxanrest polysiloxane group
Polystyrol polystyrene, PS
Polystyrolstrukturschaum structural polystyrene foam
Polysulfid polysulphide
Polysulfidbindung polysulphide linkage
Polysulfidharz polysulphide resin
polysulfidisch polysulphide

Polysulfidkautschuk polysulphide rubber
Polysulfon polysulphone, PSU
Polyterephthalat polyterephthalate
Polytetrafluorethylen polytetrafluoroethylene, PTFE
Polytetrahydrofuran polytetrahydrofuran
Polytetramethylenglykol polytetramethylene glycol
Polythiophensalz polythiophen salt
Polytitansäurebutylester polybutyl titanate
Polytrifluorchlorethylen polytrifluorochloroethylene
Polytrifluorethylen polytrifluoroethylene
Polyurethan polyurethane
Polyurethandichtungsmasse polyurethane sealant
Polyurethandispersion polyurethane dispersion
Polyurethangruppe polyurethane group
Polyurethanharnstoff polyurethane urea
Polyurethanharz polyurethane resin
Polyurethanklebdichtungsmasse polyurethane adhesive sealant
Polyurethanklebstoff polyurethane adhesive
Polyurethanlack polyester paint/varnish
Polyurethanschmelzklebstoff hot melt polyurethane adhesive
Polyurethanstruktur polyurethane structure
Polyvinyilidenfluoridpulver polyvinylidene fluoride powder
Polyvinylacetal polyvinyl acetal
Polyvinylacetat polyvinyl acetate
Polyvinylacetatleim polyvinyl acetate adhesive, PVA adhesive/glue
Polyvinylalkohol polyvinyl alcohol
Polyvinylbutyral polyvinyl butyral
Polyvinylcarbazol polyvinyl carbazole
Polyvinylchlorid polyvinyl chloride, PVC
Polyvinylchlorid hart rigid polyvinyl chloride, rigid PVC, uPVC
Polyvinylchlorid weich flexible polyvinyl chloride, flexible PVC
Polyvinylchloridacetat vinyl chloride-vinyl acetate copolymer
Polyvinylester polyvinyl ester
Polyvinylether polyvinyl ether
Polyvinyletherharz polyvinyl ether resin
Polyvinylfluorid polyvinyl fluoride
Polyvinylformal polyvinyl formal
Polyvinylidenchlorid polyvinylidene chloride, PVDC
Polyvinylidenfluorid polyvinylidene fluoride, PVDF
Polyvinylisobutylether polyvinyl isobutyl ether
Polyvinylpropionat polyvinyl propionate
Polyvinylpyrrolidon polyvinyl pyrrolidone
Polyvinyltoluol polyvinyl toluene
Polyvinylverbindung polyvinyl compound
Polyxylylenharz polyxylylene resin
Polyzucker polysaccharide
Pore pore, pinhole
Porenbeton expanded/aerated/foamed concrete
Porenbildung pinholing
porenfrei free from pinholes/pores
Porenfreiheit freedom from pinholes/pores
porig porous
porös porous
Porosität porosity
Portlandzement Portland cement
Porzellan porcelain
Porzellankugelmühle porcelain ball mill
Porzellankugeln porcelain balls
Positioniereinrichtung positioning device
positionieren to position
Positioniergenauigkeit accurate positioning
Positioniergerät position instrument/device
Positionierung positioning
positionsgenau accurately positioned
Positionsveränderung change in position
positiv positive
positiv geladen positively charged
Poster, Postersitzung poster session: *this is a relatively new phenomenon involving the presentation of a paper or product in the form of signs, text, graphs, pictures etc. fixed to boards on easels.*
Potentialdifferenz potential difference
potentialfrei zero-potential
Potentiometer potentiometer
Potentiometrie potentiometry
potentiometrisch potentiometric
Potenzansatz power law
Potenzfließgesetz power law
Potenzgesetz power law
Potenzgesetzexponent power law index
Potenzgesetzflüssigkeit power law fluid
Potenzgesetzschmelze power law melt
Potenzgesetzstoff power law fluid
Potenzgesetzverhalten power law behaviour/characteristics/flow
Potenzzahl power index
Prägeanlage embossing unit
Prägedruck embossing pressure
Prägeeigenschaften embossing properties
Prägeeinrichtung embossing unit
Prägevorrichtung embossing unit/device
Prägewalze embossing roller
Prägewerk embossing unit
Präkondensat pre-condensate
präparativ preparative
Präpolymer prepolymer
Praxisanforderungen practical requirements
Praxisbeanspruchungen practical conditions
Praxisbedingungen practical conditions
Praxisbeispiel practical example
praxisbewährt proven
praxisbezogen practice-oriented/-related
Praxiseinsatz practical use
Praxiserfahrung practical experience
Praxisergebnisse practical results
praxisgerecht meeting practical requirements
praxisnah under simulated practical conditions
praxisorientiert practice-oriented/-related

praxisrelevant practice-related
Praxisverhalten behaviour/performance under practical conditions, behaviour in practice
Praxisversuch practical test
präzipitiert precipitated
präzis precise, accurate
Präzision precision, accuracy
Präzisionsdosiereinheit precision metering unit
Präzisionsdosierpumpe precision metering pump
Preis-Durchsatzverhältnis cost-performance ratio
Preis-Leistungsindex cost-performance factor
Preis-Leistungsverhältnis cost-performance ratio/factor
Preis-Wirkungsrelation cost-benefit ratio
Preisanhebung price increase
Preisdruck downward pressure on prices
Preiserhöhung price increase
Preisgefüge price structure
preisgünstig inexpensive, low-cost, reasonably priced, economical
Preisnachlaß price reduction
Preisniveau price (level)
Preisreduzierung price reduction
Preissteigerung price increase
Preisvorteil price advantage
preiswert inexpensive, low-cost
preiswürdig inexpensive, low-cost, reasonably priced
Preiswürdigkeit reasonable/low cost
Prekondensat precondensate
Prepolymer prepolymer
Preßbedingungen pressing conditions
Preßdruck bonding/laminating pressure
Presse press
Presse, ölhydraulische hydraulic press
Preßholz compressed/densified wood
Preßplatte pressed sheet
Preßschichtstoff laminate
Preßtemperatur pressing temperature
Preßzeit pressing time
primär primary
Primärelektron primary electron
Primärion primary ion
Primärkorn primary particle
Primärpartikel primary particle
Primärradikal primary radical
Primärreaktion primary reaction
Primärstabilisator primary stabiliser
Primärstruktur primary structure
Primarteilchen primary particle
Primärweichmacher primary plasticiser
Primat priority, most important consideration/factor
primerlackiert primer coated, coated with primer
Primerlackierung application of primer, coating with primer
primerlos primerless
Primern priming, application of a primer
Primerschicht primer coat
Primerstrich primer coat
Primerung priming, application of a primer
Probe 1. sample. 2. test piece/specimen
Probeblech test panel
Probehalterung test piece support/clamp
Probekörper test piece/specimen
Probekörpergeometrie test piece geometry
Probekörperhalterung test piece support/clamp
Probekörpermitte test piece centre
Probekörperoberfläche test piece surface
Probekörpertemperatur test piece temperature
Probekörpervolumen test piece volume
Probenabmaße test piece dimensions
Probenabmessungen test piece dimensions
Probenahme sampling
Probenahmefrequenz sampling frequency
Probenaufbereitung test piece preparation
Probenbreite test piece width
Probendimensionen test piece dimensions
Probeneinlegevorrichtung test piece insertion device
Probenflüssigkeit test liquid
Probengeometrie test piece dimensions
Probenlänge test piece length
Probenmaterial sample material
Probenmengen sample quantities/amounts
Probenmitte centre of test specimen/piece
Probenoberfläche test specimen/piece surface
Probenrand edge of test specimen/piece
Probenträger test piece support/clamp
Probenvorbereitung sample preparation, test piece preparation
Probeplatte test panel
Proberaum test chamber
Probestab test piece/specimen
Probestreifen test strip
Problematik problems
problembehaftet problematic(al)
Probleme, gesundheitliche health problems
problemlos without any problem
produktabhängig product dependent, depending on the product
Produktalternative alternative product
Produktauslauf product outlet
Produktbereinigung streamlining of product range
Produktdurchsatz product throughput
Produkteigenschaften end product properties
Produkteinlauf product inlet
Produktentwicklung 1. product development. 2. product development department
Produktgruppe product group
Produktionsablauf production cycle
Produktionsabschnitt production stage
Produktionsanforderungen production requirements
Produktionsanlage production plant/line
Produktionsansatz production batch
Produktionsausfall production downtime
produktionsbedingt production related
Produktionsbedingungen production/

manufacturing conditions
Produktionscharge production batch
Produktionsdaten production data
Produktionsdatenerfassung 1. production data collection/acquisition/capture. 2. production data collecting unit
Produktionsdatenprotokoll production data record
Produktionsdatenverarbeitung production data processing
Produktionsdatum date of manufacture, production date
Produktionseinheit production unit
Produktionserhöhung increase in production, production increase
Produktionsgeschwindigkeit production rate/speed
Produktionskapazität production capacity
Produktionskontrolle production control
Produktionskosten production/manufacturing costs
Produktionskreislauf production cycle
Produktionsleistung production output/capacity
Produktionsleiter production manager
Produktionslinie production line
Produktionslos production batch
Produktionsmaschine machine, production scale machine
Produktionsmaßstab production scale
Produktionsmengen production scale quantities
Produktionspalette product range, range of products
Produktionsprogramm 1. product range. 2. production schedule/plan/programme
Produktionsprozeß production/manufacturing process
Produktionsrate production rate
Produktionsschema flow diagram
Produktionsschritte production stages
Produktionsserien, große long runs
Produktionsserien, kleine short runs
Produktionsstandort production site
Produktionsstätte plant, factory
Produktionsstillstand break in production
Produktionsstörung break in production
Produktionsstraße production line
Produktionsüberwachung production control
Produktionsunterbrechung break in production
Produktionsunterbruch break in production
Produktionsverhältnisse production/manufacturing conditions
Produktionsversuch production trial
Produktionsvolumen production volume
Produktionszahlen production figures
Produktionsziffern production figures
Produktionszustand production status
Produktionszuwachs production increase
Produktionszyklus production cycle
Produktivität productivity
Produktivitätssteigerung productivity increase, increase in productivity
Produktklasse group of products
Produktmerkblatt product data sheet
Produktpalette product range, range of products
Produktprogramm product range, range of products
Produktreihe product range, range of products
Produktsortiment product range, range of products
Produktstrom product flow
Profilometrie profilometry
Prognose forecast, prognosis
Propan propane
Propandiol propane diol
Propanol propanol
Propionaldehyd propionaldehyde
Propionsäure propionic acid
Proportionalität proportionality
Proportionalitätsgrenze proportionality limit
Proportionalitätskonstante proportionality constant
Proportionalventil proportional valve
Propylaldehyd propylaldehyde
Propylen propylene
Propylencarbonat propylene carbonate
Propylenglykol propylene glycol
Propylenglykolether propylene glycol ether
Propylenglykolmaleat propylene glycol maleate
Propylenglykolmonoalkylether propylene glycol monoalkyl ether
Propylenglykolmonomethylether propylene glycol monomethyl ether
Propylenkette propylene chain
Propylenoxid propylene oxide
Propylenoxidkautschuk propylene oxide rubber
Propylglykol propyl glycol
Protein protein
Proteingruppe protein group
proteinhaltig containing protein
Proteinkleber protein adhesive
Proteinstruktur protein structure
Proton proton
Protonenakzeptor proton acceptor
Protonendonator proton donor
Protonensäure protonic acid
Protonentransfer proton transfer
protonierfähig protonisable
protoniert protonised
Protonierung protonisation
Prototyp prototype
Prototypenfertigung prototype production
Prozedur procedure, method
Prozentanteil percentage (amount)
prozentual percentage
Prozentzahl percentage
Prozeßanalyse process analysis
Prozeßautomatisierung process automation
Prozeßbedingungen process conditions
Prozeßdaten process data
Prozeßführung process control
prozeßgeführt process controlled
prozeßgeregelt process controlled

prozeßgesteuert process controlled
Prozeßgröße process parameter/variable
Prozeßkontrolle process control
Prozeßkosten process operating costs
Prozeßlenkung process control
Prozeßoptimierung process upgrading
prozeßorgesteuert computer controlled
Prozeßparameter process parameter
Prozeßrechner process computer
Prozeßregelung process control
Prozeßschema flow diagram
Prozeßschritt process stage
Prozeßsteuerung process control
Prozeßtechnik process engineering/technology
Prozeßverarbeitung processing
Prozeßwarte control desk/panel/console
Prüfablauf test sequence
Prüfanlage test equipment
Prüfanordnung test setup
Prüfanstalt, amtliche official test establishment
Prüfapparatur test apparatus
Prüfbedingungen test conditions
Prüfbericht test report
Prüfbestimmung test specification
Prüfdaten test data
Prüfdauer test duration
Prüfdruck test pressure
Prüfeinrichtung test instrument/setup/equipment
prüfen to test, to check
Prüfergebnis test result
Prüffehler test error
Prüffeld test bed
Prüffläche test surface
Prüfflüssigkeit test liquid
Prüfformulierung test formulation
Prüfgelände test site
Prüfgerät test instrument
Prüfgeschwindigkeit testing speed
Prüfhölzer wooden test pieces
Prüfklima test atmosphere
Prüfkörper test piece/specimen
Prüfkörperentlastung removal of stress from the test piece
Prüfkraft applied load
Prüfkriterium test criterion
Prüflabor test laboratory
Prüfling test piece/specimen
Prüflösung test solution
Prüfmaschine test apparatus
Prüfmedium test medium
Prüfmethode test method/procedure
Prüfmethodik test method/procedure
Prüfmittel 1. test medium *(if liquid)*. 2. test atmosphere *(if gaseous)*. 3. test substance *(if solid)*
Prüfmodalität test procedure
Prüfnorm test standard
Prüfobjekt test specimen/piece
Prüfparameter test conditions
Prüfprogramm test programme
Prüfresultat test result
Prüfrezeptur test formulation
Prüfrichtlinie test guideline
Prüfspannung 1. test voltage. 2. test stress
Prüfstab test piece/specimen
Prüfstand test bed/rig
Prüfstandversuch test bed trial
Prüfstreifen test strip
Prüfstück test piece/specimen
Prüfsubstanz test substance, substance under test
Prüftafel test panel
Prüftemperatur test temperature
Prüftest test
Prüfung test
Prüfung, qualitätssichernde quality assurance test
Prüfungs- *see* **Prüf-**
Prüfverfahren test method/procedure
Prüfvorschrift test specification
Prüfwert test figure/result
Prüfzeit test period/duration
Prüfzeitraum test period/duration
Prüfzeugnis test certificate
Prüfzustand test status
Prüfzyklus test cycle
PSA *abbr. of* **Phthalsäureanhydrid**, phthalic anhydride
pseudoplastisch pseudoplastic, non-Newtonian
Pseudoplastizität pseudoplasticity, non-Newtonian flow
Pufferwirkung buffer effect
Pufferzone buffer zone
Pulver powder
pulverbeschichtet powder coated
Pulverbeschichtung powder coating
Pulverbeschichtungsverfahren powder coating (process)
Pulverfarbe powder coating
Pulverfliesenkleber powdered tile adhesive
Pulverform, in powdered
pulverförmig powdered, in powder form: **pulverförmiger Überzug** powder coating
Pulverfüllstoff powdered filler
Pulvergrundierung powdered primer
Pulverharz powdered resin
pulverisieren to pulverise
pulverisiert powdered
Pulverkleber powdered adhesive
Pulverkorn particle
Pulverlack powder coating
Pulverlackabfälle powder coating waste
Pulverlackbeschichtung powder coating
Pulverlackharz powder coating resin
Pulverlackierung powder coating
Pulverlackschicht powder coating
Pulverleitfähigkeit powder conductivity
Pulverpartikel powder particle
Pulverpigment powdered pigment, pigment powder
Pulverruß powdered carbon black
Pulverteilchen particle

Pulververnetzer powdered hardener/catalyst
Pulverwiderstand powder resistance
pulvrig powdery
Punktauftrag spot application *(of adhesive)*
Punktkleben spot bonding
Punktkorrosion pitting corrosion
PUR- *see* **Polyurethan-**
Putz plaster, plasterwork, stucco, rendering (coat), rendering mix
Putzfläche plaster surface
Putzmörtel rendering mix
PVAL polyvinyl alcohol
PVC hart rigid PVC, uPVC, unplasticised polyvinyl chloride
PVC-Plastisol PVC paste/plastisol
PVC-Plastisolmasse PVC paste/plastisol
PVC weich flexible PVC, plasticised polyvinyl chloride
PVDF polyvinylidene difluoride
PVK *abbr. of* **Pigmentvolumenkonzentration**, pigment volume concentration, p.v.c.
Pyraninfarbstoff pyranine dye
Pyrethrin pyrethrin
pyrogen pyrogenic
Pyrogramm pyrogram
pyrolisieren to pyrolyse
pyrolitisch pyrolytic
Pyrolyse pyrolysis
Pyrolyse-Gaschromatographie pyrolysis gas chromatography
Pyrolyseeinrichtung pyrolysis furnace
Pyrolysefragment pyrolysis fraction
Pyrolyseprodukt pyrolysis product
pyrolysieren to pyrolyse
pyrolytisch pyrolytic
Pyromellitsäure pyromellitic acid
Pyromellitsäureanhydrid pyromellitic anhydride
Pyrometer pyrometer
pyrometrisch pyrometric
Pyrrolidin pyrrolidine

Q

QS *abbr. of* **Qualitätssicherung**, quality assurance
QS-System quality assurance system
Quadratwurzel square root
Qualität quality
qualitativ qualitative
qualitativ hochwertig high-quality
Qualitätsabweichungen quality deviations
Qualitätsanalyse quality analysis
Qualitätsanforderungen quality requirements
Qualitätsansprüche quality requirements
Qualitätsaudit quality audit
Qualitätsaufzeichnungen quality assurance notes/records
Qualitätsaussagen quality statement/information
Qualitätsbeurteilung quality assessment
qualitätsbewußt quality conscious
Qualitätsbewußtsein quality awareness
Qualitätsdokumentation quality record
Qualitätseinbuße loss of quality, drop in quality
Qualitätsgarantie guaranteed quality
qualitätsgerecht high-quality
qualitätskonstant of constant quality
Qualitätskonstanz consistent/constant quality
Qualitätskontinuität constant/consistent quality
Qualitätskontrolle quality control
Qualitätskontrollsystem quality control system
Qualitätskoordinator quality assurance coordinator
Qualitätskosten quality assurance costs
Qualitätskriterium quality criterion
Qualitätsleitlinien quality assurance guidelines
Qualitätslenkung quality control
Qualitätsmaßstab quality standard
Qualitätsmerkmale quality characteristics
qualitätsmindernd quality reducing
Qualitätsminderung drop/reduction in quality
Qualitätsniveau (level of) quality
qualitätsorientiert quality orientated
Qualitätsplanung quality assurance planning
Qualitätspolitik quality assurance policy
Qualitätsprobleme quality problems
Qualitätsprozeß quality assurance process
Qualitätsprüfung quality testing/control/check
qualitätsrelevant affecting quality, relevant to quality
Qualitätsrichtlinie quality guideline
Qualitätsschwankungen quality variations
Qualitätssicherheit quality assurance
qualitätssichernde Maßnahmen quality assurance measures
Qualitätssicherung quality assurance
Qualitätssicherungsabteilung quality assurance department
Qualitätssicherungsdokumentation quality assurance documents/documentation
Qualitätssicherungsdokumente quality assurance documents
Qualitätssicherungshandbuch quality assurance manual
Qualitätssicherungsmaßnahmen quality assurance measures
Qualitätssicherungsorganisation quality assurance organisation
Qualitätssicherungsprogramm quality assurance programme
Qualitätssicherungssystem quality assurance system
Qualitätssicherungsunterlagen quality assurance documents/documentation
Qualitätssicherungszertifikat quality assurance certificate
Qualitätsstandard quality standard, standard of quality

qualitätssteigernd quality enhancing
Qualitätssteigerung improvement in quality
Qualitätssteuerung quality control
Qualitätsstrategie quality policy/strategy
Qualitätsüberprüfung quality check, check on quality
Qualitätsüberwachung quality control
Qualitätsüberwachungspaket quality control package
Qualitätsverantwortung responsibility for quality
qualitätsverbessernd quality enhancing
Qualitätsverbesserung quality improvement, improvement in quality
Qualitätsverlust loss of quality
Quanteneffekt quantum effect
quantenmechanisch quantum mechanical
quantifizieren to quantify
Quantifizierung quantification
quantitativ quantitative
quartär quaternary
Quarz quartz, silica
Quarzfeinstmehl ultra-fine silica flour
Quarzglasplatte silica glass plate
Quarzgut synthetic silica
Quarzgutmehl synthetic silica flour
Quarzkristallaufnehmer quartz transducer
Quarzkristalldruckaufnehmer quartz pressure transducer
Quarzkristallkraftaufnehmer quartz force transducer
Quarzkristallmeßwertaufnehmer quartz transducer
Quarzmehl silica flour
Quarzpulver powdered silica
Quarzsand silica sand
Quarzschiffchen silica combustion boat
Quarztiegel silica crucible
quasielastisch quasi-elastic
quasiisotrop quasi-isotropic
quasinewtonsch quasi-Newtonian
quasistatisch quasi-static
quaternär quaternary
quaterniert quaternised
quaternisierbar quaternisable
quaternisiert quaternised
Quecksilber mercury
Quecksilberdampfhochdrucklampe high-pressure mercury vapour lamp
Quecksilberdampflampe mercury vapour lamp
Quecksilberhochdrucklampe high-pressure mercury vapour lamp
Quecksilberlampe mercury vapour lamp
Quecksilbermitteldrucklampe medium-pressure mercury vapour lamp
Quecksilbersäule mercury column
quellbar swellable
Quellbarkeit swellability
Quellbeständigkeit swelling resistance
Quelldehnung swelling
Quellen swelling
Quellfestigkeit swelling resistance
Quellgrad amount of swelling
Quellmaß degree of swelling/expansion
Quellmittel swelling agent
Quellung swelling
Quellungsbeständigkeit swelling resistance, resistance to swelling
Quellungserscheinung swelling
Quellverhalten swelling characteristics
Quellvermögen swelling ability
Quellversuch swelling test
Quellwirkung swelling effect
quer transverse(ly)
Querbelastbarkeit transverse strength
Querdehnung transverse expansion
Querfestigkeit transverse strength
Querkontraktion transverse shrinkage/contraction
Querkontraktionszahl transverse shrinkage index
Querkraft transverse force
Querleimung transverse glueing
Querleimwerk transverse glueing unit
Querrichtung transverse direction
Querscherung transverse shear
Querschliff microtome section
Querschliffaufnahme microtome photograph
Querschnitt cross-section, diameter
Querschnittsänderung change in diameter
Querschnittsfläche cross-sectional area
Querschnittsübergang change in diameter
Querschnittsveränderung change in diameter
Querschrumpf transverse shrinkage
Querschwindung transverse shrinkage
Querspannung transverse stress
Querstrommembranfiltration cross-flow membrane filtration
Querzugbeanspruchung transverse tensile stress
Querzugfestigkeit transverse tensile strength
Querzugspannung transverse tensile stress
Quetschspannung compressive yield stress

R

RA-FTIR-Spektrum Raman absorption FTIR spectrum
r.F. *abbr. of* **relative Feuchte,** relative humidity
Radikal radical, group
Radikalangebot radical availability
Radikalangriff radical attack
Radikalausbeute radical yield
radikalbildend radical-forming
Radikalbildner radical former
Radikalbildung radical formation
Radikaleinfang radical interception

Radikalfänger radical interceptor
Radikalhärter free-radical hardener
Radikalinhibitor radical inhibitor
radikalisch radical
Radikalkette free radical chain
Radikalkettenmechanismus free radical (chain growth) mechanism
Radikalkettenpolymerisation free radical (chain growth) polymerisation
Radikalkonzentration radical concentration
Radikalmolekül radical molecule
Radikalpolymerisation free radical polymerisation
radikalspendend radical donating
Radikalspender radical donor
Radikalstarter radical initiator
radioaktiv radioactive
Radioaktivität radioactivity
Radiographie radiography
radiographisch radiographic
Radius radius
Radkappe hub cap
Raffination refining
Raffinerie refinery
Raffinerietechnik refinery engineering
raffiniert refined
Rahmenrezeptur starting formulation
Rahmenrichtlinie framework guideline
Rakel (doctor) knife
Rakelauftrag knife application
Rakelblatt (doctor) knife blade
Rakelmesser (doctor) knife
Rakelmesserauftrag knife application
Rakeln knife coating
Raleighstreuung Raleigh scattering
Ramanlinie Raman line
Ramanspektrometer Raman spectrometer
Ramanspektroskopie Raman spectroscopy
Ramanspektrum Raman spectrum
Ramanstreusignal Raman scattering signal
Ramanstreuung Raman scattering
Ramanstreuung, oberflächenverstärkte surface enhanced Raman scattering, SERS
Ramanstreuwirkung Raman scattering
Randabdichtung edge sealant
Randbedingungen boundary/peripheral conditions
randnah near the edge
Randwinkel contact angle
Randwinkelhysterese contact angle hysteresis
Randzone peripheral zone
Rapsöl rapeseed oil
rasch rapid, fast
raschhärtend fast curing
raschlaufend high-speed
raschtrocknend fast drying
Rasterelektronenmikroskop scanning electron microscope
Rasterelektronenmikroskopaufnahme scanning electron micrograph
Rasterelektronenmikroskopie scanning electron microscopy
rasterelektronenmikroskopische Aufnahme scanning electron micrograph
Rasterwalze halftone roller
rationalisieren to rationalise, to streamline
Rationalisierung rationalisation, streamlining
Rationalisierungsgründe economic considerations
Rationalisierungsmaßnahmen economy measures
rationell efficient, economical
rauh rough
Rauheit roughness
Rauheitsspitzen roughness peaks
Rauhigkeit roughness
Rauhtiefe roughness height
Raumdichte density
Raumfahrt 1. aerospace industry. 2. space travel
Raumfahrtindustrie aerospace industry
Raumfahrttechnik aerospace enginering
Raumfahrzeug spacecraft
Raumgewicht density
Raumkapsel space capsule
Raumklima room temperature
räumlich three-dimensional, spatial
Raumluft room atmosphere
raumsparend space saving
Raumteile parts by volume
Raumtemperatur room temperature
Raumtemperaturbereich room temperature
raumtemperaturgehärtet room temperature cured
raumtemperaturhärtend room temperature curing
Raumtemperaturhärtung room temperature curing
Raumtemperaturklebrigkeit tack at room temperature
Raumtemperaturlagerung room temperature ageing, ageing at room temperature
Raumtemperaturtrocknung drying at room temperature
raumtemperaturvernetzend room temperature curing/vulcanising
Raumtemperaturvernetzung room temperature curing/vulcanisation
Rauschpegel noise level
Rayleighstreuung Rayleigh scattering
Reagenz reagent
Reagenzglas test tube
Reagglomeration flocculation
reagierbar reactive
reagieren to react
Reaktand reactant
Reaktion reaction
Reaktion erster Ordnung first order reaction
Reaktion zweiter Ordnung second order reaction
Reaktionsablauf course of the reaction
Reaktionsbedingungen reaction conditions
Reaktionsbeginn start of the reaction

Reaktionsbereitschaft reactivity
Reaktionsbeschleuniger accelerator
Reaktionsdauer reaction time
Reaktionsende end of the reaction
Reaktionsenergie reaction energy
Reaktionsenthalpie reaction enthalpy
reaktionsfähig reactive
Reaktionsfähigkeit reactivity
reaktionsfreudig reactive
Reaktionsfreudigkeit reactivity
Reaktionsgefäß reaction vessel
Reaktionsgemisch adhesive mix
Reaktionsgeschwindigkeit reaction rate
Reaktionsgleichung (chemical) equation
Reaktionsgrad reactivity
Reaktionsharz reactive resin
Reaktionsharzbeton polymer concrete
Reaktionsharzklebstoff reactive/two-pack adhesive
Reaktionsharzmischung catalysed resin, resin-catalyst mix
Reaktionsharzmörtel polymer mortar
reaktionshemmend reaction retarding
Reaktionskinetik reaction kinetics
reaktionskinetisch reaction kinetic
Reaktionskleber reactive/two-pack adhesive
Reaktionsklebstoff reactive/two-pack adhesive
Reaktionskomponente reaction component
Reaktionslack two-pack/catalysed lacquer
Reaktionslacksystem catalysed surface coating system/formulation
Reaktionsmasse catalysed resin, resin catalyst mix
Reaktionsmechanismus reaction mechanism
Reaktionsmedium reaction medium
Reaktionsmilieu reaction environment
Reaktionsmischung adhesive mix
Reaktionsmittel 1. hardener, catalyst. 2. accelerator
Reaktionsparameter reaction conditions
Reaktionspartner reactant
Reaktionsprodukt reaction product
Reaktionsschema general reaction
reaktionsschnell fast reacting
Reaktionsschritt reaction stage
Reaktionsschrumpfung reaction shrinkage
Reaktionsschwindung reaction shrinkage
Reaktionsschwund reaction shrinkage
Reaktionsstufe reaction stage
Reaktionstemperatur reaction temperature
reaktionsträge low-reactivity
reaktionsträger less reactive
Reaktionsträgheit low reactivity
Reaktionsverhalten reaction characteristics
Reaktionsverlauf course of the reaction
Reaktionsvermögen reactivity
Reaktionsverzögerung delaying the reaction
Reaktionswärme heat of reaction
Reaktionszeit reaction time
reaktiv reactive
reaktiv härtend chemically curing
Reaktivdünner reactive thinner
Reaktivgruppe reactive group
Reaktivharz reactive resin
reaktivierbar capable of being reactivated
Reaktivierung reactivation
Reaktivierungstemperatur reactivation temperature
Reaktivität reactivity
Reaktivitätserhöhung increase in reactivity
Reaktivitätsminderung reduction in reactivity
reaktivitätssenkend reactivity reducing
Reaktivitätsunterschiede reactivity differences, differences in reactivity
Reaktivlösemittel reactive solvent
Reaktivlöser reactive solvent
Reaktivmonomer reactive monomer
Reaktivverdünner reactive thinner
Reaktivverdünnerlack paint containing reactive thinner
reaktivverdünnt thinned with reactive thinner
Reaktivweichmacher reactive plasticiser
Reaktor reactor
Realisierbarkeit feasibility
realisieren to realise, to put into practice
Realtionsharzmasse catalysed resin, resin-catalyst mix
Rechner-, rechner- *see* **Computer-, computer-**
rechtwinklig right-angled
rechtwinklige Überlappung corner lap joint
rechtwinkliger Stumpfstoß right-angle butt joint
recycelbar recyclable
recycelfähig recyclable
Recyceln recycling
Recyclat recyclate, recycled material(s), reclaim
Recycler recycling firm/contractor
Recyclierbarkeit recyclability
recyclieren to recycle
recycliert recycled
recyclingfähig recyclable
Recyclingfähigkeit recyclability
recyclingfeindlich difficult to recycle
Recyclingfirma recycling firm/contractor
recyclingfreundlich easy to recycle, easily recycled
Recyclingfreundlichkeit ease of recycling
recyclinggerecht easy to recycle
Recyclinglack recycled/reconstituted paint
Recyclingmöglichkeit recyclability, recycling option/possibility
Recyclisieren recycling
Recylingprodukte recycled products
redispergieren to redisperse
Redispersionspulver redispersible powder
Redoxinitiator redox initiator
Redoxpotential redox potential
Redoxreaktion redox reaction
Redoxsystem redox system
Reduktion reduction
Reduktionsmittel reducing agent
Reduktionspotential reduction potential
reduziert reduced

reduzierte Viskosität reduced viscosity, viscosity number
Reduzierung reduction
reemulgierbar re-emulsifiable
Reemulgierbarkeit re-emulsifiability
reemulgiert re-emulsified
Referenzgröße reference quantity
Referenzmuster reference sample
Referenzprobe reference sample
Referenzstandard reference standard
Referenztemperatur reference temperature
reflektierend reflecting
reflektiert reflected
Reflektometer reflectometer
Reflektor reflector
Reflexion reflection
Reflexionslichttaster reflected light scanner
Reflexionsspektrum reflection spectrum
Reflockulation re-flocculation
Refraktion refraction
Refraktionsindex refractive index
Refraktometer refractometer
regelbar adjustable
Regelelektronik electronic control system, electronic controls
Regelelemente controls
Regelgerät control instrument
Regelgröße controlled variable
Regelgüte control accuracy/quality
Regelinstrument control unit
Regelkreis 1. control circuit. 2. closed-loop control circuit
Regelmöglichkeiten control options
Regeln 1. control. 2. closed-loop control
Regelung, gesetzliche law, regulation
Regelungen, immissionsschutzrechtliche pollution control regulations, anti-pollution regulations
Regen rain
Regeneration regeneration
regenerierbar 1. recylable, recoverable. 2. renewable *(e.g. resources)*
regenerieren to regenerate, to reclaim, to recycle
Regenperiode 1. rainy season. 2. rainy period: **auch während längerer Regenperioden...** also during prolonged periods of rain...
Regenwasser rainwater
Regulierung control
Regulierungsmittel control agent
reibbeständig rub fast
Reibechtheit rub fastness
Reibeputz scraped finish plaster
reibfest rub fast
Reibgut millbase
Reibkraft frictional force
Reibrad abrasive wheel
Reibradverfahren Taber abrasion test
Reibstuhl rolls, roll mill
Reibtester abrasion tester
Reibung friction
Reibung, gleitende sliding friction
Reibung, innere internal friction, shear
Reibung, trockene dry friction
reibungsabhängig friction-dependent
reibungsarm low-friction
reibungsbedingt due to friction
Reibungsbeiwert coefficient of friction
Reibungsenergie frictional energy/heat
Reibungskoeffizient coefficient of friction
Reibungskraft frictional force
reibungslos 1. troublefree, smooth, without a hitch. 2. frictionless, without (any) friction
reibungsmindernd friction reducing
Reibungsverhalten frictional behaviour
Reibungsverlust frictional loss
reibungsvermindernd friction reducing
Reibungswärme frictional heat
Reibungswiderstand frictional resistance
Reibungszahl coefficient of friction
Reibverhalten frictional behaviour/properties
Reibwert coefficient of friction
Reibwiderstand frictional resistance
Reifezeit maturing period
Reinacrylat pure acrylate
Reinharz pure resin
Reinheit 1. purity. 2. cleanliness
reinigen 1. to clean *(e.g. a surface)*. 2. to purify *(e.g. a chemical compound)*
Reinigung 1. purification. 2. cleaning
Reinigung, chemische dry cleaning
Reinigungsbad cleaning bath
Reinigungsbeständigkeit resistance to cleaning
Reinigungsflüssigkeit cleaning fluid
reinigungsfreundlich easy to clean
Reinigungslösung cleaning solution
Reinigungsmaßnahmen cleaning
Reinigungsmittel cleaner
reinvioskos Newtonian
Reißarbeit fracture energy
Reißdehnung elongation at break, ultimate elongation
reißfest tear resistant
Reißfestigkeit tear strength, ultimate tensile strength
Reißfestigkeit, relative relative tear strength
Reißkraft breaking stress, ultimate tensile stress
Reißlack crackle finish
Reißlänge breaking length
Reißlast breaking stress
Reißmaschine tensile testing machine
Reißprüfung tensile test
Reißspannung yield stress
Reißversuch tensile test
Reizeffekt irritant effect
reizend irritant
Reizpotential irritant potential
Reizung irritation
Reizwirkung irritant effect
Reklamation complaint
Rekristallisation recrystallisation
Rekristallisationstemperatur recrystallising temperature

Rekristallisationsvorgang recrystallisation process
Rektifizierapparat rectifying apparatus
Rektifizierkolonne rectifying column
relativ relative
relative Dichte relative density, specific gravity
relative Dielekrizitätskonstante relative permittivity, dielectric constant
relative Luftfeuchte relative humidity
relative Luftfeuchtigkeit relative humidty
relative Reißfestigkeit relative tear strength
relative Viskosität relative viscosity
Relaxation relaxation
Relaxationsalgorithmus relaxation algorithm
Relaxationskurve relaxation curve
Relaxationsmodul relaxation modulus
Relaxationsprozess relaxation process
Relaxationsverhalten relaxation behaviour
Relaxationsvermögen ability to relax
Relaxationsversuch relaxation test
Relaxationsvorgang relaxation process
relaxieren to relax
REM 1. *abbr. of* **Rasterelektronenmikrosokop,** scanning electron microscope. 2. *abbr. of* **Rasterelektronenmikroskopie**, scanning electron microscopy
REM-Aufnahme scanning electron micrograph
Remission remission
Remissionskurve remission curve
Remissionsspektrum remission spectrum
Remissionswert remission value/figure
Renovierung renovation, restoration, refurbishment
Renovierungsarbeiten restoration work
Rentabilität profitability
Reparatur repair
Reparaturarbeiten repair work
Reparaturaufwand repair costs
reparaturfreundlich easy to repair
Reparaturkosten repair costs
Reparaturlack 1. (car) refinishing paint. 2. touching-up paint
Reparaturmörtel repair mortar
Repassivierung repassivation
Repetierbarkeit repeatability
reproduzierbar reproducible, repeatable
Reproduzierbarkeit reproducibility, repeatability
Reproduzierfähigkeit reproducibility, repeatability
reproduziergenau accurately reproducible/repeatable
Reproduziergenauigkeit accurate reproducibility/repeatability
Resinat resinate
resistent resistant
Resistenz resistance
Resolharz resol resin
Resonanz resonance
Resonanz-Ramanspektroskopie resonance Raman spectroscopy
Resonanz-Ramanstreuung resonance Raman scattering
Resonanzfrequenz resonance frequency
Resonanzintensität resonance intensity
Resonanzkurve resonance curve
Resonanzspektroskopie resonance spectroscopy
Resonanzspektrum resonance spectrum
Resonanzstreuung resonance scattering
Resorcin resorcinol
Resorcin-Formaldehydleim resorcinol-formaldehyde adhesive
Resorcinharz resorcinol resin, resorcinol-formaldehyde resin
Ressort 1. department, section. 2. responsibility
ressortfrei without specific responsibilities
Ressortleiter head of department
Ressortverteilung delegation of responsibilities
Ressourcen resources
Ressourcen, stoffliche material resources
ressourcenschonend resource-conserving
Ressourcenschonung conservation of resources
Rest 1. radical, group. 2. residue
Restacetylgehalt residual acetyl content
Restacetylgruppe residual acetyl group
Restanteile residual amounts
Restaurierung restoration
Restaurierungsarbeiten restauration work
Restbiegefestigkeit retained/residual flexural strength
Restbruchdehnung residual/retained elongation at break, residual/retained ultimate elongation
Restdehnung residual/retained elongation
Restdruck residual pressure
Restenergie residual energy
Restfarben paint residues
Restfestigkeit residual strength
Restfeuchte retained/residual moisture (content)
Restfeuchtegehalt residual moisture content
Restfeuchtigkeit residual moisture (content)
Restfunktionalität residual functionality
Resthaftkraft residual adhesive force
Restisocyanat residual isocynate
Restisocyanatgruppen residual isocyanate groups
Restklebrigkeit residual tack
restlich residual
Restlösemittel 1. residual solvent *(e.g. in a paint or adhesive film)*. 2. solvent residues *(e.g. in a drum that had contained solvent)*
Restlösemittelanteil residual solvent content
Restlösemittelgehalt residual solvent content
Restmenge residual amount
Restmonomer residual monomer
Restmonomeranteil residual monomer content. **Die geringen Restmonomeranteile...** The small amounts of residual monomer...
Restmonomergehalt residual monomer content
Restpendelhärte residual/retained pendulum hardness
Restreaktionsenthalpie residual reaction enthalpy

Restreißdehnung residual/retained elongation at break, residual/retained ultimate elongation
Restreißfestigkeit residual/retained tear strength
Restreißkraft residual/retained breaking stress, residual/retained ultimate tensile stress
Restsäuregehalt residual acid content
Restspannungen residual stresses
Reststabilität residual stability
Reststauchung residual compression
Reststoffe residues
Restwärme residual heat
Restwasser residual moisture
Restzugfestigkeit residual/retained tensile strength
Retardationsverhalten retardation behaviour
Retardationsversuch retardation test
Retardationsvorgang retardation process
Retardationszeit retardation time
Retardierungs- *see* **Retardations-**
reversibel reversible
Rezept formulation
Rezeptbestandteil formulation component
rezeptiert formulated
Rezeptierung formulation
Rezeptierungshilfe formulation aid
Rezeptur formulation
rezepturabhängig depending on the formulation
Rezepturbestandteil formulation constituent
Rezepturentwicklung developing a formulation
Rezepturgestaltung developing a formulation
Rezepturhinweis guide formulation
Rezepturkomponente formulation component
Rezepturoptimierung perfecting/upgrading a formulation
Rezepturvorschlag suggested formulation
Rezepturwechsel change of formulation
Rezepturzusammenstellung formulation
reziprok reciprocal
Reziprokwert reciprocal value/figure
rezirkulieren to recirculate
Rezykl-, rezykl- *see* **Recycl-, recycl-**
RFA *abbr. of* **Röntgenfluoreszenz-Analyse**, X-ray fluorescence analysis
RG *abbr. of* **Raumgewicht**, density
Rheogoniometer rheogoniometer
Rheogramm rheogram
Rheologie 1. rheology. 2. viscosity
Rheologieadditiv rheological additive, flow control agent
Rheologiebeeinflussung influencing of flow
Rheologieberechnung rheological calculation
Rheologiebetrachtung rheological calculation/analysis
Rheologiedaten rheological data
Rheologiehilfsmittel rheological additive, flow control agent
rheologiemodifizierend viscosity modifying
Rheologiemodifizierer rheology/flow modifier
Rheologiesteuerung control of flow (properties)
Rheologieverhalten rheological/flow behaviour
Rheologiewirkung effect on flow
rheologisch rheological
Rheometer rheometer
Rheometrie rheometry
rheometrisch rheometric
rheopex rheopectic
Rheopexie rheopexy
rhombisch rhombic
rhombusförmig rhomboid
Richtformulierung guide formulation
Richtgröße guide value
Richtkonzentration guide concentration
Richtlinie guideline
Richtrezept(ur) guide formulation
richtungsabhängig anisotropic, direction dependent
Richtungsabhängigkeit anisotropy, direction dependence
Richtungsänderung change of direction
Richtungsorientierung orientation
Richtwert guide value
Richtzahl guide value
Ricinenalkydharz dehydrated castor oil alkyd (resin)
Ricinenöl dehydrated castor oil
Ricinensäure dehydrated castor acid
Ricinoloxazolin ricinoleic oxazoline
Ricinusöl castor oil
rieselfähig free-flowing
Rieselfähigkeit free-flowing properties
Rieselverhalten free-flowing properties
Ring und Kolben-Methode ring and ball test *(for determining softening point)*
Ringöffnung ring opening
Ringöffnungsreaktion ring opening reaction
Ringschlußreaktion ring closure reaction
Ringspaltung ring opening
Ringstruktur ring structure
Risiko risk
risikobereit ready/prepared to take risks
Risikobereitschaft readiness to take risks
risikofreudig ready/prepared to take risks
Risikofreudigkeit readiness to take risks
Riß crack, tear
rißabsorbierend crack absorbing
Rißanfälligkeit susceptibility to cracking
Rißausbreitung crack propagation
Rißausbreitungsgeschwindigkeit crack growth/propagation rate
Rißausbreitungskraft crack propagation force
Rißausbreitungswiderstand crack propagation resistance
rißauslösend crack initiating
Rißauslösung crack initiation
Rißbildung cracking, crazing, checking
Rißbildungsneigung tendency to crack
Rißbildungsresistenz crack resistance
Rißerweiterungsgeschwindigkeit crack growth/propagation rate
Rißfläche crack surface
Rißfortpflanzung crack propagation
Rißfortpflanzungsgeschwindigkeit crack

propagation rate
Rißfortschritt crack propagation
rißfrei free from cracks
Rißfreiheit freedom from cracks
rißfüllend crack/gap filling
Rißfüllstoff crack/gap filler
Rißgeschwindigkeit crack growth/propagation rate
rissig cracked
Rißinitiierung crack initiation
Rißlänge crack length
Rißneigung tendency to crack
Rißrichtung crack direction
Rißstabilität crack resistance
rißüberbrückend crack spanning/bridging
rißüberdeckend crack spanning/bridging
Rißufer crack side
Rißvergrößerung crack growth
Rißverlängerung crack growth
Rißwachstum crack growth/propagation
Rißwachstumsgeschwindigkeit crack growth/propagation rate
Rißwachstumsrate crack growth/propagation rate
Rißwachstumsrichtung crack growth direction
Rißweite crack width
Rißwiderstand crack resistance
Rißzähigkeit fracture toughness/resistance
Rißzone crack zone/region
Ritzbeanspruchung scratching
Ritzfestigkeit scratch resistance
Ritzhärte scratch hardness
Rizin- *see* **Ricin-**
Roboter robot
robotergeführt robot controlled
Robotersteuerung robot control
Robotersystem robot system
roh crude
Rohcellulose crude cellulose
Rohdichte density
Rohharz crude resin
Rohöl crude oil
rohrförmig tubular
Rohrheizkörper tubular heater
Rohrleitung pipeline
Rohrsteckverbindung tubular lap joint
Rohrsteckverbindung, abgesetzte landed lap tubular joint
Rohrverbindung pipe joint
Rohrverbindung, geschäftete scarfed/tapered tubular joint
Rohrverbindung mit abgesetzter Überlappung butt lap tubular joint
Rohrverbindung mit Stumpfstoß butt tubular joint
Rohrverbindung, überlappte tubular lap joint
Rohstoff raw material, feedstock
Rohstoffbedarf raw material requirements
Rohstoffeingangskontrolle incoming raw materials control
Rohstoffhersteller raw material manufacturer
Rohstofflieferant raw material supplier
Rohstoffmaterial raw material, feedstock
Rohstoffpreis raw material price
Rohstoffressourcenschonung conservation of raw material resources
Rohstoffrückgewinnungsanlage raw material recovery plant
Rohstoffverfügbarkeit raw material availability
Rohstoffverknappung raw material shortage
Rolle roller
Rollen roller application
Rollenauftrag roller application
Rollendruckverfahren roller printing
Rollenoffsetdruck rotary offset printing
Rollenoffsetfarbe rotary offset printing ink
Rollenschälversuch climbing drum peel test
Röntgenanalyse X-ray analysis
Röntgenaufnahme X-ray photograph
Röntgenbeugung X-ray diffraction
Röntgenbeugungsbild X-ray diffraction photograph
Röntgenbeugungsdiagramm X-ray diffraction diagram
Röntgenbeugungsintensität X-ray diffraction intensity
Röntgenbeugungsmessung X-ray diffraction analysis
Röntgendiffraktometrie X-ray diffractometry
Röntgenfluoreszenz X-ray fluorescence
Röntgenfluoreszenzanalysator X-ray fluorescence analyser
Röntgenkristallstruktur X-ray crystal structure
röntgenmikroskopische Aufnahme X-ray micrograph
röntgenographisch X-ray photographic
Röntgenspektroskopie X-ray spectroscopy
Röntgenspektrum X-ray spectrum
röntgenstrahldurchlässig permeable to X-rays
Röntgenstrahlen X-rays
Röntgenstrahlendiffraktion X-ray diffraction
Röntgenstrahlendurchlässigkeit X-ray permeability
röntgenstrahlundurchlässig impermeable to X-rays
Rost 1. rust. 2. grille, grid, grating
Rostbildung rusting, rust formation
rostbraun rust brown
Rosterscheinungen signs of rusting
rostfrei free from rust
rostfreier Stahl stainless steel
Rostinhibitor rust inhibitor
Rostschutzanstrich anti-corrosive finish
Rostschutzanstrichmittel anti-corrosive paint
Rostschutzfarbe anti-corrosive paint
Rostschutzgrundierung 1. anti-corrosive primer. 2. anti-corrosive primer coat
Rostschutzlack anti-corrosive paint
Rostschutzmittel rust inhibitor, anti-corrosive agent
Rostschutzpigment anti-corrosive pigment
Rostschutzwirkung anti-corrosive effect

rot red
Rotation rotation
Rotationsachse axis of rotation
Rotationsbewegung roatary movement
Rotationsdruck rotary printing
Rotationsdruckfarbe rotary printing ink
Rotationsenergie rotational energy
Rotationsrheometer rotational rheometer
Rotationsvakuumverdampfer rotary vacuum evaporator
Rotationsviskosimeter rotational viscometer
rotbraun red-brown
Rotbuche red beech
rotgold red gold
rotierend rotating
rötlich reddish
rötlichbraun reddish-brown
Rotordrehmoment rotor torque
Rotordrehzahl rotor speed
Rotorleistung rotor efficiency/performance
Rotpigment red pigment
Rotwein red wine
Routineanalytik routine analysis
Routinekontrolle routine check
Routinemessung routine determination
Routineprüfung routine test
Routineuntersuchung routine test
Routinevorgang routine operation
RPA *abbr. of* **rheophotoakustisch**, rheophotoacoustic
RTV-1-Silicon room temperature vulcanising one-pack silicone
Rückfederung, elastische elastic recovery
Rückfluß, unter under reflux
Rückflußkühler reflux condenser
Rückführbarkeit recyclability
rückgewinnbar recoverable, reclaimable
Rückgewinnung recovery, reclamation
Rückgewinnungsleistung recovery rate
Rückgewinnungsverfahren recovery/reclamation process
Rückgrat backbone
Rückgratpolymer backbone polymer
Rückprall rebound
Rückprallelastizität rebound/impact resilience
Rückprallversuch rebound test
Rücksprunghärte rebound/impact resilience
Rückstand residue
Rückstandsbeseitigung removal of residues
rückstandsfrei free from residues
Rückstellelastizität rebound resilience
rückstellfähig resilient
Rückstellfähigkeit resilience
Rückstellkraft resilience
Rückstellung recovery *(after compression)*
Rückstellvermögen resilience
Rückstreuung back scattering
Rücktitration back titration
Rückverformung recovery
ruhende Beanspruchung static stress/load
ruhende Belastung static stress/load
Ruhezustand state of rest
Rührbedingungen stirring/mixing conditions
Rührbehälter mixing vessel
Rühren stirring
Rührer stirrer
Rührerdrehzahl stirrer speed
Rührerumdrehungszahl stirrer speed
Rührgefäß mixing tank/vessel
Rührgeschwindigkeit stirring/mixing speed
Rührorgan stirrer, mixer
Rührwelle impeller shaft
Rührwerk mixer, stirrer
Rührwerkeinrichtung mixer, stirrer
Rührwerkskugelmühle attrition mill
Rührwerksmühle attrition mill
Rührwerkzeug impeller
Rum rum
Runzelbildung wrinkling
Runzellack wrinkle-finish paint
Ruß carbon black
Rußpartikel carbon black particle
Rußpigment carbon black pigment
Rutil rutile
Rutilisierungsgrad degree of rutilisation
Rutilpigment rutile pigment
rutschfest non-skid, skidproof
Rutschfestbeschichtung non-skid coating/screed
Rutschfestigkeit non-skid properties
Rütteldichte packing density
Rüttelgewicht packing density

S

Saccharid saccharide
Saccharin saccharin
Sack sack
Safloröl safflower seed oil
Sägemehl sawdust
Sägespäne sawdust
Salicylaldoxim salicylic aldoxime
Salicylat salicylate
Salicylsäure salicylic acid
Salicylsäureester salicylate
Salpetersäure nitric acid
salpetrige Säure nitrous acid
Salz salt
Salzausblühungen efflorescence
Salzbildung salt formation
Salzlösung salt solution
Salznebelbeständigkeit salt spray resistance
Salzsäure hydrochloric acid
Salzsprühbelastung exposure to salt spray
Salzsprühbeständigkeit salt spray resistance
Salzsprühnebeltest salt spray test

Salzsprühresultat salt spray test result
Salzsprühtest salt spray test
Salzwasserlagerung immersion in salt water
SAM scanning auger microscopy
samtartig velvety
sandend sanding
sandgestrahlt sandblasted
Sandmühle sand mill/grinder
Sandstein sandstone
Sandstrahlen sandblasting
Sandwichbauweise sandwich construction
saniert restored, renovated, refurbished
Sanierung restoration, renovation, refurbishment
Sanierungsarbeiten restoration work
Sanitärbereich sanitaryware applications
Sättigung saturation
Sättigungsdampfdruck saturation vapour pressure
Sättigungsdruck saturation pressure
Sättigungsgrad degree of saturation
Sättigungskonzentration saturation concentration
Sättigungswassergehalt saturation moisture content
Sättigungszustand state of saturation
sauber clean
Säubern cleaning
sauer acid
Sauerstoff oxygen
Sauerstoffabsorption oxygen absorption
Sauerstoffangebot amount of oxygen available
Sauerstoffatom oxygen atom
Sauerstoffaufnahme oxygen absorption/take-up
Sauerstoffausschluß, unter in the absence of oxygen, under the exclusion of oxygen
Sauerstoffbarriere oxygen barrier
Sauerstoffdurchlässigkeit oxygen permeability
Sauerstoffdurchlässigkeitskoeffizient oxygen permeability coefficient
Sauerstoffempfindlichkeit oxygen sensitivity
sauerstoffhaltig oxygen containing, containing oxygen
Sauerstoffindex oxygen index
Sauerstoffinhibierung oxygen inhibition
Sauerstoffkonzentration oxygen concentration
Sauerstoffmangel lack of oxygen
Sauerstoffpermeabilität oxygen permeability
Sauerstoffplasma oxygen plasma
sauerstoffreich high-oxygen, rich in oxygen
Sauerstoffüberschuß excess oxygen
saugend absorbent
saugfähig absorbent
Saugfähigkeit absorbency
Saugpumpe suction pump
Saugvermögen absorbency
Säure acid
Säurealkohol acid alcohol
Säureamid acid amide
Säureamidgruppe acid amide group
Säureanhydrid acid anhydride
Säureanhydridhärter acid anhydride catalyst
säurearm low-acid, with a low acid content
Säurebeizen pickling
säurebeständig acid resistant
Säurebeständigkeit acid resistance
Säurechlorid acid chloride
Säuredämpfe acid fumes
Säurefänger acid interceptor
säurefest acid resistant
Säuregehalt acid content
Säuregruppe acid radical
säurehaltig containing acid
säurehärtbar acid curing
säurehärtend acid curing
Säurehärter acid catalyst
Säurehydroxylamin acid hydroxylamine
säureinstabil not acid resistant
Säurekatalysator acid catalyst
Säurekatalyse acid catalysis
säurekatalysiert acid catalysed
säurekatalytisch with an acid catalyst, acid catalysed
Säurekomponente acid component
Säurekonzentration acid concentration
säurelabil unstable in the presence of acids
säurelöslich acid soluble
säuremodifiziert acid modified
Säureradikal acid radical/group
säureresistent acid resistant
Säureresistenz acid resistance
Säurerest acid radical/group
Säurespender acid donator
säureunlöslich acid insoluble, insoluble in acids
Säurezahl acid value
Scannen scanning
Schablone template, pattern, stencil
Schadgas harmful/corrosive gas
schädigend damaging, harmful
Schädigung damage, fracture
Schädigung, thermische thermal degradation
Schädigungsarbeit fracture energy
Schädigungsenergie fracture energy
Schädigungsgrad amount of damage
Schädigungskraft fracture force
Schädigungskriterien damage criteria
schädigungslos harmless
schädlich harmful
Schadstoff harmful substance
schadstoffarm 1. low-pollutant. 2. with a low content of harmful/noxious substances
Schadstoffbelastung pollution
Schäftung scarf joint
schälbeansprucht under peel stress
Schälbeanspruchung peel stress
Schälbelastung peel stress
Schäldiagramm peel curve/diagram
Schale shell
Schälfestigkeit peel strength
Schälgeschwindigkeit peel rate
Schälkraft peel force
schallabsorbierend sound absorbent
Schallabsorption sound absorption

Schallabsorptionsgrad degree of sound absorption, amount of sound absorbed
Schallabstrahlung sound reflection
Schallamplitude sound amplitude
schallarm quiet in operation, quiet running
Schallausbreitung sound propagation
Schalldämmaß degree of sound insulation
Schalldämmasse sound insulating material
schalldämmend soundproof, sound insulating
Schalldämmplatte sound insulating panel
Schalldämmstoff sound insulating material
Schalldämmung sound insulation, soundproofing
Schalldämmungselement sound insulating element, soundproofing element
Schalldämmwert sound insulation factor
Schalldämpfung sound damping
Schalldurchgang sound transmission
Schalleitfähigkeit sound conductivity
Schallemission acoustic emission
Schallemissionsanalyse acoustic emission analysis
Schallemissionsrate acoustic emission rate
Schallenergie sonic energy
Schallfortpflanzung sound propagation
schallgedämpft soundproofed
schallgeschützt soundproofed
Schallgeschwindigkeit velocity of sound
Schallimpuls acoustic impulse
Schallisolationsvermögen sound insulation properties
schallisolierend sound insulating
Schallisoliermaterial sound insulating material
Schallpegel noise level
Schallreduzierung noise reduction
Schalöl mould oil
Schälprobekörper peel test piece
Schälspannung peel stress
Schaltkreis switching/control circuit
Schaltschema 1. circuit diagram. 2. circuitry
Schaltung circuit
Schaltung, gedruckte printed circuit
Schälung peel
Schälversuch peel test
Schälweg peel distance
Schälwert peel strength
Schälwiderstand peel strength
Schälwinkel peel angle
scharfkantig sharp edged
Schaum foam
schaumanfällig liable to foam
Schaumanfälligkeit tendency to foam
Schaumbekämpfung combating foam
Schaumbeton aerated/cellular concrete
Schaumbildung foaming
Schaumbildungstendenz tendency to foam
Schaumbläschen foam bubbles
Schaumblasen foam bubbles
schaumempfindlich liable to foam
Schaumherstellung foam production
schauminhibierend antifoam, foam inhibiting
Schaumkrater foam crater
Schaumlamelle foam lamella
Schaumlebenszeit foam life
Schäumneigung tendency to foam
Schaumprobleme foaming problems
Schaumrolle foam roller
Schaumstabilisator foam stabiliser
schaumstabilisierend foam stabilising
Schaumstoff foam
Schaumstörung foaming
Schaumstruktur foam structure
schaumunterdrückend foam suppressing
Schaumverhinderer antifoam (agent)
schaumverhindernd antifoam
Schaumverhinderung foam prevention
Schaumverhütungsmittel antifoam (agent)
Schaumvermögen foaming power
schaumzerstörend antifoam
scheibenförmig disc-shaped
scheinbar apparent
scheinbare Viskosität apparent viscosity
Scheinwerferstreuscheibe headlamp diffuser
Scheitelpunkt vertex
Schellack shellac
schematisch schematic
Scherbeanspruchung shear stress
Scherbelastung shear stress
Scherbereich shear range
Scherbewegung shear movement
Schercraze shear craze
Scherdeformation shear deformation
Scherdehnung shear strain
scherempfindlich shear sensitive
Scherempfindlichkeit shear sensitivity
scheren to shear
Scherenergie shear energy
scherentzähend pseudoplastic, non-Newtonian
Scherermüdung shear fatigue
Schererwärmung heat produced/generated through shear
Scherfeld shear field
Scherfestigkeit shear strength
Scherfestigkeit, interlaminare interlaminar shear strength
Scherfließen shear flow
Schergefälle shear rate
Schergeschwindigkeit shear rate
schergeschwindigkeitsabhängig depending on the shear rate
Schergeschwindigkeitsbereich shear rate range
Schergeschwindigkeitsgefälle shear rate/gradient
schergeschwindigkeitsunabhängig independent of the shear rate
scherinduziert shear induced
scherintensiv shear intensive
Scherkraft shear force
Scherkrafteinstellung introduction of shear forces
scherkraftreich high-shear
Schermodul shear modulus
Scherniveau shear

Scherprobekörper shear test piece
Scherprüfung shear test
Scherrate shear rate
Scherratenbereich shear rate range
Scherrateneinfluß effect of shear rate
Scherschwellfestigkeit shear fatigue strength
Scherspannung shear stress
Scherstabilität shear resistance
Scherstandfestigkeit shear strength
Scherstandvermögen shear strength
Scherströmung shear flow
scherunempfindlich unaffected by shear
Scherung shear
scherverdünnend pseudoplastic, non-Newtonian
Scherverdünnung pseudoplasticity, non-Newtonian flow/behaviour
Scherverdünnungskurve non-Newtonian flow curve
Scherverformung shear deformation, deformation due to shear
Scherverformungskraft shear deformation force
Scherverhalten shear behaviour, behaviour under shear
Scherversuch shear test
Scherzugschwellfestigkeit tensile shear fatigue strength
scheuerbeständig scrub resistant
Scheuerbeständigkeit scrub resistance
Scheuerfestigkeit scrub resistance
Scheuerresistenz scrub resistance
Scheuertest scrubbing test
Scheuerwert scrub resistance
Scheuerzyklen scrubbing cycles
Schicht 1. layer, coat(ing). 2. shift
Schicht, dampfbremsende vapour barrier
Schichtablösung (paint) film detachment
Schichtabtrennung (paint) film detachment
Schichtauftrag coating, film application
Schichtbetrieb shift operation
schichtbildend film forming
Schichtdelamination delamination
Schichtdicke film/coating thickness
Schichtdickenmesser film thickness measuring instrument
Schichtdickenmessung film thickness determination
Schichtholz plywood
Schichtlage layer
Schichtpreßstoff laminate
Schichtpreßstoffplatte laminate sheet
Schichtpreßstofftafel laminate sheet
Schichtsilikat layer-lattice silicate
Schichtstärke film/coating thickness
Schichtstoff laminate
Schichtstoffharz laminating resin
Schichtstoffplatte laminate sheet
Schiffsanstrich marine coating/finish
Schiffsanstrichsystem marine paint
Schiffsbau shipbuilding
Schiffsbauindustrie shipbuilding industry
Schiffsbewuchs marine fouling
Schiffsfarbe marine paint
Schiffskörper hull
Schiffskörperaußenfarbe marine paint
Schiffsrumpf hull
schillernd iridescent
Schimmel mould
Schimmelbefall attack by mould
Schimmelbildung mould formation
Schimmelpilz mould
Schlacke slag
Schlag impact
schlagabsorbierend impact absorbing
Schlagabsorption impact absorption
Schlagarbeit impact energy
schlagartig 1. impact-type. 2. sudden, abrupt
schlagartige Beanspruchung 1. impact stress. 2. sudden stress
schlagbeansprucht under impact stress
Schlagbeanspruchung impact stress
Schlagbelastbarkeit impact resistance
schlagbeständig impact resistant
Schlagbeständigkeit impact resistance
Schlagbiegebeanspruchung flexural impact stress
Schlagbiegefestigkeit flexural impact strength
Schlagbiegeprüfung flexural impact test
Schlagbiegestab flexural impact test piece, flexural impact bar
Schlagbiegeverhalten flexural impact behaviour
Schlagbiegeversuch flexural impact strength
Schlagbiegezähigkeit flexural impact strength
Schlagempfindlichkeit sensitivity to impact
Schlagenergie impact enervy
schlagfest impact resistant
Schlagfestigkeit impact strength/resistance
Schlaggeschwindigkeit impact speed, speed of impact
Schlaghammer pendulum
Schlagintensität impact force
Schlagkraft impact energy
Schlaglast impact stress
Schlagpendel pendulum
Schlagprüfstab impact specimen/bar, impact test piece
Schlagprüfung impact test
Schlagregen driving rain
Schlagregendichte imperviousness to driving rain
Schlagregenschutz protection against driving rain
schlagregensicher resistant to driving rain
Schlagregensicherheit resistance to driving rain
schlagresistent impact resistant
Schlagrichtung direction of impact
Schlagtest impact test
Schlagtiefung impact indentation
Schlagtiefung nach Erichsen Erichsen impact indentation
Schlagtiefungsgerät impact indentation tester
Schlagtiefungsprüfung impact indentation test
schlagunverformbar resistant to deformation

through impact
Schlagunverformbarkeit resistance to deformation through impact
Schlagverhalten impact behaviour, behaviour on impact
Schlagversuch impact test
Schlagwinkel angle of impact
schlagzäh impact resistant
Schlagzähigkeit impact strength/resistance
Schlagzähigkeitsuntersuchung impact test
Schlagzugversuch tensile impact test
Schlagzugzähigkeit tensile impact strength
Schlamm sludge
Schlämme slurry
Schlämmung slurry
Schlauchleitung pipeline
Schlaufe loop
Schlaufenkonformation loop configuration
schlechte Haftung poor adhesion
schlechter Verlauf poor flow
schlechthaftend having poor adhesion
schlechtlöslich sparingly soluble, having pooor solubility
Schleier haze
Schleierbildung blushing
Schleifband abrasive belt
schleifbar sandable, grindable
Schleifbarkeit grindability, sandability
Schleifen grinding, sanding
Schleifgewebe abrasive/emery cloth
Schleifkorn abrasive (material)
Schleifkörper abrasive/grinding wheel
Schleifleinen abrasive/emery cloth
Schleifmaterial abrasive (material)
Schleifmittel abrasive (material)
Schleifpapier sandpaper, emery paper
Schleifscheibe abrasive/grinding wheel
Schleifvlies abrasive cloth
Schleifwolle steel wool
Schleimhäute mucous membranes
schleimhautreizend irritating the mucous membranes
Schleppkraft drag force
Schlichte size
Schlichtemittel size *(in a textile context)*
Schlieren streaks
schlierenfrei free from streaks
schlierig streaky
Schliff 1. ground glass joint. 2. ground glass stopper
Schlupftendenz tendency to slip
Schlüsseleigenschaften key properties
schlüsselfertig turnkey
Schlüsselprodukt key product
Schlüsselrolle key role
Schlüsselstellung key position
Schlußfolgerung conclusion
Schlußschicht final coat
Schlußstrich top/finishing coat
schmal narrow
schmelzbar fusible
Schmelzbarkeit fusibility
Schmelzbereich melting range
Schmelzdichtstoff hot melt sealant
Schmelzdruckfarbe hot melt printing ink
Schmelze melt
Schmelzenthalpie melt enthalpy
Schmelzentropie melt entropy
schmelzfähig fusible
schmelzflüssig melted, molten
schmelzförmig melted, molten
Schmelzhaftklebstoff hot melt pressure sensitive adhesive
Schmelzhaftschicht hot melt primer coat
Schmelzharz hot melt resin
Schmelzindex melt flow index, MFI
Schmelzindexwert melt flow index, MFI
Schmelzintervall melting range
Schmelzklebefolie hot melt film adhesive
Schmelzkleber hot melt adhesive
Schmelzkleberpatrone hot melt adhesive cartridge
Schmelzklebrohstoff hot melt adhesive resin
Schmelzklebstoff hot melt adhesive
Schmelzklebstoffauftragsgerät hot melt adhesive applicator
Schmelzklebstoffgranulat granulated hot melt adhesive
Schmelzklebstoffpulver powdered hot melt adhesive
Schmelzlack hot melt varnish
Schmelzmasse hot melt compound
Schmelzpunkt melting point
Schmelzpunktanhebung melting point increase, increase in melting point
Schmelzstabilität melt stability
Schmelztemperatur melting point
Schmelztemperaturbereich melting range
Schmelzverhalten melting characteristics
Schmelzviskosität melt viscosity
Schmelzwalzenbeschichtung hot melt coating
Schmelzwärme heat of fusion
Schmiereffekt 1. lubricating effect. 2. smearing *(e.g. when polishing)*
Schmiereigenschaften lubricating properties
schmierend 1. lubricating. 2. smearing
Schmierfähigkeit lubricating properties, lubricity
Schmierfett lubricating grease
Schmierfilm lubricating film
Schmiermittel lubricant
Schmiermittelfilm lubricating film
Schmiermittelwirkung lubricating effect
Schmierschicht lubricating film
Schmierstoff lubricant
Schmiersystem lubricating system
Schmierung lubrication
Schmierwirkung 1. lubricating effect. 2. smearing *(e.g. when polishing)*
Schmirgelleinen emery cloth
schmirgeln to sand (down), to rub down
Schmutz dirt
Schmutzablagerung deposition of dirt

schmutzabweisend dirt repellent
Schmutzabweisung dirt repellency
Schmutzfleck stain
Schmutzpartikel dirt particle
Schmutzteilchen dirt particle
Schneidwerkzeug cutting tool
schnellabbindend fast setting
schnellaufend high-speed
Schnellbewitterung accelerated weathering
schnellerstarrend fast setting/solidifying
schnellfließend fast flowing
schnellhärtend fast curing/setting
Schnellhärter fast reacting hardener, high-speed hardener
Schnellhärtung accelerated cure
Schnellkleber quick action adhesive
Schnellkupplung quick action coupling
Schnellmischer high-speed mixer
Schnellrührer high-speed stirrer
schnellstabbindend very fast setting
Schnelltest accelerated test
schnelltrocknend fast drying
schnellverdunstend fast drying/evaporating
Schnittpunkt point of intersection
Schnittstelle interface
schockartig sudden
schonendes Erwärmen gentle heating
Schramme scratch
Schrotstrahlen grit blasting
Schrumpf shrinkage
schrumpfarm low-shrinkage
schrumpffrei non-shrink(ing)
Schrumpfriß shrinkage crack
Schrumpfspannung shrinkage stress
Schrumpfung shrinkage
Schub shear
Schubbeanspruchung shear stress
Schubbruchspannung shear stress at break
Schubdeformation shear deformation
Schubfestigkeit shear strength
Schubkraft shear force
Schubmodul shear modulus
Schubmodulabfall decrease in shear modulus
Schubmodulniveau shear modulus
Schubmodulverlauf shear modulus profile
Schubscherfestigkeit shear strength
Schubschwellfestigkeit shear fatigue strength
Schubspannung shear stress
Schubspannungsänderung change in shear stress
Schubspannungsspitze shear stress peak
Schubspannungssprung sudden change in shear stress
Schubspannungsverteilung shear stress distribution
schubsteif shear resistant, resistant to shear stress/forces
Schubsteifigkeit shear strength
Schubverformung shear deformation
schubweich affected by shear, not resistant to shear
Schuhabsatz heel
Schuhcreme shoe polish
Schuhfertigung shoe production
Schuhherstellung shoe production
Schuhindustrie shoe industry
Schuhklebstoff shoe adhesive
Schuhsole (shoe) sole
Schuhsolenmaterial soling material
Schulung training, instruction
Schulungsprogramm training programme
Schulungszeit training period
Schulungszentrum training centre
Schuppen flakes
schuppenförmig flake-like
schüttbar pourable
Schüttdichte apparent density
schütteln to shake
schütten to pour
schüttfähig pourable
Schüttfähigkeit pourability
Schüttgewicht apparent density
Schüttgut bulk solids, powder, loose material. *The word is used to describe any solid material which can be poured out of a bag (**schütten** = to pour)*
Schutz protection
Schutzanstrich protective coating/finish
Schutzbarriere protective barrier
Schutzbekleidung protective clothing
Schutzbeschichtung protective coating
Schutzbrille (safety) goggles
Schutzcreme barrier cream
schützend protective
Schutzfilm protective film
Schutzfunktion protective function
Schutzgas inert gas
Schutzgasatmosphäre inert gas atmosphere
Schutzhandschuhe protective gloves
Schutzhaube safety hood
Schutzhülle protective envelope
Schutzkleidung protective clothing
Schutzkolloid protective colloid
Schutzkolloideinfluß effect of protective colloid
Schutzkolloidzusatz addition of protective colloid
Schutzlack protective lacquer
Schutzlage protective layer
Schutzmaske face mask
Schutzmaßnahmen safety precautions
Schutzpigment protective pigment
Schutzrechte patent rights
Schutzschicht protective layer/coating/film
Schutzüberzug protective coating
Schutzverkleidung safety shield/cover/hood
Schutzvorrichtung safety device
Schutzvorschriften safety regulations
Schutzwirkung protective effet
Schwabbelscheibe buffing wheel
schwachpigmentiert slightly pigmented, with a low pigment content
schwachpolar slightly polar
Schwachpunkt weak spot, weakness, defect

schwachsauer slightly acid
Schwachstelle weak spot, weakness, defect
schwachstellenfrei free from defects *(e.g. a paint film)*
schwachvernetzt loosely crosslinked
Schwankungsbreite variation/fluctuation range
schwarz black
Schwarzblech steel plate
Schwarzfarbe black ink
Schwarzpigment black pigment
Schwärzung blackening
Schwebstoffe suspended matter/particles
Schwefel sulphur
Schwefeldioxid sulphur dioxide
schwefelfrei sulphur free
Schwefelkohlenstoff carbon disulphide
Schwefelsäure sulphuric acid
Schwefelverbindung sulphur compound
schweflige Säure sulphurous acid
Schweißen welding
Schweißnaht weld, seam
Schweißverbindung welded joint
Schwellbeanspruchung fatigue stress
Schwellbelastbarkeit fatigue endurance
Schwellbelastung fatigue stress
Schwelle threshold
schwellende Beanspruchung fatigue stress
schwellende Belastung fatigue stress
Schwellenwert threshold value
Schweller door sill
schwer, spezifisch heavy
schwerbenetzbar difficult to wet
Schwerbrennbarkeit flame resistance
schwerdispergierbar difficult to disperse
Schwerebeschleunigung acceleration due to gravity
Schweremballage heavy duty pack
schwerentflammbar flame resistant/retardant
Schwerentflammbarkeit flame resistance, flame retardant properties
schwerfließend poor-flow, having poor flow
schwerflüchtig low-volatility
Schwerflüchtigkeit low volatility
schwerklebbar difficult to bond/stick
schwerlöslich sparingly soluble, difficult to dissolve
Schwermetall heavy metal
schwermetallfrei free from heavy metals, not containing heavy metals
Schwermetallgehalt heavy metal content
schwermetallhaltig containing heavy metals
Schwermetallpigment heavy metal pigment
Schwermetallsalz heavy metal salt
Schwermetallverbindung heavy metal compound
Schwerpunkt centre of gravity
Schwerspat barytes
schwerzugänglich difficult to reach
Schwimmbad swimming pool
Schwimmbadfarbe swimming pool paint
Schwimmbecken swimming pool
Schwindung shrinkage
Schwindungsanisotropie anisotropic shrinkage
schwindungsarm low-shrinkage
Schwindungsdifferenzen shrinkage differences/variations
Schwindungseigenschaften shrinkage properties/characteristics
Schwindungsspannung shrinkage stress
Schwindungsuntersuchung shrinkage test
Schwindungsvorgang shrinkage process
Schwingbeanspruchung cyclic/oscillating stress
schwingbelastet subjected to vibrational stress
Schwingbelastung vibrational stress
schwingend vibrational
Schwingfestigkeit fatigue strength
Schwingfrequenz vibration frequency
Schwingschleifen honing, superfinishing
Schwingschleifer honing machine
Schwingspiel (one complete) vibration
Schwingspielzahl number of (complete) vibrations
Schwingung vibration, oscillation
Schwingungsamplitude vibration amplitude
Schwingungsbande vibration band
Schwingungsbeanspruchung vibrational stress
schwingungsdämpfend vibration damping
Schwingungsenergie vibrational energy
schwingungsresistent vibration resistant
Schwingungsverhalten fatigue behaviour
Schwingungsversuch torsion pendulum test
Schwingverhalten fatigue behaviour
Schwitzwasser condensation
Schwitzwasserbelastung exposure to condensation
schwitzwasserbeständig resistant to condensation
Schwitzwasserbeständigkeit resistance to condensation
schwitzwasserfest resistant to condensation
Schwitzwasserklima condensed moisture atmosphere
Schwitzwasserprüfung condensed moisture test
Schwund shrinkage, contraction
Schwundmaß degree/extent/amount of shrinkage
Schwundriß shrinkage crack
sechswertig hexavalent
Sedimentation sedimentation, settling out
Sedimentationsanalyse sedimentation analysis
sedimentationshemmend anti-sedimentation
Sedimentationsneigung tendency to settle (out)
Sedimentationstendenz tendency to settle (out)
Sedimentationsverhalten sedimentation characteristics/behaviour
Sedimentbildung sedimentation, settling (out)
sedimentieren to settle (out)
Seeklima marine climate
Seeluft sea air
Seewaser seawater
seewasserfest resistant to seawater
seidenglänzend silk

Seidenglanzlack silk finish paint
seidenmatt silk lustre
Seife soap
Seifenlösung soap solution
Seifenwasser soapy water
Seitengruppe side group
Seitenkette side chain
Seitenkettenpolymer branched polymer
seitenständig lateral
sekundär secondary
Sekundäreffekt side effect
Sekundärelektron secondary electron
Sekundärion secondary ion
Sekundärionenmassenspektrometrie secondary ion mass spectrometry
Sekundärkorn secondary particle
Sekundärteilchen secondary particle
Sekundenbruchteil fraction of a second
Sekundenklebstoff superglue
Sekundenschnelle, in within seconds
selbstabtastend self-scanning
Selbstbindemittel self-curing binder
selbstdichtend self-sealing
Selbstemulgierbarkeit self-emulsifying properties
selbstemulgierend self-emulsifying
selbstglättend self-levelling
selbsthaftend self-adhesive
selbsthärtend self-curing
Selbstklebeband self-adhesive tape
Selbstklebeetikett self-adhesive label
Selbstklebemasse pressure sensitive adhesive
selbstklebend self-adhesive
Selbstkleber pressure sensitive adhesive
Selbstklebeschicht self-adhesive coating
Selbstreinigungseffekt self-cleaning effect
selbstschmierend self-lubricating
selbsttragend self-supporting
selbstverlaufend self-levelling
selbstverlöschend self-extinguishing
selbstvernetzend self-curing
Selbstvernetzung self-curing
selektiv selective
SEM 1. scanning electron microscope. 2. scanning electron microscopy
SEM-Analyse scanning electron microscopy
semikristallin semicrystalline
semipermeabel semipermeable
Semipermeabilität semipermeability
semitransparent semitransparent
Senf mustard
senkrecht vertical
Senkung reduction, lowering, decrease
sensibel sensitive; *this word should never be translated as* sensible, *which has a totally different connotation.*
Sensibilisator sensitiser
sensibilisierend sensitising
Sensibilisierung sensitisation
Sensibilisierungsmittel sensitising agent
Sensor sensor, probe, transducer
sequenziell sequential
Serienfahrzeug 1. standard vehicle. 2. mass produced vehicle
Serienfertigung mass production
Seriengerät standard unit
Serienlackierung production line painting
Serienprogramm standard range
serienüblich standard
Serienversion standard model/version
SERRS surface enhanced resonance Raman spectroscopy
SERS surface enhanced Raman spectroscopy
Servoventil servo-valve
SH 1. *abbr. of* **selbsthärtend**, self-curing. 2. *abbr. of* **säurehärtend**, acid- curing
SH-Lack 1. self-curing paint. 2. acid curing paint
Shore-Härte Shore hardness
Shore-Härte A Shore A hardness
Shore-Härte D Shore D hardness
Shoreänderung change in Shore hardness
Sicherheit security, safety, reliability
Sicherheitsabstand safety margin/factor
Sicherheitsanforderungen safety requirements
Sicherheitsauflage safety directive
Sicherheitsbeiwert safety factor
Sicherheitsbestimmungen safety regulations
Sicherheitsdatenblatt safety data sheet
Sicherheitseinrichtung safety device
Sicherheitserfordernisse safety requirements
Sicherheitsfaktor safety factor
Sicherheitsforderungen safety requirements
Sicherheitsgesetze safety regulations
Sicherheitsglas safety glass
Sicherheitsgrenze safety margin
Sicherheitsgründen, aus for reasons of safety
Sicherheitshinweise safety instructions
Sicherheitskennzahlen safety data
Sicherheitskoeffizient safety factor
Sicherheitsmaßnahmen safety precautions
Sicherheitsnorm safety standard
Sicherheitsratschläge safety instructions
Sicherheitsregeln safety regulations
sicherheitsrelevant *important for safety:* **sicherheitsrelevante Daten** safety data
Sicherheitsreserve safety margin
Sicherheitsrichtlinien safety guidelines
Sicherheitsrisiko safety risk
Sicherheitsspielraum safety margin
sicherheitstechnisch safety: **sicherheitstechnische Hinweise** safety instructions
Sicherheitsvorkehrungen safety precautions
Sicherheitsvorschriften safety regulations
Sicherheitszahl safety factor
sichtbar visible
Sichtbarmachung making visible
Sichtbeton exposed concrete
Siebdruck 1. screen printing. 2. screenprint
Siebdruckfarbe screen printing ink
Siebdruckmasse screen printing ink
Siebdrucktinte screen printing ink

Siebdruckverfahren screen printing
Siebgewebe filter cloth, screen fabric
Siebpatrone cartridge filter
Siedebereich boiling range
Siedeintervall boiling range
siedend boiling
Siedepunkt boiling point
Siedetemperatur boiling point
Siedewasserlagerung immersion in boiling water
siegelbar heat sealable
Siegeleigenschaften heat sealing properties
siegelfähig heat sealable
Siegelfolie heat sealing film
Siegellack sealing varnish
Siegelschicht heat sealable coating
Siegeltemperatur heat sealing temperature
Siegelwerkzeug heat sealing instrument/tool
Signal signal
Sikkativ drier
sikkativieren to mix with drier: **Beide Harze lassen sich problemlos sikkativieren** Both resins will readily tolerate drier
sikkativiert mixed/formulated with drier
Sikkativierung mixing with drier
Silan silane
silanbehandelt silanised, silane treated
Silanendgruppe terminal silane group
Silanester silane ester
Silangruppe silane group
Silanharz silane resin
silanisiert silanised
Silanisierung silanisation
Silankette silane chain
Silanmonomer silane monomer
Silanol silanol
Silanolendgruppe terminal silanol group
silanolfunktionell silanol functional
Silanolgehalt silanol content
Silanolgruppe silanol group
Silanverbindung silane compound
silanvernetzt silane cured/crosslinked
Silanvernetzung silane cure/crosslinkage
Silanwasserstoffgruppe silane hydrogen group/radical
Silanwasserstoffrest silane hydrogen group/radical
Silazan silazane
Silber silver
silberbeschichtet silver coated
silbergefüllt silver filled
Silberkolloid colloidal silver
Silbernitrat silver nitrate
Silberoberfläche silver surface
Silberpartikel silver particle
Silberteilchen silver particle
silberüberzogen silver coated
silberweiß silvery white
Silicium silicon
Siliciumchip silicon chip
Siliciumdioxid silica
Siliciumeinheit silicon unit
siliciumfrei silicon free
siliciumhaltig containing silicon
Siliciumkarbid silicon carbide
Siliciumnitrid silicium nitride
siliciumorganisch organosilicon
Siliciumoxid silicium oxide
Siliciumscheibe silicon wafer
Siliciumwafer silicon wafer
Silicon silicone
Siliconalkydharz silicone alkyd resin
Siliconanteil silicone content
siliconarm low-silicone
Siliconat siliconate
Siliconbautenschutzmittel silicone masonry water repellent
siliconbeschichtet silicone coated
Siliconbeschichtungszusammensetzung silicone coating composition
Silicondichtstoff silicone sealant
Siliconelastomer silicone elastomer
Siliconemulsion silicone emulsion
Siliconemulsionsfarbe silicone emulsion paint
Siliconentschäumer silicone defoaming agent
Siliconfarbe silicone paint
Siliconfestharz solid silicone resin
siliconfrei silicone free
Silicongehalt silicone content
Silicongummi silicone rubber *(see explanatory note and translation example under* **Kautschuk***)*
Siliconharz silicone resin
Siliconharzemulsion silicone resin emulsion
Siliconharzfarbe silicone resin paint
Siliconharzpulver powdered silicone resin
Siliconimprägnierung silicone impregnation
siliconisiert siliconised, silicone coated
Siliconisierung siliconisation
Siliconkautschuk silicone rubber
Siliconkautschukkleber silicone rubber adhesive
Siliconkautschuklatex silicone rubber dispersion
Siliconkautschukvulkanisat vulcanised silicone rubber
Siliconklebstoff silicone adhesive
Siliconlack silicone paint
Siliconlackadditiv silicone paint additive
Siliconlösung silicone solution
siliconmodifiziert silicone modified
Siliconöl silicone fluid
Siliconpapier siliconised paper, silicone coated paper
Siliconpolymer silicone polymer
Silicontensid silicone surfactant
Silicontrennmittel silicone release agent
Silicontrennpapier silicone release/backing paper
Siliconüberzug silicone coating
Silikagel silica gel
Silikat silicate
Silikatester silicate ester
Silikatfarbe silicate paint
silikatfrei silicate-free

Silikatfüllmittel silicate filler
silikatisch siliceous
Silikon-, silikon- *see* **Silicon-, silicon-**
Silikonfett silicone grease
silikonfrei silicone-free
Silizium-, silizium- *see* **Silicium-, silicium-**
Silo silo
Siloanlage silo installation
Silofahrzeug road tanker
Silowagen road tanker
Siloxan siloxane
Siloxanalkylsulfid siloxane alkyl sulphide
Siloxanalkylthiocyanat siloxane alkyl thiocyanate
siloxanbasierend siloxane-based
Siloxanbasis, auf siloxane-based
Siloxanbaustein siloxane unit
Siloxanbindung siloxane linkage
Siloxanblock siloxane block
Siloxaneinheit siloxane unit
Siloxanemulsion siloxane emulsion
Siloxanentschäumer siloxane defoaming agent
Siloxanfarbe siloxane paint
Siloxangruppe siloxane group
Siloxanharz siloxane resin
Siloxankette siloxane chain
siloxanlöslich siloxane-soluble
Siloxanlösung siloxane solution
Siloxanöl siloxane fluid
Siloxanpolymer siloxane polymer
Siloxanrest siloxane group
siloxansubstituiert siloxane-substituted
Siloxantauchlack siloxane dipping paint
Siloxanverbindung siloxane compound
Siloxanzusammensetzung siloxane composition
Silylether silyl ether
Silylgruppe silyl group
Silylierung silylation
Silylmonomer silyl monomer
Silylverbindung silyl compound
SIMS secondary ion mass spectrometry
SIMS-Methode secondary ion mass spectrometry
Simulation simulation
simuliert simulated
sintern to sinter
Sinterofen sintering oven
Sinterpulver powder coating
Sinterung sintering
Sodalösung soda solution
Sohlenklebstoff (shoe) sole adhesive
Sojaalkyd soya bean oil alkyd
Sojaalkydharz soya bean oil alkyd (resin)
Sojabohnenöl soya bean oil
Sojafettsäure soya bean oil fatty acid
Sojaöl soya bean oil
Sojaölfettsäure soya bean oil fatty acid
Sojaoxazolin soya bean oxazoline
Solldruck required/set pressure
Sollkonzentration required concentration
Sollkurve setpoint/reference curve
Sollqualität required quality
Solltemperatur required/set temperature
Solubilisierung solubilisation
Solvatation solvation
solvatisiert solvated
Solventnaphtha solvent naphtha
Solzustand sol state
Sommermonate summer months
Sonderabfall toxic/hazardous waste
Sondereinstellung special formulation
Sondermüll toxic/hazardous waste
Sonne sun
Sonnenbestrahlung solar irradiation
Sonnenblumenöl sunflower oil
Sonneneinstrahlung solar radiation
Sonnenkollektor solar collector
Sonnenlicht sunlight
Sonnensimulation sun simulation
Sonnenspektrum solar spectrum
Sonnenstrahlung solar radiation
Sonnenstunden hours of sunshine
Sorbitanester sorbitate
Sortiment range
Spachtel 1. putty knife, trowel. 2. knifing filler, stopper, surfacer
spachtelfähig capable of being applied by knife
Spachtelmasse knifing filler, stopper, surfacer
Spalt slit, cack
Spaltkorrosion crevice corrosion
Spaltprodukt decomposition product
Spaltung 1. cleavage. 2. separation, decomposition *(in a chemical sense)*
spanabhebende Bearbeitung machining
Späne chips
Spannung 1. stress. 2. voltage, tension
Spannungs-Dehnungsdiagramm stress-strain diagram
Spannungs-Dehnungskurve stress-strain curve
Spannungs-Dehnungslinie stress-strain curve
Spannungs-Dehnungsverhalten stress-strain behaviour
Spannungs-Gleitungsverhalten stress-surface slip behaviour
Spannungs-Verformungsverhalten stress-strain behaviour
Spannungsabbau stress degradation/relaxation
Spannungsamplitude stress amplitude
Spannungsanalyse stress analysis
spannungsarm low-stress
Spannungsberechnung stress calculation
spannungsfrei stress-free
Spannungsgleichgewicht stress equilibrium
Spannungsgrenze stress limit
Spannungsinkrement stress increment
Spannungsintensität stress intensity
Spannungsintensitätsamplitude stress intensity amplitude
Spannungsintensitätsfaktor stress intensity factor
Spannungskonzentration stress concentration
Spannungskorrosion (environmental) stress

cracking
Spannungskorrosionsbeständigkeit stress cracking resistance
Spannungskorrosionsriß stress crack
Spannungskorrosionsverhalten stress cracking behaviour
Spannungskraft stress
Spannungsmaximum maximum stress
Spannungsoptik photoelasticity
spannungsoptisch photoelastic
Spannungsrelaxation stress relaxation
Spannungsrelaxationsversuch stress relaxation test
Spannungsriß stress crack
Spannungsrißanfälligkeit susceptibility to stress cracking
spannungsrißbeständig resistant to stress cracking
Spannungsrißbeständigkeit stress cracking resistance
Spannungsrißbildung, umgebungsbeeinflußte environmental stress cracking
spannungsrißempfindlich susceptible to stress cracking
Spannungsrißkorrosion stress cracking
Spannungsrißkorrosionsbeständigkeit stress cracking resistance
Spannungsrißverhalten stress cracking behaviour
Spannungsspitze stress peak
Spannungsverhältnisse stress conditions
Spannungsverlauf stress pattern
Spannungsverteilung stress distribution
Spannungswert modulus
Spannungszustand state of stress
Spannvorrichtung fixture, clamping device
Spannzange clamp
Spanplatte (wood) chipboard
Spatel spatula
SPC statistical process control
Speichermodul storage modulus
Speiseöl cooking oil
spektral spectral
Spektralbereich spectral range
Spektralenergie spectral energy
Spektralkurve spectral distribution curve
Spektralphotometer spectrophotometer
spektralphotometrisch spectrophotometric
Spektrenbereich spectral range
Spektrogramm spectrogram
spektrographisch spectrographic
Spektrometrie spectrometry
spektrometrisch spectrometric
Spektrophotometer spectrophotometer
spektrophotometrisch spectrophotometric
Spektroskopie spectroscopy
spektroskopisch spectroscopic
Spektrum 1. speectrum. 2. range
Sperreigenschaften barrier properties
Sperrfolie barrier film
Sperrgrund barrier primer coat
Sperrgrundierung barrier primer
Sperrholz plywood
Sperrholzklebstoff plywood adhesive
Sperrpigment barrier pigment
Sperrschicht barrier film/coating/layer
Sperrwirkung barrier effect
Spezialanwendung speciality application
Spezialharz speciality/special-purpose resin
Spezialkenntnisse specialised knowledge/experience
Spezialklebstoff special-purpose adhesive
Spezialmarke speciality/special-purpose grade
Spezialtype speciality/special-purpose grade
Spezifikation specification
spezifikationsgerecht according to specification
spezifisch specific
spezifisch schwer heavy
spezifische Oberfläche specific surface area
spezifische Viskosität specific viscosity
spezifische Wärme specific heat
spezifischer Durchgangswiderstand volume resistivity
spezifischer Oberflächenwiderstand surface resistivity
spezifischer Widerstand resistivity
spezifisches Gewicht specific gravity
spezifisches Volumen specific volume
sphärisch spherical
Sphärolit spherulite
Sphärolitgefüge spherulite structure
sphärolitisch spherulitic
spiegelähnlich mirror-like
spiegelhochglanzpoliert polished to a mirror finish
Spiegelung reflection
spindelförmig spindle shaped
Spinellpigment spinel pigment
Spinellstruktur spinel structure
Splitterbruch brittle failure/fracture
splittern to splinter
splittersicher splinter-proof
spreiten to spread
Spreitungsdruck spreading pressure
Spreitungskoeffizient spreading coefficient
Spreitungsvermögen spreading power
Spreitvermögen spreading power
Sprit spirit
Spritzabstand spraying distance
Spritzapplikation spray application
Spritzauftrag spray application
spritzbar sprayable
Spritzbarkeit sprayability
Spritzbedingungen spraying conditions
Spritzdruck spraying pressure
Spritzdüse nozzle
Spritzen spraying, spray application
spritzfertig ready for spraying
Spritzgeschwindigkeit spraying speed
Spritzkabine spraying booth
Spritzlack spraying paint
Spritzlackierung spray painting

Spritznebel spray mist
Spritzpistole spraygun
Spritzverfahren spraying (process)
Spritzviskosität spraying consistency
Sprödbruch brittle failure
spröde brittle
sprödelastisch brittle-elastic
sprödes Versagen brittle failure
sprödhart hard and brittle, glassy
Sprödigkeit brittleness
Sprühauftrag spray application
sprühbar sprayable
Sprühdose aerosol can
sprühen to spray
sprühgetrocknet spray-dried
Sprühlack spraying paint
Sprühnebel spray mist
Sprühpistole spraygun
Sprühsalzbeständigkeit resistance to salt spray
Sprühtrocknung spray drying
Sprühviskosität spraying viscosity/consistency
sprunghaft sudden
Spülgas purging gas
Spülmaschine dishwasher
Spuren traces
Spurenanalyse trace analysis
Spurenbestandteile traces, trace amounts
Spurenelement trace element
SQC statistical quality control
stäbchenförmig rod-like
stabförmig rod-like
stabil stable
Stabilisator stabiliser
stabilisatorarm low-stabiliser, with a low stabiliser content
Stabilisatordosierung 1. stabiliser content. 2. addition of stabiliser
stabilisatorfrei unstabilised, free from stabiliser
Stabilisatorgemisch stabiliser blend
stabilisatorhaltig stabilised, containing stabiliser
Stabilisatorkombination stabiliser blend
Stabilisatormischung stabiliser blend
Stabilisatorwirksamkeit stabiliser efficiency
Stabilisierbarkeit stabiliser tolerance
stabilisierend stabilising
Stabilisierung stabilisation
Stabilisierungsadditiv stabiliser
Stabilisierungsmittel stabiliser
Stabilisierungssystem stabiliser
Stabilisierwirkung stabilising effect
Stabilität stability, resistance
Stabilität, thermische thermal stability
Stabilitätszunahme increase in stability
Stabprobe rod-shaped test piece
Stadtgas town gas
Stahl steel
Stahl, nichtrostender stainless steel
Stahl, rostfreier stainless steel
Stahlbehälter steel tank
Stahlbeton reinforced concrete
Stahlbetonbau reinforced concrete construction
Stahlbewehrung steel reinforcement, reinforcing steel
Stahlblech steelplate, steel sheet
Stahldrahtbürste wire brush
stahlelastisch energy-elastic
Stahlelastizität energy elasticity
Stahlelektrode steel electrode
stählern steel
Stahlfolie steel foil
Stahlgestell steel frame
Stahlguß cast steel
Stahlhochbau steel structure/construction
Stahlklebung 1. bonding of steel. 2. bonded steel joint
Stahlkonstruktion steel structure
Stahlkugel steel ball
Stahlkugelmühle steel ball mill
Stahloberfläche steel surface
Stahlplatte steel sheet/panel
Stahlrohr steel pipe
Stahluntergrund steel substrate/surface
Stahlverklebung 1. bonding of steel. 2. bonded steel joint
Stahlwasserbau reinforced steel hydraulic engineering
Stahlwatte steel wool
Stahlwolle steel wool
Stammlack base, part A *(of a 2-pack paint)*
Stammlösung stock solution
Standardabweichung standard deviation
Standardanlage standard system/equipment
Standardausführung 1. standard design. 2. standard model
Standardausrüstung standard equipment
Standardbedingungen standard conditions
Standarddosierung standard amount
Standardeinrichtung standard equipment
Standardeinstellung 1. standard setting. 2. standard formulation
Standardempfehlung standard recommendation
Standardfarbenpalette standard colour range
standardisiert standardised
Standardlieferprogramm standard range
Standardmethode standard method
Standardprogramm standard range
Standardqualität standard grade
Standardrezeptur standard formulation
Standardsorte standard grade
Standardsortiment standard range
Standardtyp(e) standard grade
Standardverpackung standard pack/packaging
standfest firm, non-sag
Standfestigkeit 1. stability, rigidity. 2. non-sag properties
Standöl stand oil
Standort location
Standsicherheit stability
Standvermögen 1. stability, rigidity. 2. non-sag properties
Standversuch creep test
Standzeit 1. shelf life. 2. pot life. 3.

working/service life
Stanzbarkeit punchability
stanzen to punch, to stamp
Stanzfähigkeit punchability
Stanzlack punchable finish
Stapelanlage stacking unit
Stapelautomat automatic stacker
Stapelbarkeit stackability
Stapeleinrichtung stacking unit/device
Stapelfähigkeit stackability
Stapelfestigkeit stacking resistance
stapeln to stack
Stapelvorrichtung stacking device
starkbasisch strongly basic/alkaline
Stärke 1. thickness. 2. strength. 3. starch
Stärkederivat starch derivative
starkpolar strongly polar
starksauer strongly acid
starkvernetzt densely crosslinked
starkwandig thick-walled
starr rigid, stiff
starr-elastisch hard-elastic
Starrheit stiffness, rigidity
Startradikal initiator radical
Startreaktion starting reaction
Startreibung static friction
Startrezeptur starting formulation
Statikmischer static mixer
stationär stationary
statisch static
statische Beanspruchung static stress/load
statische Belastung static stress/load
statistisch statistical
statistische Prozeßsteuerung statistical process control
statistische Sicherheit statistical certainty
Staub dust, dirt
Staubanziehung dust/dirt attraction, attraction of dust/dirt
staubfrei free from dirt
Staubmaske face mask
Staubreste residual dirt
Staubteilchen dust particle
staubtrocken dust dry
Staubtrockenheit dust dryness
Staubtrockenzeit dust dry time
Stauchbeanspruchung compressive stress
Stauchbelastung compressive stress
Stauchbewegung movement due to compression
Stauchdruck compressive stress
Stauchdruckprüfung compression test
Stauchfestigkeit compressive strength
Stauchhärte compressive strength
Stauchhärteermüdung compressive failure
Stauchkraft compressive force
Stauchspannung compressive stress
Stauchung 1. compression. 2. compressive strain
Stauchung bei Bruch compressive strain at rupture/break
Stauchung bei Quetschspannung compressive strain at compressive yield stress
Stauchungsgeschwindigkeit rate of compression
Staudinger-Index intrinsic viscosity, limiting viscosity number
Stearamid stearamide
Stearat stearate
Stearinsäure stearic acid
Stearylalkohol stearyl alcohol
Stearylamin stearylamine
Stearylmethacrylat stearyl methacrylate
Stearylstearat stearyl stearate
Stehvermögen non-sag properties
steif stiff, rigid
Steifigkeit stiffness, rigidity
Steigung slope *(of a curve)*
steil steep
Steilabfall steep downward slope *(of a curve)*
Stein stone
Steingut earthenware
Steinkohlenteer coal tar
Steinkonservierung stone conservation
Steinrestaurierung stone restoration
Steinschlag flying stones
Steinschlagbeanspruchung impact by flying stones
steinschlagbeständig resistant to (impact by) flying stones
Steinschlagbeständigkeit resistance to (impact by) flying stones
steinschlagfest resistant to (impact by) flying stones
Steinschlagfestigkeit resistance to (impact by) flying stones
Steinschlagprüfung flying stones impact test
Steinschlagresistenz resistance to flying stones
Steinschlagschutzlack paint to protect against flying stones
Steinschlagschutzwirkung protection against (impact by) flying stones
Steinwolle mineral/rock wool
Steinzeug stoneware
Stellenwert (relative) importance
Stellmittel anti-flow additive
Stereoisomer stereoisomer
Stereokautschuk stereorubber
stereometrisch stereometric
stereospezifisch stereospecific
steril sterile
Sterilisation sterilisation
sterilisationsbeständig resistant to sterilising temperatures
Sterilisationsbeständigkeit resistance to sterilising temperatures
sterilisationsfest resistant to sterilising temperatures
Sterilisationsfestigkeit resistance to sterilising temperatures
sterilisierbar sterilisable
Sterilisierbarkeit sterilisability
sterilisierfest resistant to sterilising temperatures

Sterilisierfestigkeit resistance to sterilising temperatures
sterisch steric
sterisch gehindert sterically hindered
sterische Hinderung steric hindrance
Steueraggregat control unit
Steueranlage control system
Steueraufgabe control function
Steuerausgang control output
Steuerautomatik automatic controls
Steuerblock control unit
Steuerdaten control data
Steuereingang control input
Steuereinheit control unit
Steuereinrichtung control equipment *(plural:* controls*)*
Steuerelektronik electronic controls
Steuerelemente controls
Steuergerät control instrument
Steuergröße controlled variable
Steuerkasten control cabinet
Steuerkreis control cabinet
Steuern 1. control. 2. open-loop control
Steuerpult control desk/panel/console
Steuerschrank control cabinet
Steuersignal control signal
Steuerspannung control voltage
Steuerung 1. control. 2. control unit
Steuerungs- *see* **Steuer-**
Steuervarianten control options
Steuerventil control valve
Steuerwarte control desk/panel/console
Stichprobe 1. random test. 2. spot check
Stickoxid nitrogen oxide
Stickstoff nitrogen
Stickstoffatmosphäre atmosphere of nitrogen
Stickstoffdurchlässigkeit nitrogen permeability
Stickstoffgas nitrogen
stickstoffhaltig containing nitrogen
Stickstofflasche (liquid) nitrogen cylinder
Stickstoffstrom stream of nitrogen
Stickstoffverbindung nitrogen compound
Stiftrotor pin rotor
Stiftstator pin stator
Stillstandszeit downtime
Stöchiometrie stoichiometry
stöchiometrisch stoichiometric
Stockpunkt setting/pour point
Stoff material
Stoffdaten material constants/properties
Stoffeigenschaft material property
Stoffgröße material constant/property
Stoffkenndaten material constants/properties
Stoffkennwert material constant/property
Stoffklasse material group, group of materials
Stoffkonstante material constant
Stoffkosten (raw) material costs
Stoffkreislauf material cycle
stoffliche Ressourcen material resources
stoffliche Verwertung material recycling
stoffliche Wiederverwertung material recycling
stoffliches Recycling material recycling
Stoffrecycling material recycling
Stofftransport material transport
Stoffverhalten material performance
Stoke'sches Gesetz Stokes's Law
störanfällig liable to go wrong, liable to give trouble
Störanfälligkeit tendency to go wrong, tendency to give trouble
Störanzeige warning signal
Störanzeige, akustische audible warning signal
Störanzeige, optische visual warning signal
Störeinfluß disruptive/disturbing influence
störempfindlich liable to go wrong, liable to give trouble
störend troublesome
Störfaktor disruptive/disturbing influence
Störgröße disruptive/disturbing influence
Störmeldung warning signal
Störmeldungsanzeige alarm/warning signal
Störsignal alarm/warning signal
Störstoff contaminant, impurity
störunempfindlich unlikely to go wrong, unlikely to give trouble
Störung malfunction, disruption, disturbance, fault, interference
störungsarm almost trouble-free
Störungsfall, im if there is a fault/breakdown
störungsfrei trouble-free
störungssicher trouble-free
Störungssuche troubleshooting
Stoß 1. joint. 2. impact
stoßabsorbierend shock-absorbent
Stoßabsorption shock absorption
Stoßarbeit impact energy
stoßartige Beanspruchung impact stress
stoßbeansprucht subjected to impact (stress)
Stoßbeanspruchung impact stress
Stoßbelastbarkeit impact resistance
Stoßbelastung impact stress
stoßdämmend shock absorbent
stoßdämpfend shock absorbent
stoßelastisch impact resistant
Stoßelastizität (impact/rebound) resilience
Stoßempfindlichkeit sensitivity to impact
stoßfest impact resistant
Stoßfestigkeit impact strength/resistance
Stoßfläche impact surface
Stoßfuge butt joint
Stoßgeschwindigkeit impact speed, speed of impact
Stoßkraft impact force
Stoßprüfung impact test
stoßsicher impact resistant
Stoßstange bumper
Stoßverklebung butt joint
Stoßversuch impact test
stoßweise discontinuous(ly), intermittent(ly)
Stoßzähigkeit impact strength/resistance
Strahl 1. jet. 2. beam
Strahlanlage grit blasting unit

Strahldichtefaktor radiation density index
Strahldrucktinte ink jet printing ink
Strahlen sandblasting
Strahlen-, strahlen- *see* **Strahlungs-, strahlungs-**
strahlend radiant
strahlende Wärme radiant heat
Strahlleistung (electron) beam output
Strahlung radiation
strahlungsabsorbierend radiation absorbent
strahlungsbeständig radiation resistant
Strahlungsbeständigkeit radiation resistance
Strahlungsdichte radiation density
Strahlungsdose radiation dose
Strahlungsdosis radiation dose
Strahlungsdurchlässigkeit radiation permeability
Strahlungseinfluß effect of radiation
strahlungsempfindlich radiation-sensitive
Strahlungsenergie radiation energy
strahlungsgehärtet radiation cured
strahlungshärtbar radiation curing
Strahlungshärten radiation curing
strahlungshärtend radiation curing
Strahlungshärtung radiation curing
Strahlungsheizung radiant heaters/heating
strahlungsinduziert radiation-induced
Strahlungsintensität radiation intensity
Strahlungsoxidation radiation-induced oxidation
strahlungspolymerisierbar radiation curing
Strahlungspyrometer radiation pyrometer
Strahlungsquelle radiation source
strahlungsresistent radiation resistant
Strahlungsresistenz radiation resistance
Strahlungsstärke radiation intensity
Strahlungsverlust heat loss due to radiation, radiant heat loss
strahlungsvernetzbar radiation curable
strahlungsvernetzend radiation curing
strahlungsvernetzt radiation cured
Strahlungsvernetzung radiation curing
Strahlungszeit irradiation period
stramm stiff
Stranggranulierung strand pelletisation *(of hot melt adhesives)*
strapazierbar hard wearing
strapazierfähig hard wearing
Straßenbau roadbuilding
Straßenmarkierungsfarbe road marking paint
Straßenmarkierungsmasse road marking paint
Straßensilofahrzeug road tanker
Straßentankfahrzeug road tanker
Straßentankwagen road tanker
Straßentankzug road tanker
Streckdehnung elongation
strecken to stretch
Streckgrenze yield point
Streckgrenzfestigkeit maximum yield strength
Streckgrenzspannung maximum yield stress
Streckgrenzwert yield point
Streckspannung yield stress, tensile stress at yield
streichbar brushable
Streichbarkeit brushability
Streicheigenschaften brushability, brushing characteristics
streichen to brush
streichfähig brushable
Streichfähigkeit brushability
Streichfarbe paint (suitable) for brush application
Streichgang brushing operation/process
Streichmasse coating compound/composition
Streichpaste spread coating paste
Streichviskosität brushing consistency
Streulicht scattered light
Streuscheibe (headlamp) diffuser
Streuung variation *(e.g. of figures)*
Streuvermögen scattering power
Streuwinkel scatter angle
Streuwirkung scatter effect
Strichcode bar code
Stroh straw
strohfarben straw coloured
Stromdichte current density
Stromregelventil flow control valve
Strömung flow
Strömung, turbulente turbulent flow
Strömungsmeßgerät flowmeter
Strömungspotential streaming potential
Stromversorgung power supply
Strontiumchromat strontium chromate
Strontiumoxid strontium oxide
Struktur 1. structure. 2. texture
strukturabhängig structure-dependent
Strukturanalyse structural analysis
Strukturaufbau consistency build-up
Strukturbauteil structural component
Struktureinheit structural unit
strukturell structural
Strukturfarbe textured paint
Strukturfehler structural defect
Strukturformel structural formula
strukturiert textured
Strukturkleber structural adhesive
Strukturklebstoff structural adhesive
Strukturlack textured paint
Strukturmerkmale structural characteristics
Strukturpulverlack textured powder coating
Strukturschaum structural/integral foam
Strukturstahl structural steel
Strukturunterschied structural difference
Strukturveränderung structural change
Strukturverbindung structural joint
strukturviskos pseudoplastic, non-Newtonian
Strukturviskosität pseudoplasticity, pseudoplastic/non-Newtonian behaviour. *Literal translation of the word should be avoided since the result would be meaningless and incomprehensible to a native English or American speaker*
Strukturviskositätseffekt pseudoplasticity, pseudoplastic/non-Newtonian behaviour
Stuckgips stucco

Stückzahlen, große large numbers
Stückzahlen, kleine small numbers
stufenlos infinite(ly), stepless(ly)
stufenlos einstellbar steplessly/infinitely variable
stufenlos regelbar steplessly/infinitely variable
stufenlos verstellbar steplessly/infinitely variable
stufenweise gradually, in stages
stumpf dull, matt
Stumpfstoß butt joint
Stumpfstoß, rechtwinkliger right-angle butt joint
stumpfverklebt butt joined
Stumpfverklebung butt joint
Stundendurchsatz hourly throughput, throughput per hour
Stundenleistung hourly output, output per hour
Stützmaterial supporting material
Styrol styrene (monomer)
Styrol-Acrylatcopolymer styrene-acrylate copolymer
Styrol-Acrylatdispersion styrene-acrylate dispersion
Styrol-Acrylatmischpolymerisat styrene-acrylate copolymer
Styrol-Acrylsäurecopolymer styrene-acrylic acid copolymer
Styrol-Butadienkautschuk styrene-butadiene rubber
Styrol-Diencopolymer styrene-diene copolymer
Styrol-Diencopolymerisat styrene-diene copolymer
Styrol-Maleinsäurecopolymer styrene-maleic acid copolymer
Styrolalkyd styrene alkyd
Styrolanteil styrene content
Styrolcopolymer styrene copolymer
Styrolcopolymerisat styrene copolymer
Styroldämpfe styrene vapours
Styrolemission styrene emission
styrolfrei styrene-free
styrolhaltig containing styrene
Styrolharz styrene resin
styrolisiert styrenated
Styrollatex styrene dispersion
Styrolmischpolymer styrene copolymer
Styrolmischpolymerisat styrene copolymer
styrolmodifiziert styrene modified
Styrolmonomer styrene monomer
Styrolpolymerisat styrene polymer
Styrolreste residual styrene (monomer), styrene residues
Styrolrestgehalt residual styrene content
styrolunempfindlich unaffected by styrene
subakut sub-acute
subjektiv subjective
Submikronbereich sub-micron range
submikroskopisch sub-microscopic
Substituent substituent
substituiert substituted
Substitutionsprodukt substitute, alternative product
Substitutionsreaktion substitution reaction
Substitutionswerkstoff substitute, alternative material
Substrat substrate, surface
Substratbenetzer surface wetting agent
Substratoberfläche substrate surface
Sulfat sulphate
Sulfatasche sulphated ash
Sulfation sulphate ion
Sulfatverfahren sulphate process
Sulfenamidbeschleuniger sulphenamide accelerator
Sulfid sulphide
Sulfogruppe sulpho group
Sulfoisophthalsäure sulphoisophthalic acid
Sulfonamid sulphonamide
Sulfonat sulphonate
Sulfoneinheit sulphone group
Sulfongruppe sulphone group
sulfoniert sulphonated
Sulfonierung sulphonation
Sulfonimid sulphonimide
Sulfoniumgruppe sulphonium group
Sulfoniumsalz sulphonium salt
Sulfonsäure sulphonic acid
Sulfonsäureamid sulphonic acid amide
Sulfonsäureester sulphonate
Sulfonylaminosulfoniumverbindung sulphonyl aminosulphonium compound
Sulfophthalsäure sulphophthalic acid
Sulfosalicylsäure sulphosalicylic acid
Summenformel empirical formula
Superbenzin super-grade petrol
Superkraftstoff super-grade fuel
suspendieren to suspend
suspendiert suspended
Suspension suspension
Suspensionscopolymerisat suspension copolymer
Suspensionshilfsmittel suspending agent
Suspensionspolymerisat suspension polymer
Suspensionspolymerisation suspension polymerisation
syndiotaktisch syndiotactic
synergetisch synergetic, synergistic
Synergie synergism
Synergieeffekt synergistic/synergetic effect
Synergismus synergism
Synergist synergist
synergistisch synergistic, synergetic
Synthese synthesis
Synthesegummi synthetic rubber
Syntheseharz synthetic resin
Synthesekautschuk synthetic rubber
Syntheseverfahren synthesis
Synthetikkautschuk synthetic rubber
synthetisch synthetic
synthetisiert synthesised
Systemaudit system audit
systemisch systemic
SZ *abbr. of* **Säurezahl**, acid value

T

T-Stoß, doppelseitig verstärkter strap-supported joint
TA-Luft technical directive for the prevention of air pollution. *The best way of translating this is as* German clean air act
Tabaksamenöl tobacco seed oil
Tackverlust loss of tack
Tageslicht daylight
Taktzeit cycle time
Talkum talc
Tallöl tall oil
Tallölalkydharz tall oil alkyd (resin)
Tallölfettsäure tall oil fatty acid
Tallölharz tall oil resin
Tang seaweed
Tangente tangent
tangential tangential
Tannin tannin
Tapete wallpaper
Tapetenkleister wallpaper paste
Tarnanstrich camouflage finish
Tarnlack camouflage paint
Tastatur keyboard
Tastaturfeld keyboard
Taste key, push-button
Tastendruck, auf at the touch of a button
Tastendruck, per at the touch of a button
Tastenfeld keyboard
Tastenplatte keyboard
Taster scanner
Tauchapparatur dipping equipment
Tauchbad dip coating bath, dipping bath
Tauchbehälter dipping tank
tauchbeschichtet dip coated
Tauchbeschichtung dip coating
Tauchdauer length of immersion
Tauchen dipping
tauchgalvanisiert dip galvanised
Tauchgrundierung dipping primer
Tauchlack dipping paint
Tauchlackieren dip coating
Tauchlackierung, anodische anodic electrodeposition painting
Tauchlackierung, kathodische cathodic electrodeposition painting
Tauchlösung dipping solution
Tauchverfahren dipping (process)
Tauchvorgang dipping process/operation
Taupunkt dew point
Taupunkttemperatur dew point
TDI *abbr. of* **Toluoldiisocyanat** *or* **Toluylendiisocyanat**, toluene/toluylene diisocyanate
Technik 1. technique. 2. technology. 3. engineering
Technikum pilot plant
Technikumanlage pilot plant
Technikummaßstab pilot plant scale
Technikummaschine pilot plant machine
Technikumversuch pilot plant trial
technisch 1. technical. 2. industrial
technologisch technological
TEDMA triethylene glycol dimethacrylate
Tee tea
Teer tar
Teeraromaten aromatic tar compounds
Teerepoxyharz tar-epoxy resin
Teerfraktion tar fraction
teerhaltig containing tar
Teeröl tar oil
Teerpappe roofing felt
Teerpech tar pitch
Teerreste tar residues
teflonbeschichtet PTFE-coated
teflonisiert PTFE-coated
teilaromatisch partly aromatic
teilautomatisch semi-automatic(ally)
teilautomatisiert partly automated
Teilbereich specific area/part *(e.g. of an industry)*
teilblockiert partly blocked
Teilchen particle
Teilchenbeschaffenheit particle characteristics
Teilchendurchmesser particle size
Teilchendurchmesser, mittlerer mean particle size
Teilchenentladung particle discharge
Teilchenfeinheit particle size
Teilchenfluß particle flow
Teilchenform particle shape
Teilchenformeinfluß effect of particle shape
Teilchengröße particle size
Teilchengröße, häufigste predominant particle size
Teilchengröße, mittlere mean/average particle size
Teilchengröße, vorherrschende predominant particle size
Teilchengröße, vorwiegende predominant particle size
Teilchengrößenanalyse particle size analysis
Teilchengrößenbereich particle size range
Teilchengrößendurchmesser particle size
Teilchengrößenverteilung particle size distribution
Teilchenmorphologie particle morphology
Teilchenoberfläche particle surface
Teilchenphysik particle physics
Teilchenradius particle radius
Teilchenstruktur particle structure
Teilchenverteilung particle size distribution
Teilchenvolumen particle volume
Teilersatz partial replacement
Teilester partial ester
teilevakuiert partly evacuated
teilfluoriert partly/partially fluorinated
teilhalogeniert partly/partially halogenated
teilhydrolysiert partly hydrolysed
teilkristallin partly crystalline
teilmethyliert partly/partially methylated

teilmethyloliert partly methylolised
teiloxidiert partly/partially oxidised
teilverethert partly etherified
teilvernetzt partly crosslinked
teilverseift partly saponified
teilverzweigt partly branched
teilweise vernetzt partly crosslinked
TEM 1. transmission electron microscope. 2. transmission electron microscopy
TEM-Analyse transmission electron microscopy
TEM-Aufnahme transmission electron micrograph
Temperatur temperature
Temperatur, erhöhte elevated temperature
Temperaturabfall temperature decrease, drop/decrease in temperature
temperaturabhängig temperature-dependent, depending on the temperature
Temperaturabhängigkeit temperature dependence
Temperaturabnahme temperature decrease
Temperaturabsenkung temperature decrease, drop/decrease in temperature
Temperaturalterung heat ageing
Temperaturalterungsverhalten heat ageing behaviour/characteristics
Temperaturänderung change in temperature
Temperaturanhebung raising/increasing the temperature
Temperaturanstieg temperature increase, rise/increase in temperature
Temperaturanwendung application of heat, heating
Temperaturanwendungsbereich operating temperature range
Temperaturanzeige temperature display
Temperaturanzeigegerät temperature indicator
Temperaturaufgabe application of heat
Temperaturaufnehmer thermocouple
Temperaturausdehnungskoeffizient coefficient of expansion, expansion coefficient
Temperaturausgleich temperature adjustment
Temperaturbandbreite temperature range
temperaturbeansprucht subjected to high temperatures
Temperaturbeanspruchung exposure to high temperatures
temperaturbedingt due to temperature
Temperaturbehandlung 1. annealing, conditioning. 2. heat treatment
Temperaturbelastbarkeit thermal endurance, ability to withstand high temperatures
Temperaturbelastung exposure to high temperatures, heat exposure
Temperaturbereich temperature range
temperaturbeständig 1. heat resistant. 2. resistant to high and low temperatures
Temperaturbeständigkeit 1. heat resistance. 2. resistance to high and low temperatures: **Die Temperaturbeständigkeit reicht von -60 bis +230°C** It withstands temperatures from -60 to +230°C
Temperaturdifferenz temperature difference
Temperatureinfluß effect of temperature
Temperatureinsatzbereich operating temperature range
Temperatureinsatzgrenze maximum operating temperature
Temperatureinstellung 1. temperature setting. 2. temperature setting mechanism
Temperatureinstufung temperature rating
Temperatureinwirkung action/effect/application of heat
temperaturempfindlich heat sensitive, affected by heat
Temperaturempfindlichkeit heat sensitivity
Temperaturerfassung temperature determination/measurement
Temperaturerhöhung temperature increase
Temperaturerniedrigung temperature decrease, lowering the temperature
Temperaturfeld temperature range
temperaturfest heat resistant
Temperaturfestigkeit heat resistance
Temperaturfühler thermocouple, temperature sensor/probe
Temperaturführung 1. temperature control. 2. temperature profile
Temperaturgefälle temperature gradient/difference
temperaturgeregelt temperature-controlled
Temperaturgeschichte thermal history
Temperaturgleichgewicht temperature equilibrium
Temperaturgleichmäßigkeit even/uniform temperature
Temperaturgradient 1. temperature gradient, change in temperature. 2. temperature difference
Temperaturgrenze temperature limit
Temperaturhomogenität even/uniform temperature
Temperaturindikator temperature indicator
temperaturinduziert temperature-induced
Temperaturintervall temperature range
Temperaturkompensation temperature compensation
Temperaturkonstanz constancy of temperature
Temperaturkontrolle temperature control
Temperaturleitfähigkeit thermal diffusivity
Temperaturleitzahl thermal diffusivity
Temperaturmeßfühler thermocouple, temperature sensor/probe
Temperaturmeßgeber thermocouple, temperature sensor/probe
Temperaturmeßgerät thermometer, thermocouple, temperature measuring device
Temperaturmeßmethode method of measuring temperature
Temperaturmeßstelle temperature measuring point
Temperaturmessung temperature

measurement/determination
Temperaturmeßvorrichtung temperature measuring device
Temperaturmittelwert mean/average temperature
Temperaturniveau temperature (level)
Temperaturobergrenze upper temperature limit
Temperaturoptimum optimum temperature
Temperaturprofil temperature profile/pattern
Temperaturprogramm temperature programme
Temperaturregelgerät temperature controller, temperature control instrument
Temperaturregelkreis temperature control circuit
Temperaturregelung 1. temperature control. 2. temperature control system
Temperaturregler temperature controller, temperature control instrument
Temperaturregulierung 1. temperature control. 2. temperature control system
Temperaturrichtwert approximate temperature
Temperaturschockbeständigkeit heat shock resistance
Temperaturschreiber temperature recorder
Temperaturschwankungen temperature variations/fluctuations
Temperaturschwelle temperature limit
Temperaturschwingungen temperature fluctuations
Temperatursenkung temperature decrease
Temperatursensor thermocouple, temperature sensor/probe
Temperatursollwert required temperature
Temperatursonde thermocouple, temperature sensor/probe
Temperaturspanne temperature range
Temperaturspielraum temperature range
Temperaturspitze temperature peak, maximum temperature
Temperaturspitzenbelastbarkeit thermal endurance
Temperatursprung temperature difference
temperaturstabil thermally stable
Temperaturstabilität 1. thermal stability, heat resistance. 2. resistance to high and low temperatures
Temperaturstandfestigkeit heat resistance
Temperatursteuereinheit temperature control unit
Temperatursteuerung 1. temperature control. 2. temperature control system
Temperaturtoleranzfeld temperature tolerance range
Temperaturüberwachung 1. temperature control. 2. temperature control unit, temperature monitor
temperaturunabhängig temperature-independent, independent of the temperature
Temperaturuntergrenze lower temperature limit
Temperaturunterschied temperature difference
Temperaturveränderung change in temperature
Temperaturverlauf temperature profile, changes in temperature
Temperaturverlauf, zeitlicher time-temperature relationship
Temperaturverteilung temperature distribution
Temperaturwechsel change/fluctuation in temperature
temperaturwechselbeansprucht subjected to changing/fluctuating/alternating temperatures
Temperaturwechselbeständigkeit resistance to changing/fluctuating/ alternating temperatures
Temperaturwechselbeanspruchung exposure to changing/fluctuating/alternating temperatures
Temperaturwechselprüfung alternating temperature test
Temperaturwert temperature
Temperaturzuführung heating, application of heat
Temperaturzunahme temperature increase
Temperaturzustände temperatur conditions
Temperdauer conditioning/annealing period
Tempereffekt effect of annealing
Temperierbad constant temperature bath
temperierbar heatable
temperieren 1. to heat. 2. to control the temperature
Temperierflüssigkeit 1. heating-cooling medium, constant temperature medium, temperature control medium. 2. heating medium
Temperiergerät 1. temperature control unit. 2. heating unit
Temperierkammer constant temperature chamber
Temperierkreislauf temperature control circuit, heating-cooling circuit
Temperiermedium 1. heating-cooling medium, constant temperature medium, temperature control medium. 2. heating medium
Temperiermittel 1. heating-cooling medium, constant temperature medium, temperature control medium. 2. heating medium
tempern to anneal, to condition *(test specimens, by keeping them at a certain temperature to relieve internal stresses)*
Temperofen conditioning/annealing oven
Temperstrecke conditioning/annealing section
Tempertemperatur conditioning/annealing temperature
Temperung conditioning, annealing, post-curing
Temperzeit conditioning/annealing period
Tensid surfactant
tensidisch surfactant
Teppich carpet
Teppichauslegeware carpet tiles
Teppichboden carpeting
Teppichfliese carpet tile
Teppichklebstoff carpet adhesive
Teppichrückenbeschichtung carpet backing
Teppichrückseitenbeschichtung carpet backing
Teppichunterlage carpet underlay
Teppichunterschicht carpet backing
teratogen teratogenic
Teratogenität teratogenity

Terephthalat terephthalate
Terephthalatpolyester terephthalate-based polyester
Terephthalsäure terephthalic acid
Terephthalsäuredimethylester dimethyl terephthalate
Terephthalsäurepolyester terephthalic acid polyester
ternär ternary
Terpen terpene
Terpenharz terpene resin
Terpenkohlenwasserstoff terpene hydrocarbon
Terpentin turpentine
Terpentinöl oil of turpentine
terpolymer terpolymeric
Terpolymer terpolymer
Terpolymerisat terpolymer
Terrasse terrace
tertiär tertiary
Tertiärteilchen tertiary particle
Testbenzin white spirit
Testblech metal test piece/panel
Testdauer test duration
Testergebnis test result
Testflüssigkeit test liquid
Testlösung test solution
Testmethode test method
Testplatte test panel
Testreihe test series
tetrabasisch tetrabasic
Tetrabrombisphenol tetrabromobisphenol
Tetrabutyltitanat tetrabutyl titanate
Tetracarbonsäure tetracarboxylic acid
Tetracarbonsäureanhydrid tetracarboxylic anhydride
Tetrachlorethan tetrachloroethane
Tetrachlorethen tetrachloroethene
Tetrachlorethylen tetrachloroethylene
Tetrachlorkohlenstoff carbon tetrachloride
Tetrachlorphthalsäure tetrachlorophthalic acid
Tetrachlorphthalsäureanhydrid tetrachlorophthalic anhydride
Tetrachlorstyrol tetrachlorostyrene
Tetracyanbenzol tetracyanobenzene
Tetrafluordichlorethan tetrafluorodichloroethane
Tetrafluorethen tetrafluoroethene
Tetrafluorethenharz tetrafluoroethene resin
Tetrafluorethylen tetrafluoroethylene
Tetrafluorkohlenstoff carbon tetrafluoride
tetrafunktionell tetrafunctional
tetragonal tetragonal
Tetrahydrobenzylalkohol tetrahydrobenzyl alcohol
Tetrahydrofuran tetrahydrofuran, THF
Tetrahydronaphthalin tetrahydronaphthalene
Tetrahydrophthalsäure tetrahydrophthalic acid
Tetralin tetralin
tetramer tetrameric
Tetramer tetramer
Tetramethoxymethylmelamin tetramethoxymethyl melamine
Tetramethylsilan tetramethyl silane
Tetramethylxyloldiisocyanat tetramethylxylene diisocyanate
Tetraphenylethan tetraphenyl ethane
tetrasubstituiert tetrasubstituted
Tetrasulfid tetrasulphide
Textildruck fabric printing
Textilien textiles, fabrics
Textilimprägnierung textile impregnation
Textilindustrie textile industry
Textilschlichte textile size
Textilveredlung textile finishing
TG *abbr. of* **Thermogravimetrie**, thermogravimetry
T_g-Wert glass transition temperature
TG-Wert glass transition temperaure
TGA *abbr. of* **thermogravimetrische Analyse**, thermogravimetric analysis
TGIC-Härter triglycidyl isocyanurate hardener
TGIC-Härterkomponente triglycidyl isocyanurate hardener
theoretisch theoretical
thermisch geschädigt charred
thermisch hochbelastet subjected/exposed to high temperatures
thermisch thermal
thermisch härtend heat curing, hot setting
thermisch leitend thermally conductive
thermisch vernetzbar heat curing/curable
thermische Alterung heat ageing
thermische Analyse thermoanalysis
thermische Beanspruchung thermal stress
thermische Belastbarkeit thermal stability, heat resistance
thermische Beständigkeit thermal stability, heat resistance
thermische Bräunung brown discolouration due to overheating
thermische Eigenschaften thermal properties
thermische Leistung thermal efficiency
thermische Leitfähigkeit thermal conductivity
thermische Müllverwertung energy recovery
thermische Nutzung energy recovery
thermische Oberflächenschäden surface charring
thermische Schädigung thermal degradation
thermische Stabilität thermal stability
thermische Überanspruchung overheating
thermische Überlastung overheating
thermische Verfärbung discolouration due to charring
thermische Verwertung energy recovery
thermische Wechselbelastungen alternating exposure to heat and cold **thermische Werte** thermal properties
thermische Zersetzung thermal decomposition
thermischer Ausdehnungskoeffizient coefficient of (thermal) expansion **thermischer Längenausdehnungskoeffizient** coefficient of linear expansion
thermischer Volumenausdehnungskoeffizient

coefficient of cubical/volume expansion
thermisches Equilibrium thermal equilibrium
thermisches Härten heat curing
thermisches Recycling energy recovery
Thermoanalyse thermoanalysis
thermoanalytisch thermoanalytical
thermochrom thermochromic
Thermodiffusion thermodiffusion
Thermodraht thermocouple, temperature sensor/probe
Thermodynamik thermodynamics
thermodynamisch thermodynamic
thermoelastisch thermoelastic
Thermoelastizität thermoelasticity
thermoelektrisch thermoelectric
Thermoelement thermocouple, temperature sensor/probe
Thermofühler thermocouple, temperature sensor/probe
Thermographie thermography
Thermogravimetrie thermogravimetry
thermogravimetrisch thermogravimetric
thermogravimetrische Analyse thermogravimetric analysis
thermohärtend 1. thermosetting. 2. heat/hot curing
thermohydrolytisch thermohydrolytic
thermoinstabil thermally unstable
thermoisoliert thermally insulated
Thermokaschieranlage heat laminating plant
Thermokaschieren heat laminating
thermokonstant at constant temperature
Thermolyse thermolysis, pyrolysis
thermomechanisch thermomechanical
thermomechanische Analyse thermomechanical analysis
Thermomikroskopie thermomicroscopy
Thermooxidation thermooxidation
thermooxidativ thermooxidative
Thermoplast thermoplastic (material)
thermoplastisch thermoplastic
Thermoplastizität thermoplasticity
thermoreaktiv heat reactive
thermoreversibel thermoreversible
thermoschrumpfbar heat shrinking
thermostabil thermally stable
Thermostabilisator heat stabiliser
thermostabilisierend heat stabilising
Thermostabilisierung heat stabilisation
Thermostabilität thermal stability
Thermostat thermostat
thermostatgeregelt thermostatically controlled, thermostated
thermotrop thermotropic
Thiadiazol thiadiazole
Thiocyanatalken thiocyanate alkene
Thioester thioester
Thioether thioether
Thioetherbindung thioether linkage
Thioglykolat thioglycolate
Thioglykolsäureester thioglycolate
Thioharnstoff thiourea
Thioharnstoffharz thiourea resin
Thiohydrochinon thiohydroquinone
Thioketon thioketone
Thiokol *trade name for* polysulphide rubber
Thiokoldichtstoff polysulphide rubber sealant
Thiokolkautschuk polysulphide rubber
Thiolgruppe thiol group
Thiolverbindung thiol compound
Thioplast thioplast, polysulphide rubber
Thioverbindung thio compound
thixotrop thixotropic
thixotrop eingestellt made thixotropic
Thixotropie thixotropy
Thixotropieeffekt thixotropic effect
Thixotropiegeber thixotropic agent
thixotropierend thixotropic, imparting thixotropic properties
thixotropierendes Additiv thixotropic agent
Thixotropierharz thixotropic resin
Thixotropiermittel thixotropic agent
thixotropiert made thixotropic
Thixotropierung making thixotropic, imparting thixotropic properties
Thixotropierungsmittel thixotropic agent
Thixotropierungsverhalten thixotropic behaviour
Thixotropiewirkung thixotropic effect
thyristorgespeist thyristor controlled
thyristorgesteuert thyristor controlled
Thyristorregler thyristor control unit
Thyristorsteuerung thyristor control
TIC total ion chromatogram
tief 1. deep. 2. low *(e.g. temperature or pressure)*
Tiefbau civil engineering
Tiefdruck (roto)gravure/intaglio printing
Tiefdruckfarbe gravure printing ink
Tiefdruckgravurwalze gravure roll(er)
Tiefengrundierung penetrating primer
Tiefgarage underground car park
Tiefgrund penetrating primer
Tiefkühltruhe deep freeze
tiefschwarz deep/jet black
tiefsiedend low-boiling
Tiefsieder low-boiling solvent
Tieftemperaturadsorption low-temperature adsorption
Tieftemperaturanwendung low-temperature application
Tieftemperaturbeanspruchung exposure to low temperatures
Tieftemperaturbereich low-temperature range/region
Tieftemperaturbeständigkeit low-temperature resistance
Tieftemperaturbetrieb low-temperature operation
Tieftemperatureigenschaften low-temperature properties
Tieftemperaturen low temperatures
Tieftemperaturfestigkeit low-temperature resistance
Tieftemperaturflexibilität low-temperature

flexibility
Tieftemperaturhärtung low-temperature curing
Tieftemperaturinitiator low-temperature initiator
Tieftemperaturschlagzähigkeit low-temperature impact strength
Tieftemperaturverhalten low-temperature performance/behaviour
tieftemperaturzäh low-temperature impact resistant
Tieftemperaturzähigkeit low-temperature impact strength
Tiefung nach Erichsen Erichsen indentation
Tiefungsversuch indentation test
tierisch animal
Tinte ink
Tintenstrahldruck ink jet printing
Tintenstrahldrucker ink jet printer
Tintenstrahlschreiber ink jet printer
Titan titaniunm
Titanacetylacetonat titanium acetyl acetonate
Titanalkoxylat titanium alkoxylate
Titanat titanate
Titanatester titanate ester
Titanchelat titanium chelate
Titandioxid titanium dioxide
Titandioxidtyp grade of titanium dioxide
Titandioxidweißpigment titanium dioxide (white) pigment
Titanerz titanium ore
Titanhydroxid titanium hydroxide
Titankatalysator titanium catalyst
titanorganisch organotitanium
Titanoxidhydrat hydrated titanium oxide
Titantetrachlorid titanium tetrachloride
Titantetraisopropylat titanium tetraisopropylate
Titanverbindung titanium compound
Titanweiß titanium white
Titanylsulfat titanyl sulphate
Titration titration
titrieren to titrate
titrimetrisch by titration
TLV-Wert threshold limit value, TLV
TLVC-Wert threshold limit value ceiling, TLVC
TMA thermomechanical analysis
TMA-Wert TMA figure
Toleranz tolerance
Toleranzanforderungen tolerance requirements
Toleranzangaben tolerance details
Toleranzbänder tolerance limits
Toleranzbereich tolerance range
Toleranzbreite tolerance range
Toleranzfeld tolerance range
Toleranzgrenze tolerance limit
Toleranzüberwachungseinrichtung tolerance monitor
Toluol toluene
Toluoldiisocyanat toluene diisocyanate, TDI
toluolfrei free from toluene
toluolhaltig containing toluene
Toluolsulfonsäure toluene sulphonic acid
Toluolsulfonsäuremethylester methyl toluene sulphonate
Toluylendiisocyanat toluylene diisocyanate
Ton clay
Tönpaste tinting paste
Topfzeit pot life
Topfzeitverkürzung reduction of the pot life
Torsionsausschwingungsversuch torsion pendulum test
Torsionsbeanspruchung torsional stress
Torsionsbelastbarkeit resistance to torsional stress
Torsionsbelastung torsional stress
Torsionsklebefestigkeit torsional bond strength
Torsionsmodul torsional modulus
Torsionsmoment torsional moment
Torsionsmomentenmeßdose torsional moment transducer
Torsionspendel torsion pendulum
Torsionspendeldiagramm torsion pendulum diagram
Torsionspendelergebnis torsion pendulum test result
Torsionspendelprobe torsion pendulum test piece
Torsionspendeluntersuchung torsion pendulum test
Torsionspendelversuch torsion pendulum test
Torsionsscherbelastung torsional shear stress
Torsionsscherfestigkeit torsional shear strength
Torsionsscherversuch torsional shear test
Torsionsschwingung torsional vibration/oscillation
Torsionsschwingungsanalyse torsional vibration/oscillation analysis
Torsionsschwingungsgerät torsion pendulum (apparatus/instrument)
Torsionsschwingungsmessung torsion pendulum test
Torsionsschwingungsversuch torsion pendulum test
Torsionsspannung torsional stress
Torsionssteifheit torsional stress
Torsionsversuch torsion pendulum test
Torsionswinkel angle of torsion
Totalbruch total failure
Totalimmersionsprüfung total immersion test
Totalreflexion, abgeschwächte attenuated total reflection
Toxikant toxic substance
Toxikologie toxicology
toxikologisch toxicological
toxikologisch einwandfrei non-toxic
toxikologisch harmlos non-toxic
toxikologisch unbedenklich non-toxic
toxisch toxic
Toxizität toxicity
Toxizitätsrisiko toxicity risk
TPGDA *abbr. of* **Tripropylenglykoldiacrylat**, tripropylene glycol diacrylate
tragend load-bearing
Träger carrier, support, backing

Trägerfilm supporting/backing film
Trägerfolie supporting/backing film
Trägergewebe supporting/backing fabric
Trägerharz binder resin
trägerlos unsupported
Trägermaterial backing/supporting material, substrate
Trägermedium paint medium
Trägerpapier backing paper
Trägersubstanz paint medium
Tragfähigkeit load-bearing/-carrying capacity
trägheitsarm low-inertia
Tragstruktur load-bearing structure
Tränkanlage impregnating plant
Tränkbad impregnating bath
Tränkharz impregnating resin
Tränklack impregnating varnish
Tränklösung impregnating solution
Tränkmasse impregnating solution
Tränkmittel impregnating agent, impregnant
Tränkung impregnation
Tränkungszeit impregnating time/period
Tränkwanne impregnating bath
Transferdruckfarbe transfer printing ink
Translationsbewegung translational movement
Translationsenergie translational energy
transluzent translucent
Transluzenz translucency
Transmission transmission
transmissionselektronisch transmission electronic
Transmissionselektronenmikroskop transmission electron microscope
Transmissionselektronenmikroskopie transmission electron microscopy
transmissionselektronenmikroskopisch transmission electron microscopic
Transmissionsgrad degree of transmission
Transmissionsspektrometrie transmission spectrometry
Transmissionsspektrum transmission spectrum
transparent transparent
Transparentüberzug transparent coating
Transparenz 1. transparency. 2. clarity *(e.g.of information)*
Transparenzniveau (degree of) transparency
Transportband conveyor belt
Transportcontainer transit container
Transportgeschwindigkeit transport rate, rate of transport
Transportkosten transport costs
Transportmechanismus transport mechanism
Transportsystem transport system
Transportvorschriften transport regulations
Transveresterung transesterification
Treibgas propellant gas
Treibhauseffekt greenhouse effect
Treibmittel 1. blowing agent *(for making foam)*. 2. propellant *(e.g. for aerosol cans)*
Treibstoff fuel
treibstoffbeständig fuel resistant
Trennbeschichtung release coating
Trennbruch brittle failure/fracture
Trenneffekt non-stick/release effect
Trenneigenschaften non-stick/release properties
Trennfähigkeit non-stick properties
Trennfestigkeit bond strength, adhesion
Trennfilm release film/coating
Trennfolie release film
trennfreundlich easy-release
Trennkraft peel force
Trennlack release lacquer
Trennmittel release agent
Trennmittelreste release agent residues
Trennmittelrückstände release agent residues
Trennmittelwirkung non-stick/release effect
Trennpapier release paper
Trennsäule separating column
Trennschicht release film
Trennsubstanz release agent
Trennverfahren separating process
Trennvermögen release property
Trennwand partition wall
Treppenhaus staircase
Tri trichloroethylene
Triacrylat triacrylate
Trialkoxysilan trialkoxysilane
Trialkylformal trialkyl formal
Trialkylphosphat trialkyl phosphate
Trialkylzinnverbindung trialkyltin compound
Triallylcyanurat triallyl cyanurate
Triarylphosphat triaryl phosphate
Triarylsulfoniumsalz triarylsulphonium salt
Triazinharz triazine resin
Triazinverbindung triazine compound
Triazol triazole
tribasisch tribasic
tribochemisch tribochemical
triboelektrisch triboelectric
Tribologie tribology
tribologisch tribological
tribomechanisch tribomechanical
triborheologisch triborheological
Tributylacrylat tributyl acrylate
Tributylmethacrylat tributyl methacrylate
Tributylthioharnstoff tributyl thiourea
Tributylzinn tributyl tin
Tributylzinnfluorid tributyl tin fluoride
Tributylzinnoxid tributyl tin oxide
Tricarbonsäure tricarboxylic acid
Tricarbonsäureanhydrid tricarboxylic anhydride
Trichlorbenzol trichlorobenzene
Trichlorethan trichloroethane
Trichlorethen trichloroethene
Trichlorethylen trichloroethylene
Trichlorethylphosphat trichloroethyl phosphate
Trichlorfluormethan trichloroflurometháne
Trichlorstyrol trichlorostyrene
trichterförmig funnel-shaped
Triethanolamin triethanolamine
Triethylamin triethylamine
Triethylenglykol triethylene glycol

Triethylenglykoldimethacrylat triethylene glycol dimethacrylate
Triethylentetramin triethylene tetramine
Trifluorchlorethylen trifluorochloroethylene
Trifluornitromethan trifluoronitromethane
Trifluorpropylgruppe trifluoropropyl group/radical
Trifluortrichlorethan trifluorotrichloroethane
trifunktionell trifunctional
Triglycerid triglyceride
Triglycidylcyanurat triglycidyl cyanurate
Triglycidylisocyanurat triglycidyl isocyanurate
Triisocyanat triisocyanate
Triisopropanolamin triisopropanolamine
Trikresylphosphat tricresyl phosphate, TCP
Trimellitat trimellitate
Trimellithsäure trimellitic acid
Trimellithsäureanhydrid trimellitic anhydride
trimer trimeric
Trimerisat trimer
Trimerisation trimerisation
trimerisiert trimerised
Trimerisierung trimerisation
Trimerisierungsvorgang trimerisation reaction
Trimethylbenzophenon trimethyl benzophenone
Trimethylbenzoyldiphenylphosphinoxid trimethylbenzoyl diphenyl phosphine oxide
Trimethylolalkan trimethylolalkane
Trimethylolpropan trimethylol propane
Trimethylolpropantriacrylat trimethylolpropane triacrylate
Trimethylsilangruppe trimethyl silane group
Trimethylsilanol trimethyl silanol
Trimethylsilylester trimethylsilyl ester
Trimethylsilylgruppe trimethylsilyl group
Trinkwasser drinking water
Trioctylphosphat trioctyl phosphate
Triol triol
Triol-Polyether triol polyether
Triorganozinnverbindung triorganotin compound
Trioxyethylen trioxyethylene
Tripeptid tripeptide
Triphenylfarbstoff triphenyl dye
Triphenylmethylperoxid triphenylmethyl peroxide
Triphenylmethylradikal triphenylmethyl radical
Triphenylphosphat triphenyl phosphate, TPP
Triphenylphosphit triphenyl phosphite
Triphenylsulfoniumhexafluorarsenat triphenylsulphonium hexafluoroarsenate
Triphenylsulfoniumsalz triphenylsulphonium salt
Triphenylurethan triphenyl urethane
Tripropylenglykoldiacrylat tripropylene glycol diacrylate
Trisulfid trisulphide
trocken dry
Trockenbedingungen drying conditions
Trockenbruchwiderstand dry tear strength
trockenchemisch dry-chemical
Trockendeckfähigkeit dry hiding power
Trockendeckvermögen dry hiding power
trockene Reibung dry friction
trockener Griff dry handle
Trockenfestigkeit dry strength
Trockenfilm dry film
Trockenfilmdicke dry film thickness
Trockenfilmstärke dry film thickness
Trockengehalt dry solids content
Trockengerät drying unit
Trockengewicht dry weight
Trockengleitverschleiß wear due to dry sliding friction
trockengleitverschleißarm resistant to dry sliding friction
Trockenhaftung dry (film) adhesion
Trockenharzgehalt dry resin content
Trockenheit dryness
Trockenkanal drying tunnel
Trockenkaschieren dry laminating
Trockenklebstoff dry adhesive
Trockenlaminieranlage dry laminating plant
Trockenluft dry air
Trockenmahlverfahren dry grinding (process)
Trockenmittel 1. drier. 2. drying agent
Trockenmörtel dry mortar
Trockenofen drying oven
Trockenpigment dry pigment
Trockenpulver dry powder
Trockenputz dry plaster
Trockenquerzugfestigkeit dry transverse tensile strength
Trockenreibung dry friction
Trockenrückstand dry residue
Trockenschichtdicke dry film thickness
Trockenschichtstärke dry film thickness
Trockenschliff dry grinding
Trockenschrank drying cabinet/oven
Trockenstoff drier
Trockenstrecke drying section/tunnel
Trockensubstanz dry solids
Trockentemperatur drying temperature
Trockentunnel drying tunnel
Trockenverhalten drying characteristics
Trockenvolumen dry volume
Trockenwiderstand dry strength
Trockenzeit drying time
Trockenzustand dry state
Trockenzylinder drying cylinder
Trocknen, physikalisches physical drying
trocknend drying
Trockner drier
Trocknung drying
Trocknung, forcierte force drying
Trocknung, wärmeforcierte force drying
Trocknungsanlage drying unit
Trocknungsapparatur drying equipment
Trocknungsbedingungen drying conditions
Trocknungsbeschleuniger drier
Trocknungsbeschleunigung, zur to speed up drying
Trocknungscharakteristik drying properties/characteristics
Trocknungseigenschaften drying

properties/characteristics
Trocknungseinrichtung drying unit/equipment
Trocknungsgerät drying unit
Trocknungsgeschwindigkeit drying rate
Trocknungsluft drying air
Trocknungsmittel drier
Trocknungsphase drying stage
Trocknungsprozeß drying process
Trocknungstemperatur drying temperature
Trocknungsverhalten drying characteristics/properties
Trocknungsvermögen drying properties/characteristics
Trocknungsverzögerung delayed drying
Trocknungsvorgang drying process
Trocknungszeit drying time
Trocknungszyklus drying cycle
Tropenbedingungen tropical conditions
tropenbeständig resistant to tropical conditions
Tropenbeständigkeit resistance to tropical conditions
tropenfest resistant to tropical conditions
Tropenklima tropical conditions/climate
Tropenklimabedingungen tropical conditions
Tropentest test under tropical conditions
tropenverwendungsfähig suitable for use in the tropics *or*: under tropical conditions
Tröpfchen droplet
Tröpfchengröße droplet size
Tröpfchengrößenverteilung droplet size distribution
tropfen to drip
Tropfen 1. drop. 2. dripping
Tropfnasen runs
Tropfnasenbildung curtaining, running, sagging
Tropfpunkt, ohne non-melting, infusible
trübe cloudy, turbid
Trübung cloudiness, turbidity
Trübungspunkt cloud point
Tube tube
Tubenlack tube lacquer
Tür door
Turbinenrührer turbomixer, turbostirrer
turbulent turbulent
turbulente Strömung turbulent flow
Turbulenz turbulence
türkisblau turquoise blue
Türkischrotöl Turkey red oil
türkisgrün turquoise green
Türrahmen door frame
Türseitenteil door side panel
Tusche Indian ink
Tyndallstreuung Tyndall scattering
Typ grade *(of material)*
Typenbezeichnung product code
Typenmerkblatt product data sheet
Typenpalette range of products/grades
Typenprogramm range of products/grades
Typensortiment range of products/grades
Typenübersicht range of products/grades

U

übelriechend odourous, foul-/evil-smelling
Überanspruchung, thermische overheating
Überbrennen overstoving, overbaking
Überbrennstabilität overstoving resistance
überbrückend bridging
Überdosierung adding too much
Überdruckfarbe overprinting ink
überdurchschnittlich above average
Überempfindlichkeit hypersensitivity
Übergang transition
Übergangsbereich transition zone/region
Übergangslösung provisional/temporary solution
Übergangstemperatur glass transition temperature
Übergangszeit transitional period
Übergangszustand transitional state
Überhärtung overcuring
Überhitzung overheating
überhöht excessively high
überirdisch above ground
überlackierbar overpaintable
Überlackierbarkeit overpaintability
Überlackierechtheit overpaintability
Überlackierung overpainting
überlappend overlapping
überlappt overlapped
überlappte Rohrverbindung tubular lap joint
Überlappung overlap
Überlappung, abgeschrägte tapered lap joint
Überlappung, abgeschrägte einschnittige bevelled/tapered lap joint
Überlappung, abgeschrägte zweischnittige bevelled double lap joint
Überlappung, abgesetzte double butt lap joint
Überlappung, einschnittige simple lap joint
Überlappung, gefalzte rebated lap joint
Überlappung mit Abschrägung, abgesetzte double scarf lap joint
Überlappung, rechtwinklige corner lap joint
Überlappung, zugeschärfte einschnittige bevelled/tapered lap joint
Überlappung, zweischnittige double lap joint
Überlappungsbreite width of overlap
Überlappungsenden overlap ends
Überlappungslänge overlap length, length of overlap
Überlappungsverklebung lap joint
übermolekular supermolecular
überprüfen to check
Überprüfung check
übersättigt supersaturated
Übersättigung supersaturation
Überscherung excessive shear
Überschuß excess
überschüssig excess
Überspritzbarkeit overspray tolerance
Überspritzen overspraying
Überstreichbarkeit overpaintability

Überstreichen overpainting
Übertemperatur too high a temperature, excessive temperature
übertragbar transferrable
Übertragungskonstante propagation constant
übervernetzt overcured
Übervernetzung overcuring
Überwachung monitoring, control
Überwachungseinheit monitoring/control unit
Überwachungseinrichtung monitoring equipment
Überwachungselemente controls, monitoring devices
Überwachungskontrolleuchte pilot light
Überwachungsmöglichkeiten monitoring/control options
überwachungspflichtig requiring monitoring
Überwachungsprüfung control test
Überwachungssystem monitoring/control system
Überwachungstafel control panel
überziehen to coat
Überzug coating, finish
Überzug, pulverförmiger powder coating
Überzugsbildung film formation
Überzugseigenschaften paint film properties
Überzugsfilm coating, film
Überzugsmasse coating compound/composition, paint
Überzugsmittel coating compound/composition, paint
Überzugssystem coating system/formulation
UBS *abbr. of* **Unterbodenschutz**, underbody sealant
UBS-Masse underbody sealant
UF-Harz urea/urea-formaldehyde/UF resin
UF-Leimharz urea/urea-formaldehyde adhesive resin
UHF *abbr. of* **ultrahochfrequent**, ultra-high frequency
ultradünn ultra-thin
Ultradünnschnitt ultra-thin (microtome) section
Ultradünnschnittmikrotomie ultra-thin section microtomy
ultrafein ultra-fine
Ultrafiltration ultra-filtration
ultrahochfrequent ultra-high frequency, UHF
Ultrahochfrequenz ultra-high frequency
ultrahochmolekular ultra-high molecular weight
Ultrahochvakuum ultra-high vacuum
ultramarin ultramarine
ultrarein ultra-pure
ultrarot infrared
Ultraschall ultra-sound
Ultraschallbereich ultrasonic range
Ultraschallempfänger ultrasonic receiver
Ultraschallenergie ultrasonic energy
Ultraschallfrequenz ultrasonic frequency
Ultraschallgeber ultrasonic generator
Ultraschallimpuls ultrasonic impulse
Ultraschallmeßgerät ultrasonic measuring instrument
Ultraschallmikroskop ultrasonic/ultrasound microscope
Ultraschallprüfung ultrasonic test
Ultraschallreinigung ultrasonic cleaning
Ultraschallschwingungen ultrasonic vibrations
Ultraschallsender ultrasonic transmitter
ultraschnell ultra-fast
ultraviolett ultraviolet, UV
Ultraviolett- *see* **UV-**
Ultrazentrifugation ultra-centrifugation
Ultrazentrifuge ultracentrifuge
Umamidierung trans-amidisation
Umbruchfestigkeit cantilever strength, bending strength under two-point loading
Umdrehung rotation, revolution
Umesterung ester interchange
Umfang 1. circumference. 2. extent, range, scope
Umfangsgeschwindigkeit peripheral velocity
Umfeld environment
Umfeldbedingungen ambient conditions
Umgebung 1. environment. 2. surroundings
Umgebungsatmosphäre surrounding atmosphere
Umgebungsbedingungen ambient/environmental conditions
umgebungsbeeinflußte Spannungsrißbildung environmental stress cracking
Umgebungseinflüsse environmental influences
Umgebungsklima ambient/environmental conditions, surrounding atmosphere
Umgebungsluft surrounding atmosphere
Umgebungsmedium surrounding medium
Umgebungstemperatur ambient temperature
Umgriff throwing power *(in electrophoretic paint deposition)*
Umhüllmasse potting/embedding/encapsulating compound
Umhüllmaterial potting/embedding/encapsulating compound
Umhüllsystem potting/embedding/encapsulating compound
Umhüllung potting/embedding/encapsulation
Umkehrosmose reverse osmosis
Umlagerung transposition, rearrangement
Umlaufautoklav circulating air autoclave
Umluft circulating air
Umluftofen circulating air oven
Umlufttrockenschrank circulating air drying oven/cabinet
Umlufttrockner cirulating air dryer
Umlufttrocknung circulating air drying
Umrechnungsfaktor conversion factor
Umrechnungsformel conversion formula
Umsetzung 1. reaction. 2. conversion. 3. implementation *(e.g. of regulations)*
Umsetzungsgrad degree of conversion
Umsetzungsprodukt reaction product
Umstellung change-over
Umwandlung transformation, conversion, transition

Umwandlungsprozeß transformation process
Umwandlungspunkt transition point
Umwelt environment
Umweltanforderungen environmental requirements/demands
Umweltaspekte environmental aspects
Umweltaudit environmental audit, eco-audit
Umweltauflage environmental directive
Umweltauswirkung environmental effect, effect on the environment
Umweltbedenken environmental reservations/doubts
umweltbedenklich potentially dangerous to the environment
Umweltbedingungen environmental conditions
umweltbeeinträchtigend affecting/polluting the environment
Umweltbeeinträchtigung (environmental) pollution, pollution of the environment
Umweltbehörde environmental authority
umweltbelastend environmentally unfriendly, pollutant: **umweltbelastende Verpackungsrückstände** packaging residues which pollute the environment
Umweltbelastung environmental pollution: **Umweltbelastung der Luft** air pollution
Umweltberichterstattung environmental reporting
Umweltbericht environmental report
Umweltbetriebsprüfung environmental works audit
Umweltbewertung environmental assessment
umweltbewußt environmentally aware, environment-conscious
Umweltbewußtsein environmental awareness
umweltbezogen environmental
Umweltbilanz 1. environmental audit. 2. ecobalance, ecological equilibrium
Umweltbundesamt Federal Environment Office
Umweltcarcinogen environmental carcinogen
Umweltchemikalien chemicals in the environment
Umweltdaten environmental data
Umweltdiskussion discussion of environmental issues/problems/questions
Umwelteffekt environmental effect, effect on the environment
Umwelteigenschaften environmental properties/characteristics
Umwelteinfluß environmental impact/influence
Umwelteinwirkungen environmental influences, effects on the environment
Umweltengel *see* **Blauer Engel**
Umwelterklärung environmental declaration/statement
Umweltfarbe environment-/eco-friendly paint
Umweltfolgen environmental consequences
Umweltforderung environmental requirement/demand
Umweltforschung environmental research
Umweltfragen environmental/ecological questions/considerations
umweltfreundlich environment-friendly, environmentally friendly, ecofriendly
Umweltfreundlichkeit environment-friendliness, ecofriendliness
Umweltgedanke environmental idea/concept
Umweltgefahr environmental danger/hazard
umweltgefährdend dangerous for the environment
Umweltgefährdung danger for the environment
umweltgefährlich dangerous for the environment
umweltgerecht environment-friendly, environmentally sound/acceptable, ecofriendly
Umweltgesetz environmental law/ordinance/directive
Umweltgesetzgebung environmental legislation
Umweltgesichtspunkte environmental considerations/factors
Umweltgift environmental poison
Umweltgutachter environmental expert
Umwelthaftung environmental responsibility/liability
Umwelthaftungsgesetz environmental liability act
Umweltinformationen environmental information
Umweltinformationsgesetz environmental information law/act/directive
Umweltinitiative environmental initative
Umweltkatastrophe environmental disaster
Umweltkontamination (environmental) pollution, pollution of the environment
Umweltkrise environmental crisis
Umweltleitfaden environmental manual/handbook/guide
Umweltleitlinie environmental guideline
Umweltmanagement environmental management
Umweltmanagementsystem environmental management system
umweltneutral environment-friendly, environmentally friendly, ecofriendly.
Umweltnorm environmental standard
umweltorientiert environment-orientated
Umweltpolitik environmental politics/policy
umweltpolitisch environmental
Umweltproblem environmental problem
Umweltproblematik environmental problems
Umweltprogramm environmental programme
Umweltprüfbericht environmental test report
Umweltprüfung environmental audit/investigation/inspection
Umweltqualität environmental quality
Umweltrecht environmental legislation/law
umweltrelevant 1. environmentally important, environment-related. 2. toxic
Umweltrichtlinie environmental guideline
Umweltrisiko environmental risk
Umweltschäden environmental damage
umweltschädlich dangerous for the environment
Umweltschadstoff (environmental) pollutant
umweltschonend environment-friendly, environmentally friendly, ecofriendly

Umweltschutz environmental protection
Umweltschutzanlage pollution control equipment
Umweltschutzaspekte environmental aspects
Umweltschutzaudit environmental audit
Umweltschutzauflage environmental directive
Umweltschutzbedingungen pollution control provisos/terms/conditions
Umweltschutzbehörde environmental protection agency
Umweltschutzbemühungen environmental efforts/endeavours
Umweltschutzdaten environmental data
Umweltschützer environmentalist
Umweltschutzerfordernisse environmental requirements
Umweltschutzforschung environmental research
Umweltschutzgesetz environmental protection law/act/directive
Umweltschutzgesetzgebung environmental legislation
Umweltschutzgründe environmental reasons
Umweltschutzhandbuch environmental protection manual
Umweltschutzmaßnahmen 1. pollution control measures. 2. environmental protection measures
Umweltschutzorganisation environmentalist organisation
Umweltschutzprobleme environmental/ecological problems
Umweltschutzvorschrift environmental directive/regulation
umweltsicher environmentally safe
Umweltsicherheit environmental safety
Umweltstatistik environmental statistics
Umweltstrategie environmental strategy/policy/plan
Umwelttechnik 1. environmental technology. 2. environmental engineering
umwelttechnisch environmental
Umwelttechnologie environmental technology
umwelttechnologisch environmental
Umwelttoxikologie environmental toxicology
Umweltvergiftung environmental pollution, pollution of the environment
Umweltverhalten environmental behaviour
Umweltverhältnisse environmental conditions
umweltverschmutzend (environmentally) polluting
Umweltverschmutzung environmental pollution, pollution of the environment
umweltverträglich environmentally compatible/safe
Umweltverträglichkeit environmental impact/compatibility
Umweltverträglichkeitsprüfung environmental asssessment, EA
Umweltvorschrift environmental directive
Umweltvorsorge care for the environment
Umweltzeichen environmental symbol *(usually associated with* **Blauer Engel**, *q.v.)*
Umweltziel environmental target
unabgesättigt unsaturated
unaufgeschmolzen unmelted
unausgehärtet uncured
unbedenklich, gesundheitlich non-toxic
unbedenklich. lebensmittelrechtlich food quality, suitable for food contact applications
unbedenklich, ökologisch ecologically safe/acceptable, ecofriendly
unbedenklich, physiologisch non-toxic
unbedenklich, toxikologisch non-toxic
Unbedenklichkeit, gesundheitliche non-toxicity
Unbedenklichkeit, physiologische non-toxicity
unbedruckt unprinted
unbegrenzt unlimited, limitless
unbehandelt 1. untreated. 2. uncoated
unbelastet unstressed
unbelichtet not exposed to light
unbenetzbar unwettable
unbeschädigt undamaged
unbeschichtet uncoated
unbestrahlt non-irradiated
unbewittert unweathered
unbrennbar non-flammable
Unbrennbarkeit non-flammability
undicht leaky
Undichtheit leak
undurchdringbar impenetrable
undurchlässig impermeable
undurchsichtig opaque
Unebenheiten surface irregularities
unelastisch inflexible
unempfindlich indifferent to, unaffected by
Unempfindlichkeit resistance to, unaffected by
Unfallverhütungsvorschrift accident prevention regulation, safety regulation
ungealtert unaged, not aged
ungecoated uncoated
ungeeignet unsuitable
ungefärbt unpigmented
ungefüllt unfilled
ungehärtet uncured
ungehindert unhindered
ungekerbt unnotched
ungenau inaccurate
Ungenauigkeit inaccuracy
ungepaart unpaired
ungequollen unswelled
ungerührt unstirred
ungesättigt unsaturated
Ungesättigtheit degree of unsaturation
ungeschützt unprotected
ungestrichelt unbroken, continuous *(curve)*
ungiftig non-toxic
Ungiftigkeit non-toxicity
unglasiert unglazed
ungleichmäßig uneven
Unglück disaster
Unifarbton plain colour
Universalgrundierung general purpose primer
Universalharz general purpose resin

unkaschiert unlaminated
unkatalysiert uncatalysed
unkondensiert uncondensed
unkontrollierbar uncontrollable
unlackiert unpainted
unlegiert unalloyed
unlösbar 1. insoluble. 2. permanent *(joint)*
unlösbare Verbindung permanent joint
unlöslich 1. insoluble. 2. permanent *(bond)*
Unlöslichkeit insolubility
unmodifiziert unmodified
Unordnung disorder
unpigmentiert unpigmented
unplastifiziert unplasticised
unpolar non-polar
unpolarisiert non-polarised
unpolymerisiert unpolymerised
unporös non-porous
unregelmäßig irregular
Unreinheiten impurities
unrentabel uneconomic
unschädlich harmless
Unschädlichmachen rendering harmless
unschmelzbar infusible
Unschmelzbarkeit infusibility
unsichtbar invisible
unstabilisiert unstabilised
unsubstituiert unsubstituted
unsymmetrisch asymmetrical
Unterboden subfloor
Unterbodenschutz underbody sealant
Unterbodenschutzbeschichtung underbody sealant
Unterbodenschutzmasse underbody sealant compound
Unterbrennen understoving, underbaking
Unterdosierung adding too little
Unterdruck vacuum
Unterdruckfarbe underprinting ink
Unterdrückung suppression
unterer Grenzwert lower limit
Untergrund substrate
Untergrundoberfläche substrate surface
Untergrundvorbehandlung preparation/pretreatment of the substrate
Unterhalt upkeep, maintenance
Unterhaltskosten 1. maintenance costs. 2. running/operating costs
Unterhaltung 1. servicing, maintenance. 2. entertainment
unterhärtet undercured
Unterhärtung undercuring
unterirdisch underground, below ground
Unterkonstruktion sub-structure
Unterkühlung undercooling
Unterlage substrate
unternehmensintern in-house
Unternehmensleitung company/corporate management
Unternehmensmitarbeiter company employees
Unternehmenspolitik company/corporate policy
Unterrostung under-film rusting
Unterrostungsgeschwindigkeit under-film rusting speed
Unterrostungsmessung measurement of under-film rusting
Unterrostungsverhalten under-film rusting behaviour
Unterscheidungskriterien distinguishing features
Unterscheidungsmerkmal distinguishing feature/characteristic
Unterseite underside
Untersuchung test, investigation
Untersuchungsbericht test report
Untersuchungsergebnis test result
Untersuchungsmethode test method
Untersuchungsprogramm test programme
Untersuchungstechnik test method
Untersuchungstemperatur test temperature
Untersuchungszeitraum test period/duration
Untertagebau underground mining
untervernetzt undercured
Untervernetzung undercuring
Unterwanderung under-film/creep corrosion
Unterwanderungskorrosion under-film/creep corrosion
Unterwasseranstrich underwater coating/finish
Unterwasseranstrichstoff underwater paint
Unterwasserbeanspruchung underwater immersion/exposure
Unterwasserbereich underwater
Unterwasserfarbe underwater paint
Unterwassergranulierung underwater pelletisation *(of hot melt adhesives)*
untoxisch non-toxic
unumgesetzt unreacted
unverändert unchanged
unverdünnt undiluted
unverestert unesterified
unverletzt undamaged
unvernetzt uncured
unverrostbar rust-proof, non-rusting
unverrottbar rot-proof
unversehrt undamaged
unverseifbar unsaponifiable
unversintert unsintered
unverträglich incompatible
Unverträglichkeit incompatibility
unverzweigt unbranched
unvorbehandelt not pretreated
unvulkanisiert unvulcanised
unwirtschaftlich uneconomic
Urethan urethane
Urethanacrylat urethane acrylate
Urethanalkyd urethane alkyd (resin)
Urethanalkydharz urethane alkyd (resin)
Urethananteil urethane content
Urethanbindung urethane linkage
Urethanbindungsanteil urethane linkage content
Urethanblock urethane block
Urethanbrücke urethane bridge
Urethandispersion urethane dispersion

Urethanester urethane ester
Urethanesterrest urethane ester group
Urethanether urethane ether
Urethanethoxylat urethane ethoxylate
urethanfunktionell urethane-functional
Urethangruppe urethane group
Urethangummi urethane rubber
Urethanhaftvermittler urethane bonding/coupling agent
Urethanharz urethane resin
urethanisiert urethanised
Urethanmethacrylat urethane methacrylate
urethanmodifiziert urethane-modified
Urethanoligomer urethane oligomer
Urethanpolymer polyurethane
Urethanprepolymer urethane prepolymer
Urethanthixotropierharz thixotropic polyurethane resin
Ursprungsviskosität original viscosity
Ursprungszustand original condition
UV-Absorber UV absorbent
UV-absorbierend UV absorbent
UV-Absorption UV absorption
UV-Anteil UV content
UV-Ausrüstung UV stabiliser
UV-Belastung exposure to UV rays/radiation
UV-Beständigkeit UV resistance
UV-Bestrahlung UV irradiation
UV-durchlässig UV permeable
UV-Durchlässigkeit UV permeability
UV-Einwirkung exposure to UV rays/radiation
UV-härtbar UV curing
UV-härtend UV curing
UV-Härtung UV curing
UV-Klebstoff UV curing adhesive
UV-Lack UV-curing paint
UV-Lampe UV lamp
UV-Licht UV/ultraviolet light
UV-Lichtbeanspruchung exposure to UV/ultraviolet light
UV-Lichtquelle UV/ultraviolet light source
UV-Reaktivität UV reactivity
UV-Remission UV remission
UV-Schutz protection against UV rays/radiation
UV-Schutzmittel UV stabiliser
UV-Spektroskopie UV spectroscopy
UV-Spektrum UV spectrum
UV-stabil resistant to UV rays
UV-Stabilisator UV stabiliser
UV-Stabilität UV resistance
UV-Strahlen UV/ultraviolet rays
UV-Strahlenhärtung UV-radiation curing
UV-Strahler UV radiator
UV-Strahlung UV radiation
UV-strahlungshärtend UV-curing
UV-Strecke UV heating tunnel
UV-Streuung UV scattering
UV-vernetzend UV-curing
UV-Vernetzung UV-curing

V

V2A stainless steel
VA-Stahl stainless steel
VAE-Dispersion ethylene-vinyl acetate dispersion
Vakuum, im in vacuo
Vakuumanschluß vacuum connection
Vakuumdestillation vacuum distillation
vakuumdicht vacuum tight
Vakuumimprägnierung vacuum impregnation
Vakuumimprägnierungsanlage vacuum impregnation plant
Valenz valency
Valenzbindung valency bond
Valenzkraft valency force
Valenzschwingung valency vibration/oscillation
Valenzwinkel valency angle
Valeriansäure valeric acid
van-der-Waals-Kräfte van der Waals forces
VbF *abbr. of* **Verordnung brennbarer Flüssigkeiten**, German regulation covering flammable liquids
VC-Copolymerisat vinyl chloride copolymer
VCVAC *abbr.of* **Vinylchlorid-Vinylacetat-Copolymer**, vinyl chloride-vinyl acetate copolymer
VE-Harz vinyl ester resin
Ventil valve
veralgt covered with algae
verankern to anchor
Verankerung anchoring
verarbeitbar processable
Verarbeitbarkeit processability, ease of processing, plasticity, workability
Verarbeitung processing
Verarbeitungsbedarf required pot life
Verarbeitungsbedingungen processing conditions
verarbeitungsbereit ready for processing, ready to use
Verarbeitungseigenschaften processing characteristics
Verarbeitungseinheit processing unit
Verarbeitungsfehler processing fault
verarbeitungsfertig ready to use
verarbeitungsfreundlich easy to use
Verarbeitungsgeräte processing equipment, implements
Verarbeitungshilfe processing aid
Verarbeitungshinweise processing guidelines
Verarbeitungskonsistenz working consistency
Verarbeitungskriterien processing criteria
Verarbeitungsmerkmale processing characteristics
Verarbeitungsmittel processing aid
Verarbeitungsperiode pot life
Verarbeitungsprobleme processing problems
Verarbeitungsrichtlinie processing guideline
Verarbeitungsschwerpunkte most important processing conditions

Verarbeitungsschwierigkeiten processing difficulties
Verarbeitungsspielraum pot life
verarbeitungstechnisch processing: **verarbeitungstechnische Eigenschaften** processing characteristics
Verarbeitungstemperatur processing temperature
Verarbeitungsviskosität working consistency
Verarbeitungsvorschriften processing instructions
Verarbeitungsvorteile processing advantages
Verarbeitungszeit pot life
Verarmungsflockulation depletion flocculation
Verbesserung improvement
Verbindung 1.(chemical) compound. 2. linkage, bond
Verbindung, lösbare temporary joint
Verbindung, unlösbare permanent joint
Verbindungsfestigkeit bond strength
Verblassung fading
verblockfrei non-blocking
Verbrauch consumption
Verbrauchsdauer pot life
verbrauchsfertig ready-to-use
Verbrennungsanlage incinerator
Verbund 1. bond. 2. composite (material)
Verbundfenster sealed unit
Verbundfestigkeit bond strength
Verbundgüte bond quality
Verbundkonstruktion composite structure
Verbundmaterial composite (material)
Verbundschicht composite film/layer
Verbundsicherheitsglas laminated safety glass
Verbundsicherheitsscheibe laminated safety glass
Verbundstoff composite (material)
Verbundsystem composite system
Verbundverpackung composite pack
Verbundwerkstoff composite (material)
verdampfbar volatile
Verdampfbarkeit volatility
verdampfen to evaporate
Verdampfungsenthalpie evaporation enthalpy
Verdampfungsgeschwindigkeit evaporation rate
Verdampfungsgeschwindigkeitsanalyse evaporation rate analysis
Verdampfungsgleichgewicht evaporation equilibrium
Verdampfungsrate evaporation rate
Verdampfungsverlust heat of evaporation
Verdampfungswärme heat of evaporation
verdichten to compress, to compact
verdickend thickening
Verdicker thickener, thickening agent
Verdickeradditiv thickener, thickening agent
Verdickerharz thickening resin
Verdickung thickening
Verdickungsmittel thickener, thickening agent
Verdickungswirkung thickening effect
Verdrängungskraft displacement force
verdünnbar dilutable, thinnable
Verdünnbarkeit dilutability
verdünnen to dilute, to thin (down): **verdünnte Säure** dilute acid
Verdünner thinner, diluent
Verdünnerharz extender resin
Verdünnung dilution
Verdünnungskurve dilution curve
Verdünnungsmittel thinner
Verdünnungsverhalten dilution characteristics
Verdünnungswärme heat of dilution
verdunsten to evaporate
Verdunstung evaporation
Verdunstungsgeschwindigkeit evaporation rate
Verdunstungskälte drop in temperature due to evaporation
Verdunstungsrate evaporation rate
Verdunstungsverhalten evaporation properties/characteristics
Verdunstungsverlust evaporation loss, loss through evaporation
Verdunstungszahl evaporation index
Veredelung finishing
vereinigen to assemble, to bring together
verestert esterified
Veresterung esterification
veresterungsfähig esterifiable
verethert etherified
Veretherung etherification
Veretherungsalkohol etherifying alcohol
Veretherungsgrad degree of etherification
Veretherungsparameter etherifying conditions
Verfahrensanalyse process analysis
Verfahrensaudit process audit
Verfahrensbedingungen processing conditions
Verfahrenskontrolle process control
Verfahrensparameter processing parameter/conditions
verfahrensspezifisch processing-related
Verfahrenssteuerung process control
verfahrenstechnisch technical(lly), processing: **verfahrenstechnisch einfach** technically easy/simple; **verfahrenstechnische Bedingungen** processing conditions
Verfahrensweise procedure
Verfärbung discolouration
Verfärbung, thermische discolouration due to charring
Verfärbungsneigung tendency to discolour
verfestigen to solidify, to consolidate, to strengthen
verfestigt 1. solidified, consolidated. 2. set *(adhesive)*
Verfestigung solidifcation, setting
verfilmbar film-forming
Verfilmung film formation
Verfilmungsbedingungen film-forming conditions
Verfilmungshilfsmittel film-forming aid
Verfilmungsreaktion film-forming process
Verfilmungstemperatur film-forming temperature
verflüchtigbar volatile

verflüchtigen to evaporate, to volatilise
Verflüchtigung volatilisation, evaporation
verflüssigbar liquifiable
verflüssigt liquified
verformbar 1. plastic, flexible. 2. deformable
Verformbarkeit 1. flexibility, plasticity. 2. deformability
Verformung deformation
Verformung unter Last deformation under load
Verformungsamplitude deformation amplitude
Verformungsarbeit deformation energy
Verformungsbruch ductile fracture/failure
Verformungseigenschaften deformation characteristics/properties
Verformungsenergie deformation energy
verformungsfähig deformable
Verformungsfähigkeit 1. deformability. 2. flexibility, plasticity
Verformungsgeschwindigkeit deformation rate
Verformungsverhalten deformation behaviour/characteristics
Verformungsvermögen deformability
verfügbar available
Verfugen jointing
Verfugungsmasse 1. jointing filler/compound. 2. (tile) grout
Verfüllmasse knifing filler, stopper, surfacer
Vergießen embedding, potting
Vergiftung poisoning
vergilbend yellowing
vergilbt yellowed
Vergilbung 1. yellowing. 2. yellowness index
vergilbungsbeständig non-yellowing
Vergilbungsbeständigkeit resistance to yellowing
Vergilbungsechtheit resistance to yellowing
vergilbungsfrei non-yellowing
Vergilbungsgrad yellowness index
Vergilbungsneigung tendency to become yellow
Vergilbungszahl yellowness index
Verglasung glazing
Verglasungsmaterial glazing material
Verglasungsprofil glazing strip
Vergleichsmessung comparative determination
Vergleichsmuster reference sample/specimen
Vergleichsprobe reference sample/specimen
Vergleichsprüfung comparative test
Vergleichsrechnung comparative calculation
Vergleichsspannung comparative stress
Vergleichsuntersuchung comparative test
Vergleichsversuch comparative test
Vergleichswert comparative figure
Vergleichszahl comparative figure
Vergrößerung enlargement
Vergußmasse embedding/potting compound
Vergußmaterial embedding/potting compound
Verhalten behaviour, performance
Verhärtung hardening
Verhinderung prevention
verifizieren to verify
verkappt capped
Verkapselung encapsulation
Verkehrsabgase traffic fumes
Verkehrslast traffic load
Verkehrsschild traffic sign
Verkieselung silicification
verklebbar bondable
Verklebbarkeit bondability
Verklebeeigenschaften bonding properties/characteristics
verkleben to bond, to stick
Verklebung 1. bonded joint. 2. bonding
Verklebungsfestigkeit bond strength
Verklebungsfläche bonded area
Verklebungsmaterial adhesive
Verklebungspartner adherends, parts to be bonded
Verklebungstechnik bonding technology
Verklumpen formation of lumps
Verklumpungsneigung tendency to form lumps
verknäuelt entangled *(molecules)*
Verknäuelung entanglement *(of molecules)*
Verknüpfung linkage
Verknüpfungsreaktion crosslinking reaction
Verkochung cooking
Verkochungsbedingungen cooking conditions
verkohlen to char, to carbonise
Verkohlung charring
Verkohlungsneigung tendency to char
Verkrustung encrustation
Verlauf flow
Verlauf, schlechter poor flow
Verlauf, zeitlicher variation with time
Verlaufmittel flow control agent, flow promoter
Verlaufprobleme flow problems
Verlaufsbelag self-levelling screed
Verlaufseigenschaften flow properties/characteristics
verlaufsfördernd flow-promoting: **verlaufsfördernde Produkte** flow promoters
Verlaufshilfsmittel flow control agent, flow promoter
Verlaufsmittel flow control agent, flow promoter
Verlaufsmörtel self-levelling mortar
Verlaufsschwierigkeiten flow problems
Verlaufsstörungen flow problems
Verlaufsverbesserer flow promoter
Verlaufsverbesserung improving flow
Verlaufszusatz flow control agent, flow promoter
Verlaufverhalten flow characteristics/behaviour
Verleimbarkeit glueing/bonding characteristics
Verleimung 1. bonding, glueing. 2. bonded/glued joint, bond
Verleimungsanordnung glueing device
Verleimungsfestigkeit bond strength
Verleimungsfuge bonded/glued joint
Verleimungsmethode method of glueing
Verleimungsprobleme bonding/glueing problems
Verleimungstemperatur bonding/glueing temperature
Verlustarbeit power/energy loss

Verlustfaktor 1. dissipation factor *(electrical)*. 2. loss factor *(mechanical)*
Verlustfaktor, dielektrischer dissipation factor
Verlustfaktor, mechanischer loss factor
Verlustleistung power/energy loss
Verlustmodul loss modulus
Verluströmung leakage flow
Verlustwärme lost heat
Verlustwinkel loss angle
Verlustzahl, dielektrische loss index
Vermahlen grinding
Vermarktung marketing
Verminderung reduction, decrease
Vermischen mixing
vernachlässigbar negligible
vernetzbar crosslinkable, curable
Vernetzer curing/crosslinking agent, hardener, catalyst
Vernetzerart type of hardener/catalyst
Vernetzerkonzentration hardener/catalyst concentration
Vernetzermenge amount of hardener/catalyst
Vernetzerspaltprodukt catalyst decomposition product
Vernetzerzugabe addition of hardener/catalyst
vernetzt crosslinked, cured
vernetzt, engmaschig closely crosslinked
vernetzt, teilweise partly crosslinked
vernetzt, weitmaschig loosely crosslinked
Vernetzung crosslinkage, curing
Vernetzungsbedingungen curing conditions
Vernetzungsdichte crosslink density
vernetzungsfähig crosslinkable
Vernetzungsfähigkeit crosslinkability
Vernetzungsgeschwindigkeit curing rate
Vernetzungsgrad degree of crosslinkage
Vernetzungsharz crosslinking resin
Vernetzungshilfe curing agent, hardener, catalyst
Vernetzungskatalysator catalyst, hardener, curing/crosslinking agent
Vernetzungskomponente catalyst, hardener, curing/crosslinking agent
Vernetzungsmechanismus curing/crosslinking mechanism
Vernetzungsmethode curing method, method of curing
Vernetzungsmittel catalyst, hardener, curing/crosslinking agent
Vernetzungsphase curing/crosslinking phase
Vernetzungsprozeß curing/crosslinking process
Vernetzungspunkt crosslink point
Vernetzungsrate curing rate
Vernetzungsreaktion curing/crosslinking reaction
Vernetzungsstelle crosslink point
Vernetzungssystem curing/crosslinking agent
Vernetzungstemperatur curing temperature
Vernetzungsverhalten curing behaviour/characteristics
Vernetzungsvorgang curing/crosslinking process/reaction
Verordnung directive, ordinance, regulation
Verpackung packaging, package, pack
Verpackungsabfall packaging waste
Verpackungsband packaging tape
Verpackungsbehälter packaging container
Verpackungsbereich packaging sector/industry
Verpackungsdose can
Verpackungsfolie packaging film
Verpackungsgebiet packaging sector/industry
Verpackungsgebinde packaging container
Verpackungsindustrie packaging industry
Verpackungsklebeband adhesive packaging tape
Verpackungskleber packaging adhesive
Verpackungsklebstoff packaging adhesive
Verpackungslack can coating lacquer
Verpackungsmaterial packaging material
Verpackungsmittel packaging material
Verpackungsrohstoff packaging material
Verpackungssektor packaging sector/industry
Verpackungsstraße packaging line
verpackungstechnisch packaging
Verpackungstechnologie packaging technology
Verpackungsverordnung packaging directive/ordinance
Verpackungswerkstoff packaging material
verpilzt mouldy
Verringerung reduction
verrostet rusted
Versagen failure, breakdown
Versagen, duktiles ductile failure
Versagen, katastrophales catastrophic failure
Versagen, sprödes brittle failure
Versagenskriterium failure criterion
Versagensmechanismus failure mechanism
Versagensursache cause of failure
Verschiebung displacement
verschlauft entangled *(molecules)*
Verschlaufung entanglement *(of molecule chains)*
verschlechtert made worse, deteriorated
Verschlechterung deterioration
Verschleiß wear
verschleißanfällig susceptible to wear
Verschleißanfälligkeit susceptibility to wear
verschleißarm hard wearing, wear resistant
Verschleißarmut hard wearing properties
verschleißbedingt due to wear
verschleißbeständig wear resistant, hard wearing
Verschleißbeständigkeit wear resistance
Verschleißbetrag amount of wear
verschleißen to wear out
verschleißfest wear resistant, hard wearing
Verschleißfestigkeit wear resistance
verschleißfrei wear resistant
Verschleißgrad amount of wear
Verschleißmechanismus wear mechanism
verschleißmindernd wear resistant
Verschleißprüfung abrasion test
Verschleißrate rate of wear
Verschleißschäden damage due to wear

Verschleißschutzlage wear resistant coating/layer
Verschleißspuren signs/traces of wear
Verschleißteil part/component subject to wear
Verschleißverhalten wear characteristics
Verschleißwiderstand wear resistance
verschleißwiderstandsfähig hard wearing
Verschleißwiderstandsfähigkeit wear resistance
Verschlingung entanglement
verschlucken to swallow
verschlungen entangled
Verschmelzen fusing
verschmolzen fused
verschmutzt soiled, contaminated, polluted
Verschmutzung contamination, pollution
Verschmutzungsgefahr risk of contamination
Verschmutzungsgrad degree of contamination
Verschmutzungsneigung tendency to pick up dirt
verschneiden to blend
Verschnitt blend
verschnitten blended
Verschnittharz extender resin
Verschnittlösemittel diluent
Verschnittlöser diluent
Verschnittmittel extender
verschüttet spilt *(liquids, powders etc.)*
verseifbar saponifiable
Verseifbarkeit saponifiability
verseifend saponifying
verseift saponified
Verseifung saponification
verseifungsbeständig saponification resistant
Verseifungsbeständigkeit saponification resistance
Verseifungsfestigkeit saponification resistance
Verseifungsgeschwindigkeit saponification rate
Verseifungsgrad degree of saponification
Verseifungsrate saponification rate
Verseifungsresistenz saponification resistance
Verseifungsstabilität saponification resistance
Verseifungszahl saponification number
Versiegeln sealing
Versiegelungseffekt sealing effect
Versiegelungsmasse sealant (compound)
Versiegelungsmittel sealant (compound)
verspritzbar sprayable
Verspritzbarkeit sprayability
verspritzt sprayed
verspröden to become brittle
versprödend embrittling
versprödet embrittled
Versprödung embrittlement
Versprödungserscheinungen signs of embrittlement
Versprödungsneigung tendency to become brittle
versprödungsstabil resistant to embrittlement
Versprödungstemperatur brittleness temperature, brittle point
Versprödungstendenz tendency to become brittle
versprühbar sprayable
Versprühbarkeit sprayability
Versprühen spraying
Verstärkung reinforcement
Verstärkungsadditiv reinforcing filler
Verstärkungseffekt reinforcing effect
Verstärkungsfaser reinforcing fibre
Verstärkungsfüllstoff reinforcing filler
Verstärkungsmittel reinforcing agent
Verstärkungswirkung reinforcing effect
verstellbar adjustable
verstellbar, stufenlos steplessly/infinitely adjustable
Verstelleinrichtung adjusting mechanism
Verstopfung blockage
verstrammt stiffened *(in consistency)*
Verstrammung stiffening *(in consistency)*
Verstreichbarkeit brushability
Verstreichvorgang brushing operation: **beim Verstreichvorgang** during brushing
Versuch test, trial
Versuchsanlage experimental plant
Versuchsanordnung test setup/arrangement
Versuchsaufbau experimental setup
Versuchsauswertung interpretation of test results
Versuchsbedingungen test/experimental conditions
Versuchsbericht test report
Versuchsdaten experimental/test data
Versuchsdauer test period/duration
Versuchsdurchführung test/experimental procedure
Versuchseinrichtung test setup
Versuchsergebnis test result
Versuchsfehler experimental error
Versuchsfeld test bed
Versuchsformulierung test formulation
Versuchskessel pilot reactor
Versuchsmaßstab pilot plant scale
Versuchsmaterial test substance/material
Versuchsmenge experimental quantity
Versuchsmischer experimental mixer
Versuchsmuster test sample
Versuchsparameter test conditions
Versuchsplatte test panel
Versuchsprobe test specimen
Versuchsprodukt experimental product
Versuchsprogramm test programme
Versuchsprotokoll test record
Versuchsreihe test series
Versuchsresultat test result
Versuchsrezeptur test formulation
Versuchsschwankungen experimental variations/fluctuations
Versuchsserie test series
Versuchsstadium experimental stage
versuchstechnisch experimental
Versuchstemperatur test temperature
Versuchswerkstoff test material
Versuchswert test result
Versuchszeit test period/duration
Versuchszeitraum test period/duration

versus vs.
Verteilbarkeit dispersibility
Verteileffekt dispersing effect
verteilen 1. to disperse. 2. to distribute
Verteilung 1. dispersion. 2. distribution
Verteilungsanomalie uneven dispersion
Verteilungschromatographie partition chromatography
Verteilungsgrad degree of dispersion
Verteilungsgüte efficiency of dispersion
Verteilvorgang dispersing process
Vertiefung depression, indentation
vertikal vertical
vertikales Ausschwimmen flooding
vertikales Pigmentausschwimmen flooding
verträglich compatible
verträglich, begrenzt with limited compatibility
Verträglichkeit compatibility
Verträglichkeit, begrenzte limited compatibility
Verträglichkeitseigenschaften compatibility (characteristics)
Verträglichkeitsgrad degree of compatibility
Verträglichkeitsgrenze compatibility limit
Verträglichkeitsmacher compatibiliser
Verträglichkeitsprobleme compatibility problems
Verträglichkeitsprüfung compatibility test
Verträglichkeitsschwierigkeiten compatibility problems
Verträglichkeitsuntersuchung compatibility test
Verträglichkeitsverhalten compatibility behaviour
verträglichkeitsvermittelnd making compatible
Verträglichkeitsvermittler compatibiliser
Verunreinigung impurity, contaminant
Verunreinigungsgefahr risk of contamination
Verunreinigungsgrad degree of contamination
Verunreinigungsrisiko risk of contamination
Verweilzeit residence time
Verwendungsmöglichkeiten application possibilities
Verwendungszweck (intended) application
Verwerfung warping, deformation
verwertbar re-usable, recyclable
Verwertung recycling, recovery, reclamation
Verwertung, chemische chemical recycling
Verwertung, stoffliche material recycling
Verwertung, thermische energy recovery
Verwiegeanlage weighing equipment
Verwiegung weighing
verwindungssteif torsion resistant
verwittert weathered
verzinkt galvanised
Verzinkungsbad galvanising bath
Verzinkungsverfahren galvanising process
verzinnt tin plated
Verzinnung tin plating
Verzögerer retarding agent
Verzögerungseinrichtung delaying mechanism
verzögerungsfrei without delay, immediately
Verzug warpage
verzugsarm low-warpage
Verzugsarmut low-warpage properties
Verzugserscheinungen warping, warpage
verzugsfrei non-warping, warp-free
Verzugsfreiheit non-warping properties, freedom from distortion
Verzugsneigung tendency to warp
Verzugsspannung warpage stress
verzugsstabil warp-free
verzweigt branched
Verzweigung branching
Verzweigungsgrad degree of branching
Verzweigungsreaktion branching reaction
Verzweigungsstelle branch point
Verzweigungsstruktur branched structure
VG *abbr. of* **Verdunstungsgeschwindigkeit**, evaporation rate
Vibration vibration
Vibrationsbande vibration band
Vibrationsfrequenz vibration frequency
Vibrationszustand state of vibration
Vicat-Erweichungspunkt Vicat softening point
Vicat-Erweichungstemperatur Vicat softening point
Vicat-Wärmebeständigkeit Vicat softening point
Vicatgrad Vicat softening point
Vicatnadel Vicat indentor
Vicatstift Vicat indentor
Vicatwert Vicat softening point
Vicatzahl Vicat softening point
Vickershärte Vickers hardness
Vielseitigkeit versatility
vierbindig quadrivalent
Vierfarbendruck four-colour printing
Vierpunktbelastung four-point loading
Vierpunktbiegeversuch four-point bending test
Vierpunktbiegung four-point bending
vierwertig 1. quadrivalent. 2. quadrihydric *(if an alcohol)*
Vinylacetal vinyl acetal
Vinylacetat vinyl acetate
Vinylacetat-Ethylendispersion ethylene-vinyl acetate dispersion
Vinylacetat-Ethylencopolymerisat ethylene-vinyl acetate copolymer
Vinylacetatanteil vinyl acetate content
Vinylacetatcopolymer vinyl acetate copolymer
Vinylacetatgehalt vinyl acetate content
Vinylacetatharz polyvinyl acetate resin
Vinylacrylatharz polyvinyl acrylate resin
Vinylalkohol vinyl alcohol
Vinylbutyral vinyl butyral
Vinylbutyrat vinyl butyrate
Vinylcaprat vinyl caprate
Vinylcaproat vinyl caproate
Vinylchlorid vinyl chloride
Vinylchlorid, monomeres vinyl chloride monomer
Vinylchlorid-Vinylacetatcopolymerisat vinyl chloride-vinyl acetate copolymer
Vinylchloridcopolymer vinyl chloride copolymer
Vinylchloridcopolymerisat vinyl chloride copolymer

Vinylchloridhomopolymer vinyl chloride homopolymer
Vinylchloridpolymerisat vinyl chloride polymer
Vinylcopolymer vinyl copolymer
Vinylcopolymerisat vinyl copolymer
Vinylderivat vinyl derivative
Vinylester vinyl ester
Vinylesterharz vinyl ester resin
Vinylesterpolymer vinyl ester polymer
Vinylesterpolymerisat vinyl ester polymer
Vinylether vinyl ether
vinylfunktionell vinyl-functional
Vinylgruppe vinyl group
Vinylharz vinyl resin
Vinylharzlack vinyl (resin) paint
Vinylidenchlorid vinylidene chloride
Vinylidenchloridlatex vinylidene chloride dispersion
Vinylidenfluorid vinylidene fluoride
Vinylimidazolgruppe vinyl imadazole group
Vinylisobutylether vinyl isobutyl ether
Vinylkern vinyl core
Vinyllactam vinyl lactam
Vinyllatex vinyl emulsion
Vinyllaurat vinyl laurate
Vinyllinoleat vinyl linoleate
Vinyllinolenat vinyl linolenate
Vinylmyristat vinyl myristate
Vinylnaphthalin vinyl naphthalene
Vinylnapthenat vinyl napthenate
Vinyloleat vinyl oleate
Vinyloxyethylacrylat vinyl ethoxyacrylate
Vinylpalmitat vinyl palmitate
Vinylpolymer vinyl polymer
Vinylpolymerisat vinyl polymer
Vinylpropionat vinyl propionate
Vinylpyridin vinyl pyridine
Vinylpyrrolidon vinyl pyrrolidone
Vinylrest vinyl group/radical
Vinylseitengruppe vinyl side group
Vinylsilan vinyl silane
Vinylsilanharz vinyl silane resin
Vinylstearat vinyl stearate
vinylsubstituiert vinyl-substituted
Vinyltoluol vinyl toluene
Vinyltrialkoxysilan vinyl trialkoxysilane
Vinyltrimethoxysilan vinyl trimethoxysilane
Vinylverbindung vinyl compound
violett violet
viskoelastisch viscoelastic
Viskoelastizität viscoelasticity
viskoplastisch viscoplastic
viskos viscous
viskoses Fließen viscous flow
Viskosimeter viscometer
Viskosimetrie viscometry
viskosimetrisch viscometric
Viskosität viscosity. *There are occasions when this word is used in place of* **Konsistenz**, *in which case it should be translated as* consistency, *as in* **Applikationsviskosität**, *q.v.*
Viskosität, dynamische dynamic viscosity
Viskosität, kinematische kinematic viscosity
Viskosität, plastische plastic viscosity
Viskosität, reduzierte reduced viscosity, viscosity number
Viskosität, relative relative viscosity
Viskosität, scheinbare apparent viscosity
Viskosität, spezifische specific viscosity
Viskositätsabfall viscosity decrease/reduction, drop in viscosity
viskositätsabhängig viscosity-dependent
Viskositätsabnahme viscosity decrease/reduction, drop in viscosity
Viskositätsabsenkung viscosity decrease/reduction, drop in viscosity
Viskositätsänderung change in viscosity
Viskositätsanomalie viscosity anomaly
Viskositätsansatz viscosity equation
Viskositätsanstieg rise/increase in viscosity
Viskositätsaufbau viscosity increase
Viskositätsberechnung viscosity calculation
Viskositätsbereich viscosity range
Viskositätsberg viscosity peak
Viskositätsbestimmung viscosity determination
Viskositätsbeziehung viscosity relation(ship)
Viskositätseinstellung 1. viscosity. 2. viscosity adjustment
viskositätserhöhend viscosity-increasing
Viskositätserhöhung increase in viscosity
viskositätserniedrigend viscosity-lowering/-reducing
Viskositätserniedriger viscosity depressant
Viskositätserniedrigung viscosity reduction
Viskositätsfunktion viscosity function
viskositätsgeregelt viscosity-controlled
Viskositätsgleichgewicht viscosity equilibrium
Viskositätsgrad viscosity
Viskositätsgradient viscosity gradient
Viskositätsgrenze limiting viscosity, viscosity limit
Viskositätsindex viscosity index
Viskositätsinhomogenitäten viscosity fluctuations/variations
Viskositätskoeffizient viscosity coefficient
Viskositätskonstanz constant viscosity
Viskositätskurve viscosity curve
Viskositätsmaximum maxmum viscosity
Viskositätsmessung viscosity determination
Viskositätsminderung viscosity decrease/reduction, drop in viscosity
Viskositätsminimum minimum viscosity
Viskositätsmittel average viscosity
Viskositätsmodul viscosity modulus
Viskositätsniveau viscosity
Viskositätsprofil viscosity profile
Viskositätsregelung viscosity control
Viskositätsregler viscosity regulator
viskositätssenkend viscosity-reducing
Viskositätssenkung viscosity decrease
Viskositätssprung sudden change in viscosity

viskositätsstabil having a constant/stable viscosity
Viskositätsstreuung viscosity variation
Viskositätsunterschied difference in viscosity
Viskositätsverhalten viscosity behaviour
Viskositätsverhältnis viscosity ratio, relative viscosity
Viskositätsverlauf viscosity profile/pattern
Viskositätswert viscosity value
Viskositätszahl viscosity number
Viskositätszunahme viscosity increase
visuell visual
visuelle Beurteilung visual inspection
Vitrifizierung vitrification
Vlies 1. bonded/non-woven fabric *(plural:* non-wovens*)*. 2. mat
Vliesstoff 1. bonded/non-woven fabric *(plural: non-wovens)*
VMQ-Mischung silicone rubber mix
VOC volatile organic compound
VOC-arm low-VOC
VOC-Auflage VOC directive
VOC-Bestimmung VOC directive/regulation
VOC-Emission VOC emission
VOC-Gehalt VOC content
VOC-Niveau VOC content/level
VOC-Wert VOC content
Vol-% percent by volume, % v/v
vollaromatisch completely aromatic
Vollautomat fully automatic machine
Vollautomatik fully automatic system
vollautomatisch fully automatic
vollautomatisiert fully automated
vollelektronisch completely electronic
vollflächig over the entire area/surface *(e.g. application of adhesive)*
vollfluoriert completely fluorinated
Vollgummi solid rubber
vollhalogeniert completely/fully halogenated
vollhydriert completely hydrogenated
vollhydrolysiert completely hydrolysed
vollmethyliert completely methylated
vollmethylverethert completely methyl etherified
Vollpartikel solid particle
vollrecyclierbar completely recyclable
Vollton full colour
vollverethert completely etherified
vollvernetzt fully crosslinked
vollverseift completely saponified
Vollwärmeschutz exterior wall insulation
Volumen volume
Volumen, reduziertes effective volume
Volumen, spezifisches specific volume
Volumenabnahme volume decrease
Volumenänderung change in volume
Volumenanteil volume content
Volumenausdehnung volume/volumetric expansion
Volumenausdehnungskoeffizient coefficient of cubical/volume expansion
Volumenausdehnungskoeffizient, thermischer coefficient of cubical/volume expansion
volumenbezogen volume-related
Volumenbruch volume fraction
Volumendehnung volume expansion
Volumendilation volume expansion
Volumendosieraggregat volumetric feeder
Volumendosierung 1. volumetric feeding. 2. volumetric feeder
Volumendurchsatz volume throughput, volume flow rate
Volumeneinheit unit volume
Volumenerhöhung volume increase, increase in volume
Volumenfließindex melt volume index
Volumenfluß volume/volumetric flow
Volumenflußrate volume/volumetric flow rate
Volumengehalt volume content
volumengleich of equal volume, having the same volume
volumenintensiv bulky
Volumenkonstanz constant volume
Volumenkontraktion volume shrinkage/contraction
Volumenkonzentration volume concentration
Volumenleistung volume throughput
Volumenprozent percent by volume %v/v
Volumenregelung 1. volume control. 2. volume control unit
Volumenschrumpf volume shrinkage
Volumenschrumpfung volume shrinkage
Volumenschwund volume shrinkage
Volumenstrom volume throughput, volume/volumetric flow rate
Volumenstromregelung 1. volume control. 2. volume control unit
Volumenveränderung change in volume
Volumenvergrößerung volume increase
Volumenverkleinerung volume decrease/reduction
Volumenverminderung volume decrease/reduction
Volumenwiderstand volume resistance
Volumenzunahme volume increase
volumetrisch volumetric, by volume
Vorabinformation advance information
Vorabmischen premixing
Voranstrich primer
vorbehandelt pretreated
Vorbehandlung pretreatment
Vorbehandlungsmethode method of pretreatment
vorbeschleunigt pre-accelerated
vordispergiert predispersed
Vordispergierung predispersion
voremulgiert pre-emulsified
vorgegeben given, specified, preset, prescribed, programmed, preselected
vorgeheizt preheated
Vorgehensweise procedure
vorgeschaltet upstream
vorgrundiert primed

Vorhärtungstemperatur pre-curing temperature
vorherrschende Teilchengröße predominant particle size
Vorhersagbarkeit predictability
vorhomogenisieren to prehomogenise
Vorinformation advance information
Vorkehrungen precautions
Vorkondensat precondensate
Vorkondensation precondensation
vorkondensiert precondensed
Vorlack undercoat
Vormischung premix
vorneutralisiert preneutralised
Vornorm provisional specification
Vorprimern priming, applying a primer
Vorprodukt intermediate
Vorratsbehälter storage tank
Vorratsgefäß storage vessel/container
Vorschrift regulation, directive, ordinance, order, provision
Vorsichtsmaßnahme safety precaution/measure
Vorstrich primer, base coat
Vorstufe preliminary stage
vortemperiert preheated
Vortrockengerät predrying unit
Vortrocknung predrying
Voruntersuchung preliminary test/investigation
Vorversuch preliminary test
Vorwärmetemperatur preheating temperature
Vorwärmetemperaturbereich preheating temperature range
Vorwärmung preheating
vorzeitig premature
Vulkanisat vulcanisate
Vulkanisation vulcanisation
Vulkanisationsbeschleuniger (vulcanisation) accelerator
Vulkanisationsgeschwindigkeit vulcanising rate
Vulkanisationsmittel vulcanising agent
Vulkanisationsverhalten vulcanising behaviour
Vulkanisationsverlauf vulcanisation profile
vulkanisierbar vulcanisable
vulkanisiert vulcanised
VZ 1. *abbr. of* **Verdunstungszahl**, evaporation index. 2. *abbr. of* **Viskositätszahl**, viscosity number

W

w.w. *abbr. of* **wahlweise**, optional, if required, as required
Wabenkern honeycomb
Wachs wax
wachsartig wax-like
Wachsdispersion wax emulsion
Wachsemulsion wax emulsion
wachsend increasing
Wachsersatz wax substitute
wachshaltig containing wax
wachsmodifiziert wax modified
Wachsreste wax residues
Wägeeinrichtung weighing equipment
Wägefehler weighing error
Wägemechanismus weighing mechanism
wägen to weigh
Wahrscheinlichkeit probability
Wahrscheinlichkeitsgleichung probability equation
Walzen roller application
Walzenauftragsanlage roll application unit
Walzlack coil coating paint/enamel
Walzlackierung coil coating
Walzlackierverfahren coil coating (process)
Walzverfahren coil coating (process)
Wand wall
Wandbelag wallcovering
Wandbelagsstoff wallcovering
Wandbeschichtung wall coating
wandernd migrating
Wanderung migration
wanderungsbeständig non-migrating
Wanderungsbeständigkeit migration resistance
Wanderungsfestigkeit migration resistance
Wanderungstendenz migration tendency
Wanderungsverhalten migration behaviour
Wandfarbe wall paint
Wandfläche wall surface
Wandstärke wall thickness
Wareneingangskontrolle 1. incoming goods control. 2. incoming goods control department
Wareneingangsprüfvorschrift incoming goods test specification
Warm-Feuchtbedingungen conditions of high temperature and humidity
warmabbindend hot setting, heat curing
Wärme heat
Wärme, spezifische specific heat
Wärme, strahlende radiant heat
Wärmeabfluß heat loss
Wärmeabfuhr 1. heat dissipation. 2. cooling
Wärmeabgabe heat dissipation, dissipation of heat
Wärmeableitung heat dissipation
wärmeabsorbierend heat absorbent
Wärmeabsorption heat absorption
wärmeabstrahlend heat reflecting
wärmeaktivierbar capable of being heat activated
Wärmeaktivierung heat activation
Wärmealterung heat ageing
Wärmealterungsbeständigkeit heat ageing resistance
Wärmealterungsverhalten heat ageing behaviour/properties
Wärmealterungsversuch heat ageing test
Wärmealterungswerte heat ageing properties
Wärmeanwendung application of heat

Wärmeaufbau heat build-up
Wärmeausdehnung thermal expansion
Wärmeausdehnungskoeffizient coefficient of thermal expansion
Wärmeausdehnungskoeffizient, linearer coefficient of linear expansion
Wärmeausdehnungsverhalten thermal expansion behaviour
Wärmeausdehnungszahl coefficient of thermal expansion
Wärmeaushärtung heat cure/curing
Wärmeaustauscher heat exchanger
Wärmebeanspruchung thermal stress, exposure to high temperatures
wärmebehandelt heat treated
Wärmebehandlung 1. heat treatment. 2. heat ageing
wärmebelastbar heat resistant, resistant to high temperatures
wärmebelastet heated, subjected to high temperatures
Wärmebelastung exposure to heat, exposure to high temperatures
wärmebeständig heat resistant
Wärmebeständigkeit heat resistance
Wärmedämmeigenschaften thermal insulating properties
wärmedämmend heat/thermally insulating
Wärmedämmputz heat/thermal insulating plaster
Wärmedämmstoff heat/thermal insulating material
Wärmedämmung thermal insulation
Wärmedämmverbundsystem composite thermal insulation system
Wärmedämmvermögen thermal insulation properties
Wärmedauerbelastung long-term exposure to high temperatures
Wärmedauerfestigkeit long-term heat resistance
Wärmedehnung thermal expansion
Wärmedehnungszahl coefficient of thermal expansion
Wärmedissoziierung thermal dissociation
Wärmedurchgang heat transfer/transmission
Wärmedurchgangskoeffizient heat transfer coefficient, coefficient of heat transmission
Wärmedurchgangszahl heat transfer coefficient
Wärmedurchlaßkoeffizient heat transfer coefficient, coefficient of heat transmission
Wärmedurchlaßwiderstand heat transmission resistance
Wärmedurchschlag thermal breakdown
Wärmeeinbringung application of heat
Wärmeeinfluß effect of heat
Wärmeeinwirkung heat exposure, action/effect of heat
wärmeempfindlich heat sensitive, affected by heat
Wärmeenergie thermal energy
Wärmeentwicklung heat evolution, evolution of heat
Wärmeentwicklungsgeschwindigkeit heat evolution rate, rate of heat evolution
Wärmeentzug removal of heat
wärmeerweichbar heat softenable
wärmeerzeugend heat-producing
Wärmeerzeugung production of heat
wärmefest heat resistant
Wärmefestigkeit heat resistance
Wärmefluß heat flow
wärmeforciert trocknend force drying
wärmeforcierte Trocknung force drying
wärmeformbeständig heat resistant
Wärmeformbeständigkeit 1. deflection temperature, heat distortion temperature. 2. heat resistance
Wärmeformbeständigkeit nach ISO R 175 deflection temperature under load according to ISO R 175
Wärmeformbeständigkeit nach Martens Martens heat distortion temperature
Wärmeformbeständigkeit nach Vicat Vicat softening point
wärmeformstabil heat resistant
Wärmefühler thermocouple
wärmegedämmt thermally insulated
wärmehärtbar heat setting/curing
wärmehärtend heat setting/curing
Wärmehärtung heat curing
Wärmehaushalt 1. heat content. 2. temperature. 3. heat balance
Wärmeinhalt heat content, enthalpy
wärmeintensiv very/extremely hot
Wärmeisolation thermal insulation
Wärmeisoliereigenschaften heat insulating properties
wärmeisolierend heat insulating
Wärmeisolierung thermal insulation
Wärmekapazität heat capacity
wärmelabil unstable at high temperatures
Wärmelagerung heat ageing
wärmeleitend heat/thermally conductive
Wärmeleiter conductor of heat
wärmeleitfähig thermally conductive
Wärmeleitfähigkeit thermal conductivity
Wärmeleitung thermal conductivity
Wärmeleitungsvermögen thermal conductivity
Wärmeleitzahl thermal conductivity
Wärmemenge amount of heat
Wärmemenge, freigesetzte amount of heat released
Wärmenachbehandlung annealing
Wärmenest heat accumulation
Wärmeofen oven
Wärmeoxidation thermal oxidation
Wärmequelle heat source
wärmereaktiv heat reactive
wärmereflektierend heat-reflecting
Wärmereserve heat reserve
Wärmerückgewinnung heat recovery
Wärmerückgewinnungsanlage heat recovery plant

Wärmerückgewinnungsgrad degree of heat recovery
Wärmeschock thermal shock
Wärmeschockbeständigkeit thermal shock resistance
Wärmeschockverhalten thermal shock resistance
Wärmeschrank drying oven/cabinet
Wärmeschrumpfung heat shrinkage
Wärmeschutz heat/thermal insulation
Wärmeschwingung temperature fluctuation
wärmesensitiv heat sensitive
Wärmesicherheit heat resistance
Wärmesinterung heat sintering
Wärmespannung thermal stress
Wärmespektrum heat spectrum
Wärmesperre heat barrier
wärmestabil heat resistant
Wärmestabilisator heat stabiliser
wärmestabilisiert heat stabilised
Wärmestabilität thermal stability
Wärmestand thermal stability
wärmestandfest heat resistant
Wärmestandfestigkeit heat resistance
Wärmestau heat build-up, accumulation of heat
Wärmestrahler radiant heater
Wärmestrahlung heat radiation
Wärmetauscher heat exchanger
Wärmetönung heat effect, heat produced, change in temperature
Wärmeträger heat carrier, heat transfer medium
Wärmeträgermedium heat transfer medium
Wärmeträgermittel heat transfer medium
Wärmetransfer heat transfer
Wärmetransport heat transport
Wärmetrocknung heat drying
Wärmeübergang heat transfer
Wärmeübergangskoeffizient heat transfer coefficient
Wärmeübergangszahl heat transfer coefficient
Wärmeübertragung heat transfer
Wärmeübertragungsmedium heat transfer medium
Wärmeübertragungsmittel heat transfer medium
wärmeunempfindlich unaffected by heat
wärmeverfilmbar capable of forming a film at high/elevated temperatures
Wärmeverfilmung film formation at high/elevated temperatures
Wärmeverhalten high temperature behaviour/performance
Wärmeverlust heat loss
wärmevernetzbar heat curable/curing
wärmevernetzend heat curing
wärmevernetzt heat cured
Wärmevernetzung heat curing
Wärmeverteilung heat distribution
wärmezerstörbar likely to be broken down by heat
Wärmezufuhr 1. heating, application of heat. 2. heat input
Warmfestigkeit heat resistance
warmhärtbar heat/hot curing
warmhärtend heat/hot curing
Warmhärtung heat/hot curing
Warmkleben hot bonding
Warmlagerung heat ageing
Warmlagerungstemperatur heat ageing temperature
Warmlagerungsversuch heat ageing test
Warmlagerungszeit heat ageing period
Warmluft hot air
Warmluftgebläse hot air blower
Warmlufttrockner hot air dryer
Warmschmelzklebstoff hot melt adhesive
wärmstabil heat resistant
Warmwasser warm/hot water
Warmzugversuch high-temperature tensile test
Warnlampe warning light
Warnleuchte warning light
Warnsignal alarm/warning signal
Warnzeichen warning sign
Wartezeit waiting time/period
Wartezeit, geschlossene closed assembly time
Wartezeit, offene open assembly time
Wartung servicing, maintenance
Wartungsansprüche maintenance requirements
Wartungsanweisungen servicing/maintenance instructions
Wartungsarbeit maintenance work
wartungsarm low-maintenance, requiring little maintenance
wartungsfrei maintenance-free, requiring no maintenance
Wartungsfreiheit freedom from maintenance, requiring no maintenance
wartungsfreundlich easy to service/maintain
Wartungskosten maintenance/servicing costs
Wartungsmaßnahmen servicing, maintenance
Wartungstechniker service engineer
waschbeständig wash resistant
Waschbeständigkeit washing resistance
waschfest wash resistant
Waschmaschine washing machine
Waschmittel detergent
Waschprimer wash/etch primer
Waschwert wash resistance
Wasser water
Wasser-in-Öl-Dispersion water-in-oil emulsion, W/O emulsion
Wasser-in-Öl-Emulsion water-in-oil emulsion, W/O emulsion
Wasser-Zementfaktor water-cement ratio
Wasser-Zementwert water-cement ratio
Wasserabdunstrate evaporation rate of water
Wasserabdunstung evaporation of water
Wasserabgabe loss of water
wasserabsorbierend water absorbent
Wasserabsorption water/moisture absorption
wasserabstoßend water repellent
wasserabwaschbar water washable

wasserabweisend water repellent
Wasserabweisung water repellency
Wasseradsorption water adsorption
wasseraktivierbar capable of being water activated
Wasseranteil water content, amount of water
wasserarm with a low moisture content
Wasseraufnahme water/moisture absorption
Wasseraufnahmefähigkeit ability to absorb water
Wasseraufnahmegeschwindigkeit water absorption rate
Wasseraufnahmekapazität water absorptive capacity
Wasseraufnahmekoeffizient coefficient of water absorption
Wasseraufnahmevermögen water absorptive capacity
wasseraufnehmend water absorbent
Wasserbad waterbath
wasserbasierend water-based, waterborne
Wasserbasis, auf water-based, waterborne
Wasserbasisanstrich water-based/waterborne paint
Wasserbau hydraulic engineering
Wasserbäumchen water treeing *(an electrical phenomenon)*
Wasserbauwerk hydraulic structure
Wasserbeize water-based (wood) stain
wasserbelastbar *implying that something - e.g. a paint film - can be brought in contact with water:* **Die beschichteten Flächen sind bereits einige Stunden nach dem Auftrag voll wasserbelastbar.** The coated surfaces are water resistant already a few hours after application of the paint
Wasserbelastung immersion in water, exposure to water
Wasserbesprühung spraying with water
wasserbeständig water resistant
Wasserbeständigkeit water resistance
Wasserdampf water vapour
Wasserdampfatmosphäre water vapour atmosphere
Wasserdampfdiffusion water vapour diffusion
Wasserdampfdiffusionswiderstand water vapour diffusion resistance
Wasserdampfdiffusionswiderstandsfaktor coefficient of water vapour diffusion resistance
Wasserdampfdruck water vapour pressure
wasserdampfdurchlässig water vapour permeable, permeable to water vapour
Wasserdampfdurchlässigkeit water vapour permeability
Wasserdampffestigkeit water vapour resistance
wasserdampfgesättigt saturated with water vapour
Wasserdampfpartialdruck partial water vapour pressure
Wasserdampfpermeabilität water vapour permeability
Wasserdampfpermeation water vapour permeation/diffusion
Wasserdampfregenerierung water vapour regeneration
Wasserdampfsättigungsdruck saturation water vapour pressure
Wasserdampfsperre water vapour barrier
Wasserdampfstabilität water vapour resistance
Wasserdampfteildruck partial (water) vapour pressure
wasserdampfundurchlässig water vapour impermeable, impermeable to water vapour
wasserdicht waterproof, watertight
Wasserdichtigkeit water impermeability, impermeability/imperviousness to water
wasserdispergierbar water dispersible
wasserdispergiert dispersed in water
Wasserdispersionslack emulsion paint
Wasserdruck water pressure
wasserdurchlässig water permeable
Wasserdurchlässigkeit water permeability
Wassereindringverhalten water penetration behaviour
Wassereindringzahl water penetration factor/index
Wassereinlagerung immersion in water
Wassereinwirkung immersion in water
wasserempfindlich affected by water
Wasserempfindlichkeit sensitivity to water
wasseremulgierbar water emulsifiable
Wasserentmischung separation of water *(e.g. in an emulsion paint)*
wasserfest water resistant
Wasserfestigkeit water resistance
wasserfrei anhydrous, free from water
wasserfreundlich hydrophilic
Wassergehalt water/moisture content
wassergekühlt water cooled
Wasserglas waterglass, sodium silicate
wasserhaltig 1. water-based, waterborne*(paints, adhesives)*. 2. aqueous, containing water
Wasserhaushalt water/moisture content
wasserhell water-white
wasserhemmend water retardant
Wasserinhalt water/moisture content
wasserklar water-white
Wasserklarlack waterborne clear varnish/lacquer
Wasserlack waterborne/water-based paint
Wasserlacksystem waterborne paint
Wasserlagerung immersion in water
Wasserlagerungsversuch water immersion test
wasserlöslich 1. water soluble. 2. water miscible *(e.g. solvents)*
Wasserlöslichkeit water solubility
Wassermenge amount of water
wassermischbar water miscible
Wassermolekül water molecule
Wasserpermeabilität water permeability
Wasserphase aqueous phase
wasserquellbar water swellable
Wasserquellbarkeit water swellability

wasserredispergierbar water-redispersible
wasserreduzierbar 1. water-thinnable. 2. waterborne, water-based *(paint or adhesive)*
wasserresistent water resistant
Wasserresistenz water resistance
Wasserretention water retentivity
Wasserringgranulierung water-cooled die face pelletisation *(of hot melt adhesives)*
Wasserrückhaltevermögen water retentivity
Wasserrückstand residual moisture
wassersaugend water absorbent
Wassersaugfähigkeit water absorbency
Wassersauggeschwindigkeit water absorption rate
Wassersiedepunkt boiling point of water
wassersperrend water repellent/impermeable
Wasserstand water level
Wasserstoff hydrogen
wasserstoffaktiv hydrogen-active
Wasserstoffakzeptor hydrogen acceptor
Wasserstoffatom hydrogen atom
Wasserstoffbindung hydrogen bond
Wasserstoffbindungskräfte hydrogen bond forces
Wasserstoffbrücke hydrogen bridge
Wasserstoffbrückenbildung hydrogen bridge formation
Wasserstoffbrückenbindung hydrogen bridge bond
Wasserstoffion hydrogen ion
Wasserstoffperoxid hydrogen peroxide
wasserstoffrei hydrogen-free
Wasserstoffsuperoxid hydrogen peroxide
Wassertropfen droplet of water
Wasserumlauftemperiergerät circulating water temperature control unit
Wasserumwälzeinheit water circulating unit
wasserundurchlässig water-impermeable, watertight
Wasserundurchlässigkeit water impermeability
wasserunempfindlich unaffected by water
Wasserunempfindlichkeit indifference to water, water resistance, resistance to water
wasserunlöslich water insoluble
wasserunverträglich water-immiscible, immiscible with water
wasserverdünnbar 1. water-thinnable. 2. waterborne, water-based *(paint or adhesive)*
Wasserverdünnbarkeit water dilutability
wasserverdünnt 1. diluted with water. 2. waterborne, water-based *(paint or adhesive)*
Wasserverschmutzung water pollution
wasserverträglich water-miscible, miscible with water
Wasserwiderstand water resistance
Wasserzudosierung addition of water
Wasserzugabemenge amount of water added or: to be added
wäßrig 1. aqueous *(solution)*. 2. water-based, waterborne *(paint or adhesive)*
WDD *abbr. of* **Wasserdampfdurchlässigkeit**, water vapour permeability
Wechselbeanspruchung cyclic stress
Wechselbelastung cyclic stress
Wechselbelastung, thermische alternating exposure to heat and cold
Wechselbeziehung interrleation
Wechselbiegefestigkeit flexural fatigue strength
Wechselbiegeversuch flexural fatigue test
Wechselfeldspannung a.c. voltage
Wechselfestigkeit fatigue strength, endurance limit
Wechselkartusche exchangeable cartridge
Wechselklima changing climatic conditions
Wechselklimabeanspruchung exposure to changing climatic conditions
Wechselklimalagerung ageing under changing climatic conditions
Wechselklimaprüfung testing under changing climatic conditions
Wechselknickversuch folding endurance test
Wechsellast cyclic load/stress
wechselnde Beanspruchung cyclic loading/stress
wechselnde Belastung cyclic loading/stress
Wechselspannung alternating/a.c. voltage
Wechselspannungserzeuger a.c. voltage generator
Wechselstrom alternating current, a.c.
Wechseltorsion alternating torsion
wechselweise alternately
Wechselwirkung interaction
Wechselwirkungsenergie interactive energy
Wechselwirkungskräfte interactive forces
Weglänge distance
weich soft, flexible
Weich-PVC flexible/plasticised PVC
weichelastisch flexible, pliable
weichflexibel flexible, pliable
weichgemacht plasticised
Weichharz soft resin
Weichheitsgrad softness, flexibility
Weichhholz soft wood
weichmachend plasticising, softening
Weichmacher plasticiser
Weichmacheranteil plasticiser content
Weichmacheraufnahme plasticiser absorption
Weichmacherbeständigkeit plasticiser resistance
Weichmacherdämpfe plasticiser vapours
Weichmachereinfluß effect of plasticiser
Weichmacherextraktion plasticiser extraction
weichmacherfest plasticiser resistant
Weichmacherflüchtigkeit plasticiser volatility
weichmacherfrei unplasticised
Weichmacherfreiheit freedom from plasticisers
Weichmachergehalt plasticiser content
Weichmachergemisch plasticiser blend
weichmacherhaltig plasticised, containing plasticiser
Weichmachermigration plasticiser migration
Weichmacherresistenz plasticiser resistance
Weichmacherrückhaltevermögen plasticiser

retentivity
Weichmacherverlust plasticiser loss
Weichmacherwanderung plasticiser migration
Weichmachung plasticisation
Weichmachung, äußere external plasticisation
Weichmachung, innere internal plasticisation
Weichmachungsmittel plasticiser
Weichmachungsvermögen plasticising capacity
weichpastös paste-like
Weichstahl mild steel
Weichwerden softening
weiß white
Weißblech tinplate
Weiße whiteness
Weißgrad whiteness
Weißheitsgrad whiteness
Weißkraft whiteness
Weißlack white paint/enamel
weißlich whiteish
Weißpigment white pigment
weißpigmentiert (pigmented) white
Weißpunkt 1. minimum film forming temperature. 2. powder point *(of an emulsion)*
Weiterbildung further education
weiterentwickelt upgraded, improved
Weiterentwicklung improvement, further development, improved version
Weiterreißen tear propagation
Weiterreißfestigkeit tear propagation resistance
Weiterreißkraft tear propagation force
Weiterreißversuch tear propagation test
Weiterreißwiderstand tear propagation resistance
Weiterverdünnen further dilution
weitgefächert wide-ranging, widespread
weitgehend löslich largely soluble
weitmaschig vernetzt loosely crosslinked
weitreichend far-reaching
Welle 1. wave. 2. shaft
Wellenlänge wavelength
Wellenlängenbereich wavelength range
Wellenzahl wave number
weltweit worldwide
Wendelmischer helical mixer
Werksanlage plant
werkseitig in-plant
Werkshalle factory shed
Werksleitung works management
Werkstatt workshop
Werkstätte workshop
Werkstoff material
Werkstoffabtrag material wear/abrasion
Werkstoffanisotropie material anisotropy
werkstoffbezogen material-related
Werkstoffdaten material constants
Werkstoffeigenschaften material properties
Werkstoffermüdung material fatigue
Werkstoffkenndaten material constants
Werkstoffkenngröße material constant/property
Werkstoffkombination material combination
werkstoffmechanisch mechanical *(properties)*
Werkstoffprüflabor material testing laboratory
Werkstoffprüfung material testing
werkstoffspezifisch material-related, specific to the material
Werkstoffverbund composite (material)
Werkstoffversagen material failure/breakdown
Werkstück workpiece
Wert 1. figure, value. 2. property
Werte, elektrische electrical properties
Werte, mechanische mechanical properties
Werte, thermische thermal properties
Werteabfall deterioration of properties
Wertebild (general) properties
Werteniveau (general) properties
Wertigkeit valency
Wertsiegel quality seal
Wertstoff re-usable/recoverable material
Wertstoffrecycling recycling of usable materials
Wetter weather
Wetterbeanspruchung exposure to the elements
Wetterbedingungen weather conditions
Wetterbelastung weathering
wetterbeständig weather resistant
Wetterbeständigkeit weathering resistance
wetterecht weather resistant, weatherproof
Wetterechtheit weathering fastness *(of pigments)*
wetterfest weather resistant, weatherproof
Wetterfestigkeit weathering resistance
wetterstabil weather resistant, weatherproof
Wetterstabilität weathering resistance
Wetterverhältnisse weather conditions
widersprüchlich contradictory
Widerstand resistance
Widerstand, elektrischer electrical resistance
Widerstand, spezifischer resistivity
Widerstand zwischen Stöpseln insulation resistance
widerstandsbeheizt resistance heated
Widerstandsdraht resistance wire
Widerstandserhöhung increase in resistance
widerstandsfähig resistant, durable
Widerstandsfähigkeit resistance
Widerstandsfühler resistance thermocouple
Widerstandsheizband resistance band heater
Widerstandsheizelement resistance heater
Widerstandsheizkörper resistance heater
Widerstandsheizung resistance heater
Widerstandstemperaturfühler resistance thermocouple
Widerstandsthermometer resistance thermometer
wiederanfeuchtbar remoistenable
wiederaufbereiten to reprocess, to reclaim, to recondition
Wiederaufbereitung reprocessing, reclamation, reconditioning
Wiederaufheizzeit reheating time, time required for reheating
Wiederaufschmelzen remelting
wiederaufwärmen to reheat
Wiedererwärmung reheating

Wiedergewinnung recovery
wiederholbar repeatable, reproducible
Wiederholbarkeit repeatability, reproducibility
wiederholgenau repeatable, reproducible
Wiederholgenauigkeit repeatability, reproducibility
wiederlösbar temporary *(joint)*
Wiederverarbeitbarkeit recyclability
Wiederverarbeiten reprocessing, recycling
wiederverschließbar re-sealable
wiederverwendbar recyclable, re-usable
Wiederverwendbarkeit recyclability, re-usability
Wiederverwendung recycling, re-use
wiederverwertbar recyclable, re-usable
Wiederverwertung recycling
Wiederverwertung, energetische energy recovery
Wiederverwertung, stoffliche material recycling
Wind wind
Windgeschwindigkeit wind velocity/speed
Windlast wind load/force
Windrichtung wind direction
Windschutzscheibe windscreen
winkelförmige Nutverbindung slip recessed joint
Winkelgeschwindigkeit angular velocity
Winkelschälversuch T-peel test
Wintermonate winter months
Wirbelbett fluidised bed
Wirbelschicht fluidised bed
Wirbelschichtpyrolyse fluidised bed pyrolysis
Wirbelschichtreaktor fluidised bed reactor
Wirbelschütten fluidised pouring
Wirbelsintergerät fluidised bed coater
Wirbelsintern fluidised bed coating
Wirbelstrom eddy current
Wirbelwirkung vortex effect
Wirkkonzentration active concentration
Wirkmechanismus mode of action
Wirkprinzip mode of action
wirksam effective
wirksame Oberfläche effective surface
Wirksamkeit effectiveness
Wirkstoff active substance/ingredient
Wirksubstanz active substance/ingredient
Wirkungsmechanismus mode of action
wirkungsstark powerful
Wirkungsweise mode of action
Wirtschaft 1. management. 2. trade, industry, business. 3. economy, economic system, sector of the economy
wirtschaften to manage
wirtschaftlich 1. economic(al). 2. economically efficient, profitable. 3. industrial, commercial. 4. financial
Wirtschaftlichkeit cost-effectiveness, economic advantages/viability/ aspects, profitability, efficiency: *since this word covers a wide range of ideas associated with productivity, high outputs, low costs etc. one can often expand the translation, e.g.* **Ein weiterer Beitrag zur Erhöhung der Wirtschaftlichkeit** another contribution towards greater efficiency and increased profits; **der Wirtschaftlichkeit wegen** for economic reasons
Wirtschaftlichkeitsbetrachtung profitability study, cost-benefit study
Wirtschaftlichkeitsaspekte economic aspects, economic points of view
Wirtschaftlichkeitsgründe economic reasons/considerations
Wirtschaftlichkeitsüberlegungen economic considerations
Wirtschaftsklima economic climate
Wirtschaftspotential economic potential
Wirtschaftsrezession economic recession
Wirtschaftszweig sector of the economy
Wirtsmedium host medium
wischen to wipe
wischfest rub resistant
Wismutanteil bismuth content
Wismutphosphat bismuth phosphate
Wismutsalz bismuth salt
Wismutvanadat bismuth vanadate
Witterung weather, wind and weather, the elements
Witterungsbedingungen weather conditions
witterungsbeständig weather resistant
Witterungsbeständigkeit weathering resistance
Witterungsbeständigkeitseigenschaften weathering characteristics
Witterungseigenschaften weathering characteristics
Witterungseinfluß effect of weathering *plural:* weathering influences
witterungsfest weather resistant
Witterungsschutz protection against the weather/elements
witterungsseitig on the exposed side
witterungsstabil weather resistant
Witterungsstabilität weathering resistance
Witterungsverhalten weathering characteristics/behaviour
Witterungsverhältnisse weather conditions
Wolframcarbid tungsten carbide
Wollastonit wollastonite
WOM Weather-Ometer
WU-Beton water-impermeable concrete
Wunderkleber superglue
würfelförmig cubical, cube-shaped
WVS *abbr. of* **Wärmedämmverbundsystem**, composite thermal insulation system

Xanthan xanthan
Xanthon xanthone

Xenonbogen xenon arc
Xenonbogenstrahler xenon lamp
Xenonbogenstrahlung xenon radiation
Xenonstrahlung xenon radiation
Xenotest xenotest
XLD crosslink density
Xylenol xylenol
Xylol xylene

YI-Wert yellowness index

Z

zäh 1. tough. 2. viscous
Zäh-Sprödbruchübergang ductile-brittle failure transition point
Zäh-Sprödübergang rubber-glass transition
Zähbruch ductile fracture/failure, tough fracture/failure
Zähelastifizierung imparting high impact strength
zähelastisch ductile, tough and resilient
zähfließend viscous
zähflüssig viscous
Zähflüssigkeit high-viscosity
zähhart hard and tough
Zähigkeit 1. toughness, strength. impact strength/resistance. 3. viscosity
Zähigkeit, kinematische kinematic viscosity
Zähigkeitsverhalten 1. strength characteristics. 2. viscosity characteristics
Zähigkeitsverlust loss of strength
Zähigkeitswert 1. toughness. 2. viscosity
Zahlenbeispiel numerical example
Zahlenmittel number average
Zahlenwert 1. numerical value. 2. figure
Zahnradpumpe gear pump
Zahnspachtel spreader/spreading comb
Zahntraufel spreader/spreading comb
zähsteif tough and rigid
zähviskos high-viscosity, viscous
Zaun fence
Zeit time
Zeit-Biegewechselfestigkeit flexural fatigue strength
Zeit, offene open assembly time
zeitabhängig time-dependent
Zeitabhängigkeit time dependence
Zeitabstand interval
Zeitaufwand (amount of) time required/needed
zeitaufwendig time consuming
Zeitbedarf (amount of) time required
Zeitbruchkurve creep rupture curve
Zeitdauer time, interval, period of time
Zeitdehnlinie creep curve
Zeitdehnspannung creep stress
Zeitdehnverhalten creep behaviour
Zeiteinfluß effect of time
Zeiteinheit unit of time
Zeitfaktor time factor
Zeitfestigkeit fatigue endurance
zeitintensiv time consuming
Zeitintervall interval
zeitlich time-dependent/-related, over a period
zeitlich steuerbar time-controlled
zeitlicher Temperaturverlauf time-temperature relationship
zeitlicher Verlauf variation with time
Zeitmaßstab time scale
zeitraffend accelerated *(e.g. tests)*
Zeitraffversuch accelerated test
zeitraubend time consuming
Zeitraum period, interval
Zeitschalter time switch
Zeitschaltuhr time switch
Zeitspanne period, interval
Zeitspannung creep stress
Zeitspannungslinie creep curve
zeitsparend time-saving
Zeitstandbeanspruchung creep stress
Zeitstandbiegefestigkeit flexural creep strength
Zeitstandbiegeversuch flexural creep test
Zeitstandbruch creep fracture/failure
Zeitstanddiagramm creep diagram
Zeitstanddruckverhalten compressive creep behaviour
Zeitstanddruckversuch compressive creep test
Zeitstandfestigkeit creep strength
Zeitstandfestigkeitskurve creep strength curve
Zeitstandfestigkeitslinie creep strength curve
Zeitstandfestigkeitsschaubild creep diagram
Zeitstandkurve creep curve
Zeitstandprüfung creep test
Zeitstandschaubild creep diagram
Zeitstandverhalten creep behaviour, long-term behaviour/performance
Zeitstandversuch creep test
Zeitstandzugbeanspruchung tensile creep stress, long-term tensile stress
Zeitstandzugbelastung tensile creep stress, long-term tensile stress
Zeitstandzugfestigkeit tensile creep strength
Zeitstandzugprüfung tensile creep test
Zeitstandzugverhalten tensile creep behaviour
Zeitstandzugversuch tensile creep test
Zeitsteuerung time control
zeitunabhängig independent of time

Zeitungsdruckfarbe newspaper printing ink
Zeitwechselfestigkeit endurance/fatigue limit
Zeitwechselfestigkeitsversuch fatigue test
Zelle cell
Zellkern cell nucleus
Zellstoff cellulose, wood pulp
Zellstoffindustrie cellulose/wood pulp industry
Zellstoffproduktion cellulose/wood pulp production
Zellulose- *see* **Cellulose-**
Zement cement
Zementadditiv cement additive
Zementestrich cement screed
Zementmatrix cement matrix
Zementmischung cement mix
Zementmörtel cement mortar
Zementputz cement plaster
Zementschlamm laitance
Zementschlämme laitance
Zementschlempe cement slurry
Zementspachtel cement surfacer
Zementstein 1. hardened cement. 2. artificial stone
Zementverflüssiger cement plasticiser
zementverträglich cement-compatible
Zementzumischung addition of cement
Zentralatom central atom
Zentrifugalkraft centrifugal force
Zentrifugaltrockner centrifugal dryer
Zentrifugation centrifugation
Zentrifuge centrifuge
zentrifugieren to centrifuge
zentrisch centric
zerbrechen to break
zerbrechlich fragile
Zerfall decomposition, degradation
zerfallen to break up, to dissociate, to disintegrate, to decompose
Zerfallgeschwindigkeit decomposition rate
Zerfallsprodukt decomposition product
Zerfallsprozeß decomposition process
Zerfallsreaktion decomposition reaction
Zerfallstemperatur decomposition temperature
zerkleinern to shred, to cut up, to break down
zerplatzen to burst *(e.g. bubbles in a paint film)*
Zerreißfestigkeit tear strength, ultimate tensile strength
Zerreißkraft tensile strength at break
Zerreißmaschine tensile testing machine
Zerreißprobe tensile (test) specimen
Zerreißprüfmaschine tensile testing machine
Zerrüttungsprüfung destructive test
Zerrüttungsuntersuchung destructive test
zerschnitten cut up, shredded
zersetzt decomposed
Zersetzung decomposition, degradation
Zersetzung, thermische thermal decomposition
Zersetzungsbereich decomposition temperature range
Zersetzungserscheinungen signs of decomposition
Zersetzungsgeschwindigkeit decomposition rate
Zersetzungsmechanismus degradation mechanism
Zersetzungsprodukt decomposition product
Zersetzungspunkt decomposition temperature
Zersetzungsreaktion decomposition reaction
Zersetzungstemperatur decomposition temperature
Zersetzungstemperaturbereich decomposition temperature range
zerspringen to crack
zerstäubt atomised
Zerstäubung atomisation
Zerstäubungsdruck atomising pressure
zerstörbar destructible
zerstörend destructive *(test)*
zerstört destroyed
Zerstörung destruction, breakdown
zerstörungsfrei non-destructive *(test)*
Zerteilung breaking down *(e.g. of agglomerates)*
Zertifizierung certification
Zetapotential zeta potential
Ziegel brick, tile
Ziegelmauerwerk brickwork
ziegelrot brick red
Ziegelwand brick wall
Zielgröße target quantity
Zielsetzung purpose, aim, objective
Zielwert target figure/value
Zimmertemperatur room temperature
Zink zinc
Zinkblech zinc sheet
Zinkborat zinc borate
Zinkchlorid zinc chloride
Zinkchromat zinc chromate
Zinkchromatgrundierung 1. zinc chromate primer. 2. zinc chromate primer coat
Zinkferrit zinc ferrite
Zinkgehalt zinc content
Zinkgelb zinc yellow
zinkhaltig containing zinc
Zinkhydrophosphat zinc hydrophosphate
Zinknitrat zinc nitrate
Zinkoctoat zinc octoate
Zinkoxid zinc oxide
Zinkoxidtyp grade of zinc oxide
Zinkoxychlorid zinc oxychloride
Zinkphosphat zinc phosphate
Zinkphosphat-Eisenoxidgrundierung zinc phosphate-iron oxide primer
Zinkphosphatierung zinc phosphating
Zinkpigment zinc dust
zinkplattiert galvanised
Zinkprimer zinc-rich primer
Zinkpulver zinc dust
zinkreich zinc-rich
Zinksalz zinc salt
Zinkseife zinc soap
Zinksilikat zinc silicate
Zinkstaub zinc dust

Zinkstaubbeschichtung zinc-rich coating
Zinkstaubfarbe zinc-rich paint
Zinkstaubformulierung zinc-rich paint
Zinkstaubgrund zinc-rich primer
Zinkstaubgrundbeschichtung zinc-rich primer coat
Zinkstaubgrundierung 1. zinc-rich primer. 2. zinc-rich primer coat
Zinkstaubpartikel zinc dust particle
Zinkstaubpigment zinc dust
Zinkstaubprimer zinc-rich primer
Zinkstaubsystem zinc-rich paint
Zinkstearat zinc stearate
Zinksulfid zinc sulphide
Zinktetraoxychromat zinc tetraoxychromate
Zinkverbindung zinc compound
Zinkweiß zinc white
Zinn tin
Zinnbasis, auf tin-based
Zinnbutyl butyl tin
Zinncarboxylat tin carboxylate
Zinndioxid tin dioxide
zinnfrei tin-free
Zinnkatalysator tin catalyst
zinnorganisch organotin
Zinnoxid tin oxide
Zinnstabilisator tin stabiliser
zinnstabilisiert tin stabilised
Zirkonacetylacetonat zirconium acetylacetonate
Zirkonat zirconate
Zirkondioxid zirconium dioxide
Zirkonoxid zirconium oxide
Zirkonoxychlorid zirconium oxychloride
Zirkonpropionat zirconium propionate
Zirkonsalz zirconium salt
Zirkonsilikat zirconium silicate
zitronengelb lemon yellow
Zitronensäure citric acid
Zitronensäureester citrate
Zr-Salz zirconium salt
Zubehör ancillary equipment, accessories
Zubehörteile ancillary equipment, accessories
zudosieren to add, to incorporate
Zudosierung addition, incorporation
Zufallsergebnis chance result
Zug tension
Zug-Dehnungsdiagramm tensile stress-elongation curve
Zug-Dehnungskurve tensile stress-elongation curve
Zug-E-Modul tensile modulus of elasticity
Zug-Elastizitätsmodul tensile modulus of elasticity
Zug-Kriechmodul tensile creep modulus
Zug-Kriechversuch tensile creep test
Zug-Zeitstandfestigkeit tensile creep strength
Zugabemenge amount added
Zugarbeit tensile energy
zugbeansprucht under tensile stress
Zugbeanspruchung tensile stress
Zugbeanspruchungsrichtung direction of tensile stress
zugbelastet under tensile stress
Zugbelastung tensile stress
Zugbruchlast tensile breaking stress
Zugdeformation tensile deformation
Zugdehnung elongation
Zugdehnung bei Streckgrenze elongation at yield
Zugdehnungsversuch tensile test
Zugdose tensile force transducer
Zugeigenspannung internal tensile stress
Zugeinrichtung tensile test instrument
zugeschärfte einschnittige Überlappung bevelled/tapered lap joint
Zugfestigkeit tensile strength
Zugfestigkeit bei Bruch tensile strength at break, ultimate tensile strength
Zugfestigkeitsprüfer tensile testing machine
Zugfestigkeitswert tensile strength (figure)
Zuggeschwindigkeit tensile testing rate/speed
zügige Beanspruchung dynamic stress
zügige Belastung dynamic stress
Zugkraft tensile force
Zugkraftaufnehmer tensile force transducer
Zuglast tensile stress
Zugluft draught
Zugmaschine tensile testing machine
Zugmodul tensile modulus
Zugprobe tensile specimen/bar, tensile test piece
Zugprobekörper tensile specimen/bar, tensile test piece
Zugprobestäbchen tensile specimen/bar, tensile test piece
Zugprüfmaschine tensile testing machine
Zugprüfstab tensile specimen/bar, tensile test piece
Zugprüfung tensile test
Zugrichtung direction of tensile stress
Zugscherbeanspruchung tensile shear stress
Zugscherbelastung tensile shear stress
Zugscherfestigkeit tensile shear strength
Zugscherprobe tensile shear test piece
Zugscherprobekörper tensile shear test piece
Zugscherprüfling tensile shear test piece
Zugscherprüfung tensile shear test
Zugscherspannung tensile shear stress
Zugscherung tensile shear
Zugscherversuch tensile shear test
Zugschwellast repeated tensile stress
Zugschwellbelastung repeated tensile stress
Zugschwellbereich region of repeated tensile stress
Zugschwellfestigkeit tensile fatigue strength
Zugschwellversuch tensile fatigue test
Zugschwingungsversuch tensile fatigue test
Zugspannung tensile stress
Zugspannung bei Streckgrenze tensile stress at yield
Zugspannungs-Dehnungskurve tensile stress-elongation curve
Zugspannungscraze tensile craze

Zugspannungsmaximum maximum tensile stress
Zugspannungsspitze tensile stress peak
Zugstab tensile specimen/bar, tensile test piece
Zugverformung tensile deformation, deformation under tensile stress
Zugverformungsrest tension set
Zugverhalten tensile behaviour
Zugversuch tensile test
Zuluft incoming air
Zumischharz blending resin
Zumischung addition
Zunderschicht scale
Zündpunkt ignition point
Zündquelle source of ignition
Zündtemperatur ignition point
Zündversuch ignition test
zurückgewonnen recovered
Zusammenballen agglomeration
Zusammenbau assembly
zusammenfließen to coalesce, to flow together
zusammenfügen to join
zusammengeflossen coalesced
Zusammenhaltekraft cohesive force
Zusammenlaufen coalescence
zusammenschmelzen to fuse (together)
Zusammensetzung composition, constitution
Zusammenstoß collision
Zusatz 1. addition. 2. additive
Zusatzbindemittel additional binder
Zusatzdispergiermittel extra dispersing agent
zusatzfrei additive-free, free from additives
Zusatzmenge amount (to be) added, amount
Zusatzmittel additive, filler
Zusatzprüfung additional test
Zusatzstoff additive, filler
Zuschlagstoff additive, filler
Zustandsänderung change of state
Zustandsform physical form
zweiachsig biaxial
zweibasisch dibasic
zweibindig divalent
zweidimensional two-dimensional
Zweikomponenten-Polyurethanbeschichtung two-pack polyurethane paint
Zweikomponentenanstrichfarbe two-pack paint
Zweikomponentenanstrichmittel two-pack paint
Zweikomponentendecklack two-pack top coat
Zweikomponentendichtstoff two-pack sealant
Zweikomponentenepoxidharzlack two-pack epoxy paint
Zweikomponentenflüssigsilikonkautschuk two-pack liquid silicone rubber
Zweikomponentengrundierung two-pack primer
Zweikomponentenkleber two-pack adhesive
Zweikomponentenklebstoff two-pack adhesive
Zweikomponentenklarlack two-pack clear varnish
Zweikomponentenlack two-pack paint
Zweikomponentenlacksystem two-pack paint
Zweikomponentenmischung two-pack mix
Zweikomponentenpolyurethandecklack two-pack polyurethane paint
Zweikomponentenpolyurethanlack two-pack polyurethane paint
Zweikomponentensystem 1. two-pack formulation. 2. two-pack paint. 3. two- pack adhesive
Zweikomponentenurethansystem two-pack polyurethane formulation/system/paint
zweikomponentig two-pack
Zweikreissystem dual circuit system
Zweiphasengemisch two-phase system
Zweiphasenmorphologie two-phase morphology
Zweiphasenstruktur two-phase structure
Zweiphasensystem two-phase system
zweiphasig two-phase
zweischichtig two-coat
Zweischichtsystem two-coat paint/system
zweischnittige Laschung double strap/butt-strap joint
zweischnittige Überlappung double lap joint
zweiseitig on both sides, double sided
Zweistufenhärter two-stage hardener
Zweistufenprozeß two-stage process
zweistufig two-stage
Zweitlack recycled/reconstituted paint
Zweitopfsystem two-pack system/formulation
zweitwichtigst second most important
Zweiwellenschneckenkneter twin screw compounder
zweiwertig 1. bivalent, divalent. 2. dihydric *(if an alcohol)*
Zwischenanstrich undercoat
Zwischenfraktion intermediate fraction
Zwischenlagenhaftung inter-layer adhesion
zwischenlagern 1. to store temporarily. 2. to mature, to age
zwischenmolekular intermolecular
Zwischenphase intermediate phase
Zwischenprodukt intermediate (product)
Zwischenraum interstice
Zwischenreaktion intermediate reaction
Zwischenschicht intermediate layer
Zwischenschichthaftung inter-layer adhesion
Zwischenshorehärte intermediate Shore hardness
Zwischenstadium intermediate stage
Zwischenstufe intermediate stage
Zwischentermin intermediate target date
Zwischenwert intermediate figure
Zwitterion zwitterion
Zwitternatur zwitter character
Zyklenprüfung cyclic test
zyklisch cyclic
Zyklisierung cyclisation
Zyklus cycle
Zyklusprüfung cyclic test
Zykluszeit cycle time
zylinderförmig cylindrical
zylindrisch cylindrical